ELECTRON DEFICIENT BORON AND CARBON CLUSTERS

Dedicated to William N. Lipscomb on his 70th Birthday

ELECTRON DEFICIENT BORON AND CARBON CLUSTERS

Edited by

George A. Olah
University of Southern California
Los Angeles, California

Kenneth Wade
University of Durham
Durham, England

Robert E. Williams
University of Southern California
Los Angeles, California

A WILEY-INTERSCIENCE PUBLICATION

JOHN WILEY & SONS

New York • Chichester • Brisbane • Toronto • Singapore

In recognition of the importance of preserving what has
been written, it is a policy of John Wiley & Sons, Inc. to
have books of enduring value published in the United
States printed on acid-free paper, and we exert our best
efforts to that end.

Library of Congress Cataloging in Publication Data:
Electron deficient boron and carbon clusters/George A. Olah, Kenneth
  Wade, Robert E. Williams.
      p.   cm.
    "A Wiley-Interscience publication."
    Includes bibliographical references.
    ISBN 0-471-52795-5
    1. Boranes. 2. Carboranes. I. Olah, George A. II. Wade,
  Kenneth. III. Williams, Robert E.
  QD181.B1E43 1990
  546'.671—dc20
                                                   90-41326
                                                        CIP

Printed in the United States of America

10 9 8 7 6 5 4 3 2 1

CONTRIBUTORS

Michael Buehl, Universitat Erlangen-Nurnberg, Erlangen, West Germany

F. Albert Cotton, Texas A & M University, College Station, Texas

Thomas P. Fehlner, University of Notre Dame, Notre Dame, Indiana

Thomas C. Flood, University of Southern California, Los Angeles, California

Keith Fuller, California State University, Los Angeles, Los Angeles, California

Norman N. Greenwood, University of Leeds, Leeds, England

Russel N. Grimes, University of Virginia, Charlottesville, Virginia

Narayan S. Hosmane, Southern Methodist University, Dallas, Texas

Sang O. Kang, University of Pennsylvania, Philadelphia, Pennsylvania

John A. Maguire, Southern Methodist University, Dallas, Texas

D. Michael P. Mingos, University of Oxford, Oxford, England

George A. Olah, University of Southern California, Los Angeles, California

Thomas P. Onak, California State University, Los Angeles, Los Angeles, California

Paul v. R. Schleyer, Universitat Erlangen-Nurnberg, Erlangen, West Germany

Sheldon G. Shore, Ohio State University, Columbus, Ohio

Larry G. Sneddon, University of Pennsylvania, Philadelphia, Pennsylvania

Kenneth Wade, University of Durham, Durham, England

David J. Wales, University Chemical Laboratories, Cambridge, England

Robert E. Williams, University of Southern California, Los Angeles, California

D. P. Workman, Ohio State University, Columbus, Ohio

In January 1989 the Loker Hydrocarbon Research Institute held one of its biannual research symposia; the topic selected was Electron Deficient Clusters. The lively symposium and its discussions contributed much to focus the views of the participants on their mutual interest in cluster chemistry, particularly on electron deficient clusters of boron and carbon. It was subsequently suggested that it would be worthwhile to have the relevant contributions enlarged and published as chapters in a volume. We agreed to edit the book and invited some additional contributors to round out the topic. We are dedicating the book to Bill Lipscomb on the occasion of his 70th birthday. He has been the leading exponent in advancing the concept of multicenter two electron bonding so essential to cluster chemistry and has pioneered these concepts in boron chemistry. His colleagues and friends wish him the very best and look forward to his continuing contribution to chemistry for many years to come.

GEORGE A. OLAH
KENNETH WADE
ROBERT E. WILLIAMS

Los Angeles, California
Durham, England
November 1990

Mookie, the Olah's cocker spaniel, enlightens
an otherwise blank page.

(photo by Mark Sassaman)

▰▰▰▰ CONTENTS

DEDICATION TO "THE COLONEL"

This volume evolved from a symposium on Electron Deficient Clusters held at the Loker Hydrocarbon Research Institute in January 1989. The organizers of this symposium were wise enough to first get all of the participants out to USC and have us contribute to, and thoroughly enjoy, a delightful three days before they broached the heavy subject of publishing the talks. I initially reacted to this with that extreme lightness of heart given only to those who have the pleasure of watching others do the noble work without having to do any themselves. I had been in the happy position of giving an after dinner talk of such consummate insubstantiality that even my own mother would have declared it unsuitable for publication.

I was not, however, to get off so easily. The organizers had the happy inspiration of dedicating the volume to one of the participants, namely, our esteemed colleague, Bill Lipscomb, on (or about) the occasion of his seventieth birthday. When they suggested that I might wish to write the dedicatory introduction, I found that I too would have to labor in the vineyards, but with the saving grace of performing a labor of love.

I may or may not have known Bill Lipscomb longer than anyone else at this conference, but it has certainly been for a long time. It is hard to believe that the thirty years since he arrived at Harvard, where I first met him, have flown by so fast. He has filled these years with a parade of stunning scientific spectaculars and I will recall a few of them for you presently.

Actually, I got to see and hear Bill Lipscomb about a year before he became a Harvard professor, because I attended a lecture he gave in the guise of a seminar speaker. The insiders knew that this was not just your run of the mill seminar but that he was being sized up for a job offer. Needless to say, his lecture left no doubt that he was the man they should hire—or almost none. Bill has always been possessed of an exuberant love of telling a good joke (or making an outrageous pun) and as I remember, he could not resist the urgings of the imp of the perverse (as Poe calls it) on this important occasion. The Harvard faculty member who introduced him was thanked for doing it so well with the story of the good angel. The good angel kissed one man on the forehead and he became a reigning mathematical maven, kissed another on the fingers and he became a piano virtuoso, and so forth. Then, came the punch line: "I don't know where the good angel kissed professor X, but he is certainly a marvelous chairman." Rumor among us young lowlifes had it that while the audience at large enjoyed this innocent ribaldry, some of the Harvard faculty were, like Queen Victoria, "not amused." However, to Harvard's everlasting good name, Bill got the job and the rest, as they say, is history.

For the enlightment of those who have been in a deep coma for the last thirty years, let me mention a few highlights of the Lipscomb scientific legacy. From such a wealth of material (well over 500 publications) any such short list is arbitrary and subjective. Having said that, I must begin, as an inorganic chemist, by citing Bill's work on boron hydrides, with all its implications for valence theory in general, as his premier accomplishment. There were some people in Stockholm who where also of this opinion. When he began his work on these compounds in about 1950 (having previously mastered the necessary low-temperature crystallographic techniques), there were clues to the potentialities of the field strewn about, but not being perceived nor appreciated. Bill did that wonderful thing that great scientists do: he looked where others had looked, but he saw what they had not seen. In particular, he could see the scope and importance of the field and he had the breadth of talent (and the capacity for hard work) to see that these were realized. He pioneered not only the field but the philosophy of the multifaceted attack: structure, theory, other physical measurements, and even synthesis (no mean feat for a physical chemist).

The theoretical work on the boron hydrides was done at two levels, one involving so-called topological concepts that give a valuable conceptual picture of what holds these substances in their remarkable molecular shapes, and the other involving the use of accurate numerical computations. The latter effort evolved over the years into a more general program of research into molecular quantum mechanics, from which many valuable results not necessarily pertaining to boron compounds have been obtained.

As an inorganic chemist and writer of textbooks in that field, I must point out that Bill's publication list bristles with other important contributions to the study of inorganic structures and bonding. His early work on mercury amides, on $(NO)_2$, on crystalline HF, on Roussin's black salt, on $(PCF_3)_5$, and many others I could mention were critical and timely contributions. These by themselves would make a publication list in which a lesser man could take great pride.

It was not long after his arrival at Harvard that Bill's latent interest in protein structure became an active one; he was on the lookout for a good one to work on. When a friend of mine at Harvard Medical School asked me if I wouldn't like to work on carboxypeptidase, I said "no, I'm not ready to tackle anything as demanding as that, but I know someone who is." It was at a very pleasant dinner in our apartment at MIT (where my wife and I were "houseparents" in a student residence) for the Stockmayers, the Vallees and the Lipscombs that "man met molecule" so to speak. I think there is no doubt that carboxypeptidase was an excellent choice in view of its structural and chemical properties and its broad biochemical interest, and it has indeed proved to be a classic of structural enzymology. However, it was a tough job, perhaps the toughest one since hemoglobin, given the unsophisticated state of the art of protein crystallography at the time. Nevertheless, in 1963 impressive progress was reported and by 1965 a complete 6A map was in hand. A few years later this had been carried to atomic resolution. Of the four enzyme structures first carried to atomic resolution (lysozyme, *staph* nuclease, ribonuclease, and carboxypeptidase), carboxypeptidase probably presented the greatest technical difficulties, especially with the preparation of good heavy atom derivatives.

While continuing to go at full throttle on his existing programs in molecular quantum mechanics, boron chemistry, and carboxypeptidase, in 1968 Bill took on a new, major challenge, aspartate transcarbamylase, a large regulatory enzyme with subunits, likely to display instructive allosteric effects. It will come as no surprise that this too has been brought off with skill and panache, and has proved to be another landmark in understanding at the atomic level how enzymes manage to be the phenomenal catalysts that they are.

As a steady basso continuo to all of these major themes, Bill and his students produced a steady stream of significant contributions in other areas. Perhaps the most notable of these is the calculation done with Russell Pitzer of the rotational barrier in ethane, which showed that the attainable accuracy of ab initio calculation was sufficient to estimate this barrier to within about 1%.

Oscar Levant observed that "the pun is the lowest form of humor—when you don't think of it first." Bill has been thinking of them first for as long as I have known him. Way back in 1963 at a party to felicitate me on an award I had just received he told me that while he was delighted by the choice, I should not suppose that he was on the "cottonpickin" committee. Then, again, when my colleague Dave Shoemaker became a consultant to what was then called the Esso Corporation, he reminded Dave that "honey bees make honey but Esso bees make money."

For his bountiful contributions to so many areas of chemistry (if not for his puns), Bill has been bountifully recognized. Besides the Nobel Prize which he received in 1976 he has received an impressive list of honorary degrees, a denumerable infinity of lectureships and honorary memberships in many learned societies. He is the only person I know to have had a mineral named in his honor. I hope it gleams brightly enough.

The best news of all about Bill, as we could all see at USC in January, is that he is in terrific shape, mentally and physically. He hasn't mellowed at all. He takes his work as seriously as ever, and he is still at the cutting edge of several important fields. No doubt we shall have the pleasure of seeing him still in fine fettle at a meeting ten years from now—provided *we* are still spry enough to get there.

On behalf of all of the participants in this volume I wish you a happy birthday and continued good health.

F. A. COTTON
May 1989

ELECTRON DEFICIENT BORON AND CARBON CLUSTERS

CHAPTER 1

General Concepts and Definitions

G. A. OLAH, K. WADE and R. E. WILLIAMS

In planning this book, our intention was to bring together topics of common interest in the area of cluster chemistry. The term "electron deficient" is used in the title to emphasize that our particular interest is in systems in which electrons are in relatively short supply. However, as the terms "electron deficient" and indeed "cluster" are used in more than one chemical sense these days, it appeared both appropriate and necessary to explain the sense in which we ourselves use them, and to explore briefly not only the value of such terms but also the value of other related terms commonly used in cluster chemistry though rarely defined.

The term "electron deficient" has at least three different senses in which it has been used in connection with covalent inorganic or organic systems. In organic chemistry, it is commonly used as meaning "center for nucleophilic attack" in referring to carbon atoms bearing electron-withdrawing substituents. Secondly, it is also used in referring to compounds containing coordinatively unsaturated atoms, such as the carbon atoms of carbenium ions R_3C^+ or the metal atoms of 16-electron transition metal complexes, which can accommodate an extra pair of electrons and thus act as Lewis acids. The third usage is as a label for molecules or parts thereof in which there are too few electrons to allow the bonding to be described exclusively in terms of 2-center electron pair bonds (2c2e bonds) [1, 2]. It is in this third sense that it is used in our discussions. The boron and carbon cluster compounds on which this book focuses are those in which the molecular skeleton is held together by fewer skeletal bond pairs than the number of points of contact between skeletal atoms.

The use of the term in this last sense enables us to distinguish cluster compounds like the boron hydrides and certain carbocations from "electron precise" systems which contain exactly the right number of electrons to give each pair a 2-center bonding role, as in methane or hydrocarbons in general, and indeed from "electron rich" systems like ammonia, water or Lewis bases in general containing nonbonding (lone pair) electrons [3, 4].

To draw such a distinction might seem to place undue emphasis on the Lewis 2-center bond concept, an inspired development in bonding theory when introduced in 1916 [5], but which some argue has outlived its usefulness. We feel this is emphatically not the case. Since a 2c2e bond between two atoms requires

Electron Deficient Boron and Carbon Clusters, Edited by George A. Olah,
Kenneth Wade, and Robert E. Williams.
ISBN 0-471-52795-5 © 1991 John Wiley & Sons, Inc.

each atom to contribute an atomic orbital to form the bond, then the total number of valance shell atomic orbitals provided by the atoms of an electron precise molecule must equal the total number of valence shell electrons it contains. If the number of valence shell electrons exceeds the number of valence shell orbitals, then the molecule will be electron rich. If the number of valence shell orbitals exceeds the number of valence shell electrons, then the molecule will be electron deficient (it might alternatively be justifiably referred to as "orbital rich", which might sound less pejorative than "electron deficient", though inviting confusion with "Lewis acid").

It should be stressed that borane-type clusters are *not* electron deficient in the sense that they need more electrons in order to hold together. They contain precisely the right numbers of valence electrons for the structures they adopt. Adding more electrons causes them to adapt more open structures, which are more compactly *clustered*, than those of electron precise systems. In an electron deficient compound, atoms group together (typically by forming triangular arrangements) in a way that allows one pair of electrons to be shared between more than two atoms, typically in 3-center electron pair bonds, so as to compensate for the shortage of electrons. This leads to a phenomenon that may be referred to as "hyper-coordination" [2], in which an atom is surrounded by (and in direct bonding contact with) a number of atoms that exceeds the number of its own valence shell orbitals. For example, the carbon atom in the methanonium ion, CH_5^+ (through which protolysis of methane occurs in highly acidic media) [6] is hypercoordinate in that it is bonded directly to five hydrogen atoms but uses only four valence shell atomic orbitals. Its carbon atom can be regarded as sp^3 hybridized, using three hybrid orbitals to form normal 2c2e CH bonds and the fourth hybrid orbital to form a 3-center 2 electron (3c2e) CH_2 bond to the remaining two hydrogen atoms [7]:

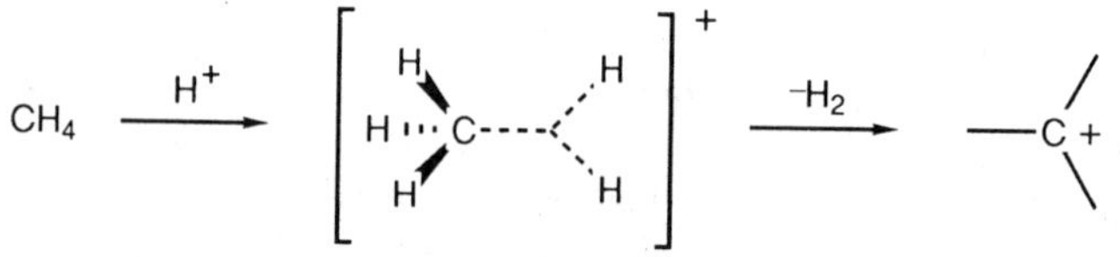

The carbon atom of CH_5^+ is thus tetravalent but 5-coordinate, like the boron atom of the isoelectronic neutral transient intermediate BH_5 through which protolysis of borohydride anion BH_4^- proceeds [8]:

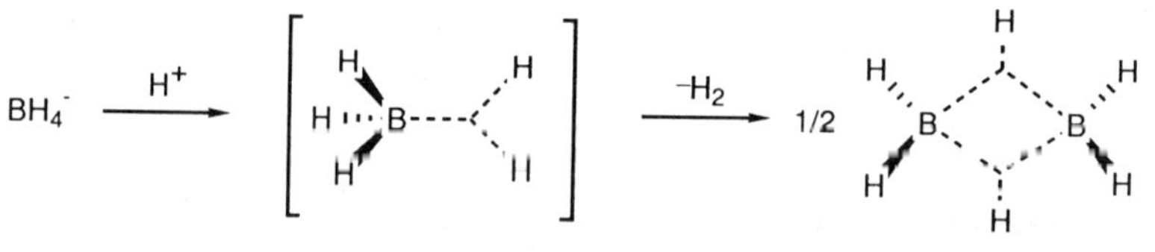

The product of this last reaction, diborane (B_2H_6) is the archetypal electron deficient molecule whose existence first alerted chemists to the fact that molecules could exist with fewer bond pairs than 2-center bonding contacts. With eight atoms and only six bond pairs, it was evident from its formula alone, long before its actual structure was known, that its bonding could not be described solely in terms of 2c2e bonds, since a minimum of seven such bonds would have been needed (n atoms need at least n-1 2-center links to hold together).

It was to explain the bridged structure of diborane [9–11] and in particular its hypercoordinated bridging hydrogen atoms, that Longuet-Higgins [12] invoked the idea of 3c2e B H B bonds which, together with 3c2e **BBB** bonds, were then used extensively by Lipscomb [13, 14] to rationalize the structures of the series of higher boranes whose intricate and unprecedented structural chemistry he himself established by meticulous X-ray crystallographic work. Examples of the way that 2c2e **BB** and **BH** bonds together with 3c2e **BHB** and **BBB** bonds can be used very effectively to rationalize the bond networks in some representative boranes are shown in Figure 1.1 (further discussion of such localized bond schemes is to be found in Chapter 3).

In working out what networks of 2- and 3-center bonds might be used to

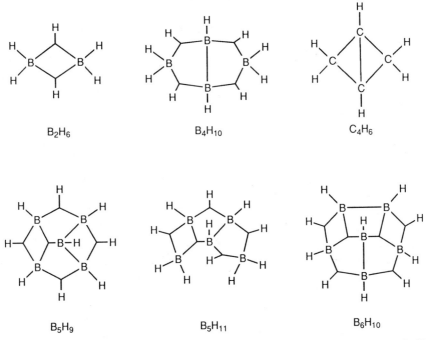

Figure 1.1. Two- and three-center bond networks of the boranes B_2H_6, B_4H_{10}, B_5H_9, B_5H_{11}, and B_6H_{10}; bicyclobutane C_4H_6 (isoelectronic with B_4H_{10}) is also shown.

rationalize known borane structures or predict those that were unknown, Lipscomb simplified the problem by excluding from his discussion the electron pair and hydrogen atom involved in the outward-pointing (*exo*) BH bond each boron atom formed. He therefore represented borane structures as having formulae B_pH_{p+q}, that is, as clusters of p BH units each contributing three atomic orbitals, and held together by the electrons they contributed (two apiece) supplemented by the q electrons from the *endo* hydrogen atoms which, whether bridging or terminal, lay on the same pseudo-spherical surface as the boron atoms. He thus effectively introduced the concept of *skeletal* or *framework electrons*, the numbers and distribution of which have subsequently proved of enormous interest to cluster chemists [1–4, 15–22]. Lipscomb himself assigned the $2p + q$ skeletal electrons of boranes B_pH_{p+q} to s BHB, t BBB, y BB and x BH bonds. Subsequent workers have focussed on the relationship between the way the three-dimensional shapes of cluster molecules reflect the total numbers of skeletal electrons they contain. The skeletal boron atoms of boranes, indeed the skeletal atoms of borane-type cluster compounds in general, define most or all of the vertices of an n-vertex deltahedron where the number of skeletal electron pairs is $(n + 1)$. The familiar three-dimensional shapes of typical borane-type clusters [23, 24] are shown in Figure 1.2, which shows how they reflect the numbers of skeletal electron pairs, progressively opening up from *closo* (p atoms, $p + 1$ pairs) to *nido* (p atoms, $p + 2$ pairs) to *arachno* (p atoms, $p + 3$ pairs) as the excess of skeletal bond pairs over skeletal atoms increases from one to two to three.

From Figure 1.2 one can see that, as the cluster shapes open up from *closo* to *nido* to *arachno*, the numbers of links between skeletal atoms that have to be accounted for by the available skeletal bond pairs fall dramatically (see also Table 1.1). The parent *closo* n-vertex polyhedra all have more edges ($3n$-6), that

TABLE 1.1. Numbers of Skeletal Atom Contacts (Polyhedron Edges) in *Closo-*, *Nido-*, and *Arachno*-Borane-Type Clusters.

Parent Polyhedron	Number of skeletal electron pairs	Number of skeletal atom contacts		
		Closo	*Nido*	*Arachno*
Trigonal bipyramid	6	9	5,6	3
Octahedron	7	12	8	4,5
Pentagonal bipyramid	8	15	10,11	5–7
D_{2d} dodecahedron	9	18	13–14	8–11
Tricapped trigonal prism	10	21	16,17	11–13
Bicapped square antiprism	11	24	19,20	14–16
C_{2v} octadecahedron	12	27	19–21	16–19
Icosahedron	13	30	25	20,21

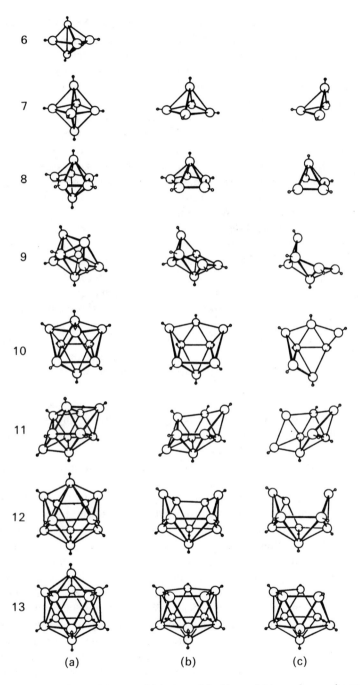

Figure 1.2. The deltahedral shapes of (a) *closo-* (b) *nido*, and (c) *arachno*-carboranes and related clusters; S = number of skeletal electron pairs. The endo hydrogen atoms of *arachno*-species have been omitted for clarity.

is, points of bonding contact between skeletal atoms, than skeletal bond pairs. However, among *nido* species, tetrahedral clusters held together by six skeletal bond pairs are not electron deficient, but electron precise, while the smaller *arachno* clusters include many species not formally electron deficient unless bridging ligands (such as the bridging hydrogen atoms of B_4H_{10}) are present. Tetraborane(10) is electron deficient because of those bridging hydrogen atoms. Its isoelectronic analogue, bicyclobutane, C_4H_6, (Figure 1.1) is electron-precise, containing eleven electron pairs (seven of which may formally be treated as skeletal pairs) to assign to six 2c2e CH bonds (four *exo*, two *endo*) and five 2c2e CC bonds. The family of borane-type structures thus extends into the electron precise, indeed electron-rich, arena, and it would be unhelpful arbitrarily to restrict discussion only to electron deficient systems, though the latter are our prime concern in the present volume.

So far, we have avoided defining the term "cluster", though we have pointed out that the phenomenon referred to as clustering—unusually close aggregation of a group of atoms, in which some become hypercoordinated—is a characteristic of electron deficient systems. A cluster compound may be defined as one containing or consisting of a group of atoms in which each is directly bonded to all or most of the remainder. However, this definition effectively requires one to regard the cross-polyhedral bonding in the *closo* borane anions $B_nH_n^{2-}$ and carboranes $C_2B_{n-2}H_n$ (Figure 1.2) as direct bonding, and runs into difficulties if one is dealing with some of the larger close-packed metal clusters that have been isolated recently. Perhaps a better definition is that a cluster compound is one containing a finite group of atoms held together (at least to a significant extent) by bonds directly between those atoms, even though some other atoms may be associated intimately with the cluster [25]. Generally (though not always), cluster compounds contain deltahedral, deltahedral fragment, or close-packed arrangements of their skeletal atoms.

There remains the question as to whether a cluster compound is necessarily electron deficient, an issue touched on in our comparison above of B_4H_{10} (a cluster compound) with bicyclobutane C_4H_6 (not normally regarded as a cluster compound, though organic chemists tend to use the term "cluster" less frequently than organometallic or inorganic chemists anyway). It seems appropriate to regard tetranuclear systems containing tetrahedral arrangements of their skeletal atoms as cluster compounds, and indeed, triangular trinuclear systems are generally regarded as cluster compounds—their skeletal atoms are arranged as compactly as possible. However, by no means all tetrahedral tetranuclear systems are electron deficient. With four skeletal electron pairs, the boron subhalide B_4Cl_4 is electron deficient in its B_4 core, even though the chlorine ligands are electron rich, whereas tetrahedrane, C_4H_4 and tetraphosphorus P_4, are not electron deficient since each contains six skeletal electron pairs. Likewise, among triangular trinuclear clusters, some (like $B_3H_8^-$) are electron deficient, whereas others (like cyclopropane C_3H_6) are not. Among higher nuclearity clusters, octahedral clusters of the borane type (i.e., with seven skeletal pairs), such as $B_6H_6^{2-}$ or the related carborane, $C_2B_4H_6$, are electron deficient systems, as indeed is the niobium halide system $Nb_6Cl_{12}^{2+}$ (whose octahedral Nb_6 core is held together by eight 3c2e Nb_3 face bonds—the doubly bridging chlorine ligands span the edges of the octahedron). By contrast, the molybdenum halide cluster $Mo_6Cl_8^{4+}$ is electron precise; it contains enough electrons to allocate a pair to each of its twelve Mo–Mo edge 2c2e bonds (its eight μ_3 chlorine ligands bridge the faces of the Mo_6 octahedron) [26, 27]. Cluster compounds are thus not necessarily electron deficient, and electron deficient compounds are not necessarily cluster compounds.

In focussing our attention on whether the skeleton of a particular molecule may be electron deficient, we have already met examples (B_4Cl_4, $Nb_6Cl_{12}^{2+}$) in which an electron deficient skeleton bears electron rich substituents which do not necessarily release electron density into the core. One fascinating aspect of cluster chemistry is the manner in which the polyhedral electron deficient skeletons of many borane-type cluster molecules can survive intact in an electron rich environment. The high connectivity of the skeletal atoms compensates for the relatively low electron density in the skeletal bonds.

A further interesting aspect of borane-type clusters is the extent to which their BH units are capable of replacement by other units that can contribute two electrons and three orbitals for cluster bonding. The units that replace the BH units of the parent species need not be isoelectronic, but can be isolobal [28] with the former; the orbitals they provide must have similar energies, extensions in space, and lobal characteristics to those of a BH unit, but do not have to be identical with them (Figure 1.3 shows how the $p_zd_{z^2}$, p_xd_{xz}, and p_yd_{yz} hybrid

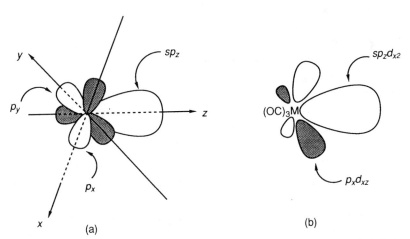

Figure 1.3. The orbitals that (a) BH or CH units and (b) $Fe(CO)_3$ or $Co(\eta^5\text{-}C_5H_5)$ units can use for skeletal bonding in clusters. (a) The radially oriented sp hybrid and tangentially oriented p AOs of a BH or CH unit. (b) The radially oriented d_{z^2} and tangentially oriented pd hybrid AOs of an $Fe(\eta^5\text{-}C_5H_5)$ unit (the tangential p_yd_{yz} orbital is not shown).

orbitals of a pyramidal $Fe(CO)_3$ unit have lobal characteristics that resemble those of a BH unit). It was the demonstration by Hawthorne and his co-workers [29–31] that various metal units can be incorporated in polyhedral borane clusters which led to the explosive growth of mixed metal-boron cluster chemistry, the development of the isolobal concept [28] and its subsequent use in many other areas of chemistry. A key contributor in these developments was Hoffmann [28, 32] who earlier had collaborated with Lipscomb in unravelling the intricacies of borane cluster bonding [33]. The conceptual wheel thus turned full circle, and indeed continues to revolve. Principles developed for 3c2e (or multicenter) bonding in boranes were found equally applicable carboranes, metalloboranes, and more recently to hypercarbon chemistry (including that of hydrocarbons) [2].

REFERENCES

1. K. Wade, *Electron Deficient Compounds*, Nelson, London, 1971.
2. G. A. Olah, G. K. S. Prakash, R. E. Williams, L. D. Field and Wade, K. *Hypercarbon Chemistry*, Wiley, New York, 1987.
3. D. M. P. Mingos, *Nature (London) Phys. Sci.*, **236**, 99 (1972).
4. D. M. P. Mingos, *Chem. Soc. Rev.*, **15**, 31 (1986).
5. G. N. Lewis, *J. Am. Chem. Soc.*, **38**, 762 (1916).
6. G. A. Olah, G. Klopman and Schlosberg, R. H., *J. Am. Chem. Soc.*, **91**, 3261 (1969).
7. W. A. Latham, W. J. Hehre and Pople, J. A., *Tetrahedron Lett.*, **91**, 2699 (1970).
8. G. A. Olah, P. W. Westerman, Y. K. Mo and Klopman, G., *J. Am. Chem. Soc.*, **94**, 7859 (1972).
9. W. Dilthey, *Z. Angew. Chem.*, **34**, 596 (1921).
10. R. P. Bell and Longuet-Higgins, H. C., *Proc. R. Soc.*, **183**, 357 (1945).
11. W. C. Price, *J. Chem. Phys.*, **16**, 894 (1948).
12. H. C. Longuet-Higgins, *J. Chem. Phys.*, **46**, 275 (1949).
13. W. N. Lipscomb, *Adv. Inorg. Chem. Radiochem.*, **1**, 117 (1950).
14. W. N. Lipscomb, *Boron Hydrides*, Benjamin, New York, 1963.
15. K. Wade, *Chem. Commun.*, 791 (1971).
16. K. Wade, *Adv. Inorg. Chem. Radiochem.*, **18**, 1 (1976).
17. B. F. G. Johnson (Ed.), *Transition Metal Clusters*, Wiley, Chichester, 1982.
18. R. N. Grimes (Ed.), *Metal Interactions with Boron Clusters*, Plenum, New York, 1982.
19. A. J. Stone, *Polyhedron*, **3**, 1299 (1984).
20. M. McPartlin and Mingos, D. M. P., *Polyhedron*, **3**, 1321 (1984).
21. M. Moskowits (Ed.), *Metal Clusters*, Wiley, New York, 1986.
22. W. N. Lipscomb, *Inorg. Chem.*, **18**, 2328 (1979).
23. R. E. Williams, *Inorg. Chem.*, **10**, 210 (1971).
24. R. E. Williams, *Adv. Inorg. Chem. Radiochem.*, **18**, 67 (1976).
25. F. A. Cotton, *Q. Rev. Chem. Soc.*, **20**, 389 (1966).

26. F. A. Cotton and Haas, T. E., *Inorg. Chem.*, **3**, 10 (1964).

27. S. F. A. Kettle, *Theoret. Chim. Acta*, **3**, 211 (1965).

28. M. Elian, M. M.-L. Chen, D. M. P. Mingos and Hoffmann, R., *Inorg. Chem.*, **15**, 1148 (1976).

29. M. F. Hawthorne, D. C. Young and Wegner, P. A., *J. Am. Chem. Soc.*, **87**, 1818 (1965).

30. M. F. Hawthorne, D. C. Young, T. D. Andrews, D. V. Howe, R. L. Pilling, A. D. Pitts, M. Reintjes, L. F. Warren Jr. and Wegner, P. A., *J. Am. Chem. Soc.*, **90**, 879 (1968).

31. G. B. Dunks and M. F. Hawthorne, in: *Boron Hydride Chemistry* (E. L. Muetterties, Ed.), Academic Press, New York, 1975, pp. 383–430.

32. R. Hoffmann, *Angew. Chem. Int. Ed.*, **21**, 711 (1982).

33. R. Hoffman and Lipscomb, W. N., *J. Chem. Phys.*, **36**, 2179, 3489 (1962); **37**, 2872 (1962).

CHAPTER 2

Geometrical Systematics of *Nido*-Carboranes, -Polyboranes and -Carbocations; Dominance of Aperture Dependent Electron Distribution and Charge Smoothing

R. E. WILLIAMS

Electron Deficient Boron and Carbon Clusters, Edited by George A. Olah, Kenneth Wade, and Robert E. Williams.
ISBN 0-471-52795-5 © 1991 John Wiley & Sons, Inc.

2.1. INTRODUCTION

Electron deficient, electron precise, and electron rich compounds are usually assumed to be mutually exclusive categories and it is presumed that a given compound will fall within only one of these three categories.

For our purposes, an electron rich compound has at least one lone pair of electrons on a skeletal atom (a 1c2e lone pair) functioning as a "terminal group". Compounds such as carbenes, amines, and alcohols, wherein lone pairs are associated with the carbons, nitrogens or oxygens, are examples of electron rich compounds.

Electron precise compounds are those that have just enough valence electrons to form two center two electron bonds (2c2e bonds) and all skeletal atoms are coordinatively saturated. Compounds such as hydrocarbons, olefins, alkynes, etcetera, are typical of this category.

Electron deficient molecules contain too few electrons to coordinatively saturate all skeletal atoms involving solely 2c2e bonds. Electron deficient compounds are of two kinds; the first type incorporates solely 2c2e bonds and simply leaves certain atoms coordinatively unsaturated. Carbenium cations, R_3C^+ and trialkl- or trihaloboranes (R_3B, X_3B) are examples of electron deficient coordinatively unsaturated compounds; the tricoordinate central carbons and borons have access to less than eight valence electrons.

The second type of electron deficient compound involves multicenter bonding, that is, 3c2e bonds or higher $\geqslant$4c2e bonds. Multicenter bonding allows skeletal atoms access to eight valence electrons; such electron deficient compounds are coordinatively saturated. Compounds of this second type include diborane, B_2H_6 which involves two 3c2e bonds, and trimethylaluminum dimer, Me_6Al_2 (wherein the aluminums and methyls are associated by two 3c2e bonds).

In light of three more or less mutually exclusive definitions, consider the *arachno*-compound, RSB_2H_5 [1], which incorporates features of all three categories, that is, electron rich, electron precise, and electron deficient (see Figure 2.1).

In the preferred structure of RSB_2H_5, a lone pair of electrons (1c2e) is located on sulfur; that portion of the molecule may be considered as electron rich. The sulfur is associated with the two borons via two electron precise 2c2e bonds, an

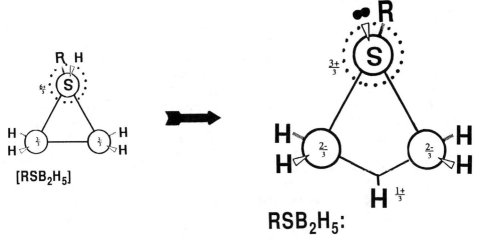

Figure 2.1. Charge smoothing selects structure of RSB_2H_5.

electron precise region. Opposite the sulfur atom, there is an electron deficient moiety; that is, a bridge hydrogen which is bonded to two borons via one 3c2e bond. In the case of RSB_2H_5, all three types of architectural features are incorporated in close proximity.

The hypothetical removal of the proton from the bridged position and relocating it upon the lone pair of the sulfur (see Figure 2.1) would create a completely electron precise molecule (all 2c2e bonds). Both the electron deficient and electron rich regions would be eliminated. In carrying out this hypothetical proton transfer, however, large charge separations (dipoles), would result. The sulfur would become strongly positively charged while the two borons would become negatively charged.

"Charge smoothing" is defined as the driving force that minimizes such dipoles and causes the partially electron deficient, partially electron rich structure to be preferred in the case of RSB_2H_5. The charge smoothing influence is apparently stronger than any driving force that might favor electron precise 2c2e bonds at the expense of 3c2e bonds or lone pairs (1c2e). *It is suggested that charge smoothing is an extremely important structural influence in fluxional electron deficient compounds.*

Carboranes, polyboranes and carbocations may be subdivided into the *closo*, *nido*, *arachno*, and *hypho*-catagories. Examples of *nido*-compounds and how they are structurally derived from *closo*-configurations will be considered here, emphasizing those with even numbers of skeletal atoms. Comprehensive discussions of *arachno-* and *hypho-* categories will be given elsewhere, but the general concepts will become apparent in this chapter.

2.2. BACKGROUND

2.2.1. Early Studies of Polyboranes

Prior to the investigations of Alfred Stock [2] in the 1920s and early 1930s, all of the speculative structures of the hydrides of boron were incorrect. Stock identified tetraborane-10, B_4H_{10}, pentaborane-9, B_5H_9, pentaborane-11, B_5H_{11} (see Chapters 4 and 6), hexaborane-10, B_6H_{10}, probably hexaborane-12, B_6H_{12} and decaborane-14, $B_{10}H_{14}$, although he could only speculate about their structures. Stock noted that these compounds subscribed to two series of empirical formulae, B_nH_{n+4} and B_nH_{n+6} which he labeled the hydrogen-rich and hydrogen-poor boron hydrides.

The early 1950s were watershed years in polyborane chemistry in that Lipscomb and co-workers [3] and Kasper, Lucht and Harker [4] and co-workers began elucidating the structures of the higher polyboranes (see Figure 2.2). In contrast to related hydrocarbons, these new three-dimensional structures were difficult to depict in only two dimensions.

In the 1950s and early 1960s, the concept was enshrined that the icosahedron was the "sacred template" of polyboranes and that all were constructed of networks of boron atoms disposed about the vertices of a *regular* icosahedron, except pentaborane-9 which was noted to be a fragment of a *regular* octahedron.

Lipscomb [5] and Longuet-Higgins [6] initiated theoretical studies to determine if polyboranes with completed icosahedral and octahedral structures would be capable of existence and concluded that the dianions, $B_6H_6^{2-}$ and $B_{12}H_{12}^{2-}$, should be stable (subsequently confirmed); see Figure 2.3.

The relatively electron rich 4-connected vertices (4k vertices) of the octa-

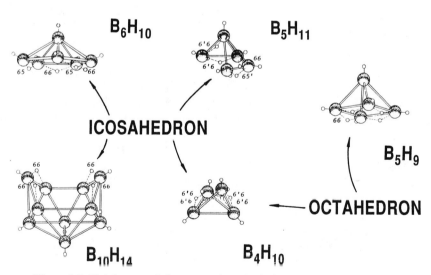

Figure 2.2. Polyborane skeletons are icosahedral (octahedral) fragments.

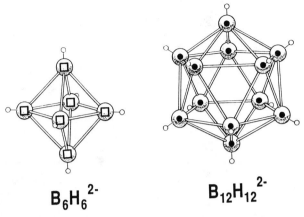

$$B_6H_6^{2-} \qquad\qquad B_{12}H_{12}^{2-}$$

Figure 2.3. Structures of $B_{12}H_{12}^{2-}$ and $B_6H_6^{2-}$.

hedron are identified by open squares while the less electron rich 5-connected vertices (*5k*) in the icosahedron are identified by small solid circles (more below). In Figures 2.2 to 2.60, the "sticks" in all ball and stick molecular illustrations signify connections only between atoms (or vertices) not electron pairs or bonds (2c2e or 3c2e, etc.).

2.2.2. Early Studies of Carboranes

Concurrently with the calculations of Lipscomb [5] and Longuet-Higgins [6], Williams et al. concluded (based upon ^{11}B-NMR spectra) that they had isolated two new octahedral compounds of empirical formula $C_2B_4H_6$ [7]; see Figure 2.4.

One isomer each of two other new compounds were also discovered, that is, $C_2B_3H_5$ [7, 8] and $C_2B_5H_7$ [9, 10] (also displayed in Figure 2.4). It was noted that the carbons were located in the low coordinate vertices of 1,5-$C_2B_3H_5$ and 2,3- or 2,4-$C_2B_5H_7$ although higher coordinate vertices were available in both cases. Of the two isomers of $C_2B_4H_6$, the 1,6-isomer seemed to be more stable (less reactive) than the 1,2-isomer. It was assumed [7] (confirmed by Onak et al. [10]) that the carbons-adjacent isomer, 1,2-$C_2B_4H_6$, was produced initially (from the antecedents B_5H_9 and C_2H_2) and that the 1,2-$C_2B_4H_6$ isomer subsequently rearranged into the more stable 1,6-$C_2B_4H_6$ isomer.

It followed that if the most stable isomers had (1) *nonadjacent carbons* in (2) *lowest coordinated sites* then the lone isomer of $C_2B_5H_7$ should be the 2,4-$C_2B_5H_7$ isomer (confirmed by Beaudet et al. [11, 12]). Both carbon-location trends in 2,4-$C_2B_5H_7$ are driven by charge smoothing forces, that is, the carbons with one more positively charged proton per nucleus (than the boron nuclei) "sought out" and occupied the lowest coordinated electron rich vertex sites. Over 15 years later, Schaeffer et al. discovered the less stable carbons-adjacent

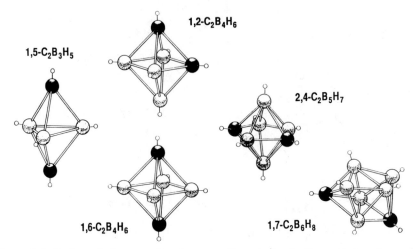

Figure 2.4. Carbons in low coordination nonadjacent sites in the earliest carboranes.

isomer, $2,3$-$C_2B_5H_7$ [13] and Sneddon et al. recently demonstrated that $2,3$-$C_2B_5H_7$ rearranges into $2,4$-$C_2B_5H_7$ upon heating [14].

The globular configurations of the first small *closo*-carboranes foreshadowed the generality that the preferred polyhedral shapes for all $C_2B_nH_{n+2}$ *closo*-carboranes would be the *most spherical* alternatives.

The next larger *closo*-carborane, $C_2B_6H_8$, was considered and the *most spherical* 8-vertex polyhedral structure with *nonadjacent carbons in low coordination sites* was anticipated. Several years later, $1,7$-$C_2B_6H_8$ was synthesized and found to have the expected structure [15] (see Figure 2.4).

Several groups independently discovered the icosahedral *closo*-carboranes $1,2$- [16–19], $1,7$- [20, 21], and $1,12$-$C_2B_{10}H_{12}$ [22] while Hawthorne and co-workers discovered the intermediately sized *closo*-carboranes, $C_2B_9H_{11}$ [23–25], $1,6$- and $1,10$-$C_2B_8H_{10}$, and $C_2B_7H_9$ [26] (see Figure 2.5).

Throughout this manuscript, we will be discussing various deltahedra. Deltahedra, by definition are polyhedra that have triangular (III-gonal) facets exclusively but may have vertices of various connectivities, k. For example (see Figure 2.3), an octahedron has six 4-connected, $4k$, vertices, an icosahedron, twelve $5k$ vertices, and so on (terminal connectivities are not counted); both the octahedron and icosahedron are deltahedra. *Nido*-deltahedral fragments have one or more square (IV-gonal), pentagonal (V-gonal) or hexagonal (VI-gonal) open faces produced by the removal of $4k$, $5k$, and $6k$ vertices, respectively.

It is a property of the geometry of the *closo*-$B_nH_n^{2-}$ deltahedra that the most spherical deltahedra have the least varied or most uniformly connected vertices which in turn leads to the least varied and most uniform negative charge distribution, again attributable to charge smoothing forces. The least connected vertices in $B_nH_n^{2-}$ deltahedra have the highest electron density and negative charge (details below), thereby hypothetically "attracting" the positively charged

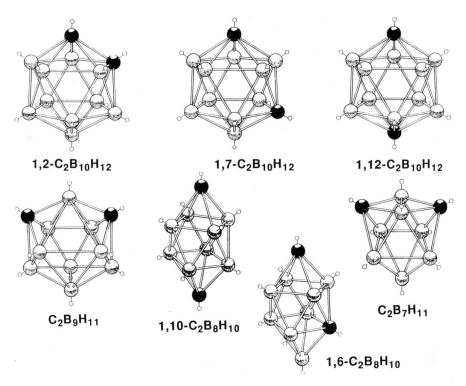

1,2-$C_2B_{10}H_{12}$ 1,7-$C_2B_{10}H_{12}$ 1,12-$C_2B_{10}H_{12}$

$C_2B_9H_{11}$ 1,10-$C_2B_8H_{10}$ $C_2B_7H_{11}$

1,6-$C_2B_8H_{10}$

Figure 2.5. Remaining $C_2B_nH_{n+2}$ *closo*-carboranes.

(more electronegative) carbons when two CH vertices are substituted for two BH^- vertices to produce the *closo*-$C_2B_nH_{n+2}$ compound (carbon has one more nuclear proton than boron); again charge smoothing is the driving force.

Neighboring similarly charged vertices repel each other so that charge smoothing tends to favor carborane isomers (see Figures 2.4 and 2.5) wherein the carbons are separated, especially if nonadjacent sites of the same or lower connectivity, k, are available.

2.2.3. Fluxionality Among the Polyboranes and Carboranes

One of the unique features of polyborane chemistry (compared to hydrocarbon chemistry) is that, for a given empirical formula, there is, almost without exception, one and only one structure. As an example, in the center of Figure 2.6 is the only structure that has ever been observed for B_6H_{10}. It seems that for all known simple polyboranes, there is only one stable structure [27] with the possible exception of *arachno*-B_9H_{15}.

In contrast, there are many alternative structures for the analogous neutral hydrocarbon of the empirical formula C_6H_{10}; a large number of alkenes, alkynes, and cyclic compounds come to mind. Why would the empirical formula

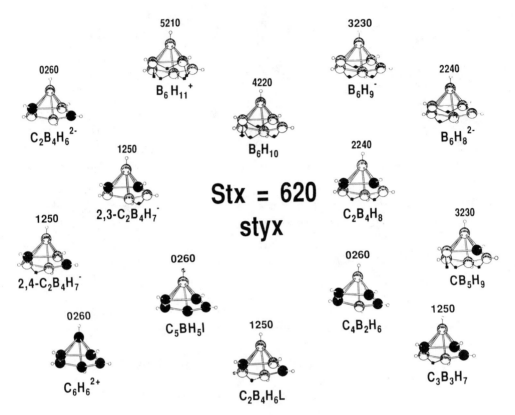

Figure 2.6. The polyborane, carborane, carbocation continuum exemplified by compounds isoelectronic and isostructural with *nido*-B_6H_{10}.

B_6H_{10} yield only one structure where its "carbon cousin" C_6H_{10} exists in many configurations?

The one structure per empirical formula phenomenon is a consequence of the incorporation of one or more multicenter 3c2e bonds in the polyboranes which tend to allow fluxionality. Many polyboranes are fluxional at room temperature or below, others are fluxional in solution and most are fluxional at elevated temperatures. Spontaneous hydrogen exchange is common (tautomerism [27]) and exchange of location between skeletal borons is much more facile within polyboranes than is exchange of carbons within neutral hydrocarbons.

The borons in alkylboranes tend to "swing along" carbon chains (brachiation) via intermediates involving 3c2e bonds in the hydroboration rearrangement [28]. Given the inherently fluxional nature of polyboranes, the thermodynamically most stable conformer is almost always accessible from any less stable analog; charge smoothing is the dominant process in selecting that "one" preferred structure for each empirical formula.

A CH unit is isoelectronic with both a BH^- group and a neutral BH–H– unit (the –H– represents a bridge hydrogen). When substitution of carbon for boron takes place, the number of multicenter bonds decreases and numerous isomers involving both carbon and boron are observed. Many of these are displayed in Figure 2.6, but even when just two 3c2e bonds remain, they all reflect the fundamental pentagonal pyramid structure characteristic of B_6H_{10} [29, 30] even when there are no boron atoms, for example, $C_6Me_6^{2+}$ [31].

2.2.4. From Lipscomb-styx to "Chop-Stx"

Above the various molecules in Figure 2.6 are six sets of identifying numbers that vary between 0260 and 5210. These numbers are the Lipscomb-styx [32] numbers which identify the numbers of various kinds of skeletal architectural features within the various compounds. For example, the 4220 printed above the molecule B_6H_{10} in Figure 2.6 reveals that the molecule incorporates four BHB 3c2e bonds (four bridge hydrogens), two BBB 3c2e bonds, two BB 2c2e bonds and zero endo hydrogens (no BH_2 groups); exoskeletal terminal groups are ignored. In contrast, the carbocation $C_6H_6^{2+}$ (styx = 0260) has zero bridging hydrogens, two CCC 3c2e bonds, six CC 2c2e bonds and zero CH_2 groups. The structures of $C_6H_6^{2+}$ and B_6H_{10} are very similar, but the styx numbers (0260 and 4220 differ greatly) do not reflect their kinship and are difficult to commit to memory.

Fortunately the styx system can be simplified easily. The second and fourth styx numbers are *always* 2 and 0 (see Figure 2.7). The first and third styx

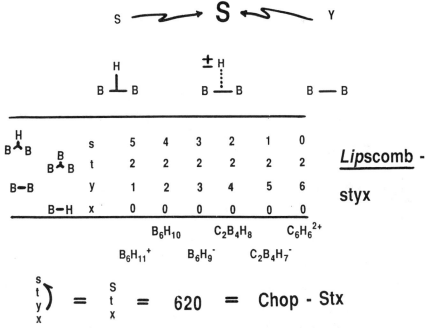

Figure 2.7. Conversion of Lipscomb-styx to Chop-Stx.

numbers (representing **BHB** and **BB** architectural features) *always* equal 6 when added together. If we hypothetically ignore whether a bridge hydrogen is "perched" on a BB bond or not, then the first (BHB) and third (BB) numbers, the s and y of styx, may be added together and identified as S in Stx [33, 34]. The Stx number then becomes 620 which we name the "Chop-Stx" number as we have "chopped off" the bridge hydrogens. The single Stx number, 620, characterizes every compound related to B_6H_{10} in Figure 2.6.

In Figure 2.7, the mechanics of changing all "Lip-styx" numbers to Chop-Stx numbers are summarized. In this reduction of the six styx numbers (0260 to 5210) to the single Stx number (620), there is one bit of information that is seemingly "lost" and that is how many of the "BB" bonds (the S in Chop-Stx) have hydrogens perched on them (the BHB bridge hydrogens or the s in Lip-styx).

Actually, no information is lost in converting from Lip-styx to Chop-Stx. Any styx or Stx number is absolutely meaningless unless associated with an empirical formula and the Stx number together with the empirical formula, automatically reveals the number of bridge hydrogens. For example, the Stx number of B_6H_{10}, that is, 620, denotes six **BB** bonds ($\pm$ bridge hydrogens), two **BBB** 3c2e bonds and zero **BH$_2$** groups. The empirical formula, B_6H_{10} "says" that there are "four more" hydrogens than borons while the 0 in 620 "says" none (of the "four more" hydrogens) are utilized to produce BH_2 groups; it follows that all "four more" hydrogens are bridge hydrogens, that is, the 6 in 620 must reflect four BHB bonds and two BB bonds.

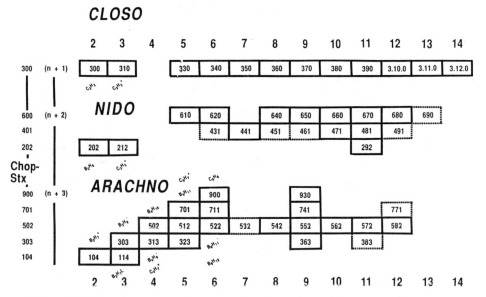

Figure 2.8. Chop-Stx systematics categorize the entire polyborane, carborane, carbocation continuum [33, 34].

In the case of $C_2B_4H_8$ [35, 36] (where Stx is also 620), there are six BB (or BC or CC) bonds ($\pm$ bridge hydrogens), two BBB (or BBC or BCC) bonds and zero BH_2 (or CH_2) groups. Since the eight hydrogens in the empirical formula exceed the skeletal carbons plus borons by two, and there are no BH_2 (or CH_2) groups, the two excess hydrogens *must* be bridge hydrogens (BHB, BHC or CHC). The easily confused Lip-styx numbers (4220 versus 2240 in the cases of B_6H_{10} and $C_2B_4H_8$) can thus be converted to the easily remembered Chop-Stx number (620 in both cases), with absolutely no loss of information (see Figure 2.6).

The usefulness of converting from the styx nomenclature to the Stx convention is illustrated in Figure 2.8.

Figure 2.8 cross-correlates the Chop-Stx numbers for all *closo-*, and *nido-* and *arachno-*families [37–39] of compounds; the same may be done for the *hypho-* and *klado-* [40] families as well. The sums of the Stx numbers are equal to the Wade electron count numbers [41, 42]. In the middle of Figure 2.8 is the number 620 which represents not only B_6H_{10}, but *all* of its isoelectronic analogs as displayed in Figure 2.6. In this chapter, we will focus our attention primarily on the 620, 640, 660, and 680 families of *nido-*compounds and will illustrate how they may be derived from the 300 series of *closo-*compounds.

2.3. EMPIRICAL FORMULAE, DELTAHEDRA-DELTRAHEDRAL FRAGMENT SYSTEMATICS

2.3.1. Geometrical Relationships

In 1971, Williams [38] published the geometrical systematics illustrated in Figure 2.9. Prior to 1970, the various polyboranes and carboranes had been identified as belonging to two classes [37], that is, *closo-* (a corruption of *clovo-* Gk. cage) and *nido-* (Gk. nest) compounds because of their closed and more open configurations.

The compounds labeled as *nido-*species [37] before 1970 were actually the two kinds identified by Stock [2] prior to 1933. Therefore, in 1971, the "two kinds of *nido-*compounds" were subdivided [38] into the *nido-* (nest) and *arachno-* (Gk. web) classes of compounds. Stock's "hydrogen poor" compounds, B_nH_{n+4}, inherited the label *nido-*compounds, open nest-like, while his "hydrogen rich" series, B_nH_{n+6} become labeled *arachno-*compounds [38] as they were more open or web like. Later, an even more "hydrogen rich" family of compounds, $[B_nH_{n+8}]$, were discovered by Shore which became labeled *hypho-* compounds (Gk. net) [39].

It was noted [38] (see Figure 2.9) that (1) the skeletons of *nido-*B_5H_9 and *nido-* B_6H_{10} could be derived by the removal of one highest coordinated vertex from the "one-vertex-larger" *closo-*octahedron and *closo-*pentagonal bipyramid respectively, (2) in the same fashion, the 10-vertex *nido-*$B_{10}H_{14}$ could be derived from the most spherical *closo-*11-vertex deltahedron, and (3) the 11-vertex *nido-* skeletons could be obtained from the *closo-*12-vertex-icosahedron. If all of these known 5-, 6-, 10-, and 11-vertex *nido-*structures could be derived by the removal

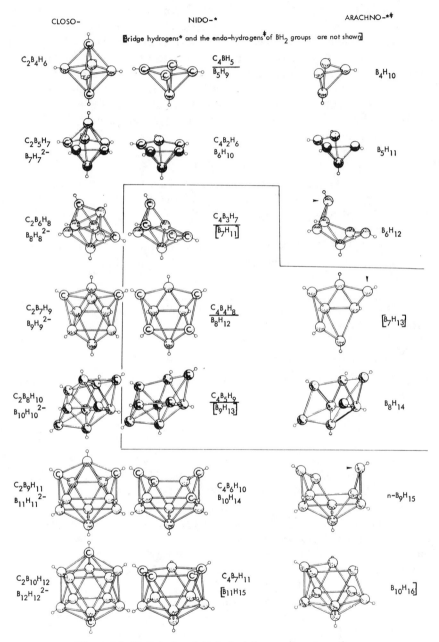

Figure 2.9. Deltahedra-deltahedral fragment systematics.

of one highest coordinated vertex from the most spherical *closo*-deltahedra, it was reasoned that probably (4) the "missing" 7-, 8-, and 9-vertex *nido*-configurations could be similarly predicted [38].

It was also noted that the known 4-, 5-, 6-, 9-, and 10-vertex *arachno*-species could all be derived by the removal of one additional high coordination edge-vertex from the "one-vertex-larger" *nido*-configurations; it was concluded that the missing 7- and 8-vertex *arachno*-structures could probably be generated in like fashion [38].

This "island" of missing configurations (the boxed area in Figure 2.9) consisted of three *nido*-compounds with 7-, 8-, and 9-vertices and two *arachno*-compounds with 7- and 8-vertices.

Eighteen years later, examples of all three predicted configurations for the *nido*-structures of Figure 2.9 and one of the two *arachno*-structures have been confirmed.

Following Williams' publication [38] (Figure 2.9), Wade [41, 42] (and later Rudolph [43]) recognized that the geometry/empirical formulae relationships of Figure 2.9 could be extended beneficially by changing from simple empirical formula categories [38] to skeletal electron counting. This relationship is useful not only to relate the various classes of polyboranes, carboranes, and car-bocations as Williams had done, but moreover could also be extended to encompass most metallaorganics, transition element clusters and carbides, borides, etcetera.

2.3.2. Skeletal Electron Counting

In Figure 2.10 are listed three classification systems for polyborane, carborane, and carbocation clusters, (1) the empirical formulae categorizations pioneered by Stock [2] and elaborated by Parry [44], (2) the electron pair, bond-type representations of Lipscomb (i.e., the styx/Stx systematics [32]) and (3) the

EMPIRICAL FORMULAE			BOND TYPES	ELECTRON COUNTING
STOCK	PARRY (EDWARDS)		LIPSCOMB	WADE (RUDOLPH)
<1933	<1959		<1954	1971
		"CAPO"		n
"poor" ⌐	$B_nH_n(BH)_3$	CLOSO	styx (Stx)	n+1
B_nH_{n+4} ⌐	$B_nH_n(BH_3)_2$	NIDO	"	n+2
B_nH_{n+6} ⌐	$B_nH_n(BH_3)_3$	ARACHNO	"	n+3
"rich" ⌐	$B_nH_n(BH_3)_4$	HYPHO	"	n+4

Figure 2.10. Relationships of empirical formulae, bond types, and skeletal electron counting.

underlying skeletal electron counting of Wade [41,42] and elaborated by Mingos.

There are subtle differences between the empirical formula relationships, Lipscomb's styx/Stx formulation [32] (which, when the styx numbers are totaled, are equivalent to the number of skeletal electron pairs in each class), and Wade's totaling of these numbers [41] ($n + 1, n + 2$, etc., where n also equals the number of nonhydrogen skeletal atoms).

The empirical formula relationships and styx/Stx formalism [38] are adequate if one is restricted solely to the carborane, polyborane, carbocation continuum, but become excessively complicated when transition element and other heteroelement groupings are involved. Wade's skeletal electron pair counting, on the other hand, is almost universal and allowed the empirical formula-geometrical systematics of Figure 2.9 to be extended into many additional areas of chemistry [41, 42]. Many important theoretical extensions to the geometrical and electron counting systematics have been made by Stone et al. [45], Mingos et al. [46], Gimarc et al. [47], and King et al. [48] (see Chapters 3 and 5).

The three missing *nido*-structures of Figure 2.9 are discussed below.

2.3.3. Three "Missing" *Nido*-Structures

The predicted 9-vertex *nido*-configuration [38] was soon confirmed by Huffman and Streib [49] and is displayed in Figure 2.11 as $C,C'-Me_2C_2B_7H_9$. However, a full decade passed before the second "missing configuration" was reported and almost two decades before examples of all three were confirmed. In this chapter, terminal alkyl groups will be ignored as their identification tends to obscure the cluster patterns; the latter 9-vertex compound will hereinafter be identified as $C_2B_7H_{11}$, caveat emptor.

In 1981, Sneddon [50] published the first example of the predicted *nido*-8-vertex configuration, $S(CpCo)_2B_5H_7$, see Figure 2.11. Sneddon et al. [51] also reported the first example of a 7-vertex *nido*-compound, $C_2B_5H_8^-$, with the predicted skeletal configuration shown in Figure 2.11.

In Figure 2.9, the preferred locations for the carbon atoms were indicated [38] in the most favored isomers of many yet-to-be-discovered *nido*-carboranes based upon low coordination sites and driven by charge smoothing as extrapolated from the smallest *closo*-carboranes (see Figure 2.4).

2.3.4. Carbon Locations

In both Figures 2.11 and 2.12, are illustrated compounds wherein these carbon locations were subsequently confirmed.

Based on pentaborane-9, B_5H_9, Williams [52] predicted that $C_5H_5^+$ should have the structure shown in Figure 2.12. Later, Huffman [53] independently predicted the same $C_5H_5^+$ configuration and subsequently, Masamune [54] reported the synthesis and structure of $1,2-Me_2C_5H_3^+$, the dimethyl derivative of $C_5H_5^+$ shown in Figure 2.12.

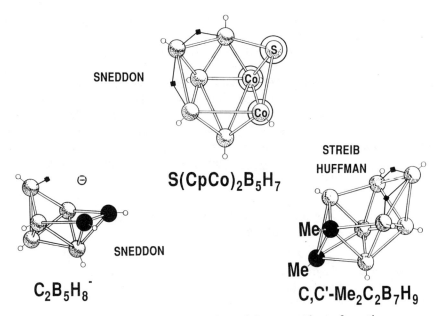

SNEDDON

$S(CpCo)_2B_5H_7$

STREIB
HUFFMAN

SNEDDON

$C_2B_5H_8^-$

Me
Me

$C,C'-Me_2C_2B_7H_9$

Figure 2.11. First confirmed 7-, 8-, and 9-vertex *nido*-configurations.

Binger [55] and Onak [56, 57] discovered derivatives of the compound $C_4B_2H_6$ while Hermanek, Plesek, Stibr and co-workers discovered alkyl derivatives of $C_4B_7H_{11}$ [58] and $C_4B_6H_{10}$ [59].

Williams [39] expanded the original systematics [38] and based upon the knowledge that bridge hydrogen's needs for electron rich, low coordinated environments are greater than carbons and that a boron-Lewis base group was isoelectronic with a CH group, it was predicted [39] (1) where the Lewis base should be located in $5\text{-}L\text{-}B_{10}H_{12}$, (2) where the carbon should be placed in its isoelectronic analog, $5\text{-}CB_9H_{13}$, and (3) what the structure of $6,9\text{-}C_2B_8H_{10}^{2-}$

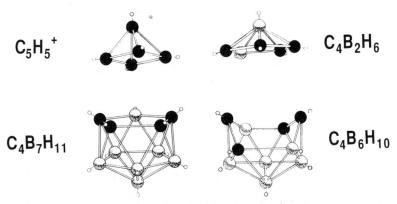

$C_5H_5^+$

$C_4B_2H_6$

$C_4B_7H_{11}$

$C_4B_6H_{10}$

Figure 2.12. Confirmed carbon-location isomers.

should be. Hermanek, Plesek and Stibr et al. later confirmed these structures [60, 61], see Figure 2.13.

Williams incorrectly suggested [38] *nido*-B_8H_{12} and its congener *nido*-$C_4B_4H_8$ should have pentagonal open faces (see Figure 2.9 and left side of Figure 2.14). It was thought that crystal packing forces might have "opened" their structures to less favorable configurations with hexagonal open faces and that they probably had structures with V-gonal open faces in solution.

The probable explanation [39] for why *nido*-B_8H_{12} has the more open structure (upper right-hand diagram in Figure 2.14) and why $C_4B_4H_8$ [62, 63] also has the more open structure led, inter alia, to the recognition of the effects discussed in this chapter.

2.3.5. *Nido*-Configurations; V-Gonal versus VI-Gonal Open Faces; Unsegregated Evaluation

As illustrated in Figure 2.9, *nido*-compounds were usually anticipated [38] to have deltahedral fragment structures with V-gonal open faces because the highest connected vertex in most precursor *closo*-deltahedra was usually a *5k* vertex. In the case of *nido*-10-vertex species, the precursor *closo*-11-vertex-deltahedron had a *6k* vertex, thus the VI-gonal open face was expected for *nido*-decaborane type compounds.

In this discussion (Section 2.3.5 and Figure 2.15), an attempt to organize all *nido*-compounds under one simple set of patterns independent of composition of matter is undertaken, that is, the compounds will not be segregated and treated as separate groups. In a subsequent discussion (Section 2.6.7 and Figure 2.61), it will be seen whether advantages accrue from treating heteroatom derivatives separately, for example, considering thiaboranes and azaboranes as groups separate and distinct from carboranes, polyboranes, and carbocations.

In Figure 2.15, the various sized *nido*-core-clusters are listed. Along the left-hand side, we have also listed the corresponding Chop-Stx numbers discussed earlier in this chapter. Plotted across the top are listed the alternative open faces.

If one follows the dotted line from top to bottom in Figure 2.15, it is seen that V-gonal open faces (identified by the large plus signs or crosses) are seemingly

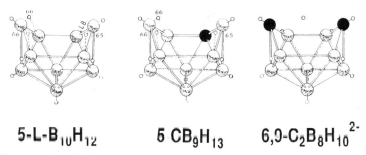

$$5\text{-}L\text{-}B_{10}H_{12} \qquad 6\ CB_9H_{13} \qquad 6,9\text{-}C_2B_8H_{10}{}^{2-}$$

Figure 2.13. Additional confirmed carbon and Lewis base location isomers.

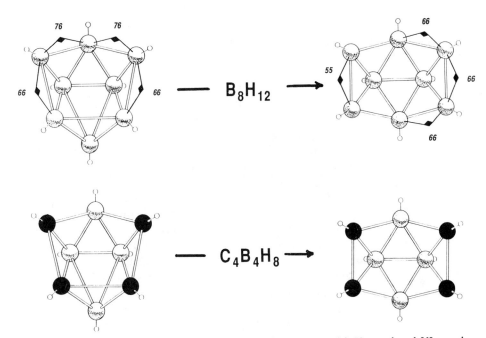

Figure 2.14. Alternative structures for B_8H_{12} and $C_4B_4H_8$ with V-gonal and VI-gonal open faces.

NIDO-CONFIGURATIONS

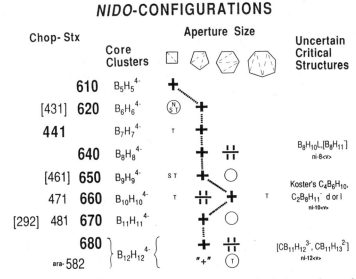

Figure 2.15. Nonsegregated cross-correlation between *nido*-skeleton size and aperture size.

preferred for most of the *nido*-core-clusters. This dotted line identifies the configurations that reflect the predicted geometrically preferred *nido*-configurations in Figure 2.9. The configuration for the *nido*-compound, B_8H_{12} (with the VI-gonal open face), was known [32] (indicated by a double cross in Figure 2.15), but was not included in Figure 2.9 as it was incorrectly [38] thought to be an aberrant structure caused by crystal packing forces. Koster [64] has possibly produced at least one *nido*-10-vertex carborane with a V-gonal open face, also identified by a double cross in Figure 2.15.

In light of recent research by Grimes et al. [65–67] three or four compounds probably isoelectronic with $[B_{12}H_{12}^{4-}]$ may be added. Anticipated structures are identified by a large cross and a large double cross in Figure 2.15. Two aberrant structures are represented by a small T within a circle and a small cross in quotes reflecting the latter's questionable status. T, N, and S in Figure 2.16 stand for transition element group, nitrogen, and sulfur, respectively.

It is seen that sufficiently large *nido*-species, that is, larger than $[B_5H_5^{4-}]$, have alternative configurations with V-gonal open faces and those larger than $[B_7H_7^{4-}]$ seemingly *could* have alternative structures with VI-gonal open faces.

Several lines of investigation led to the belief that there are simply no electronically acceptable configurations with VI-gonal open faces for either the 9-vertex or 11-vertex *nido*-carboranes or polyboranes, hence the open circles in Figure 2.15.

The fragments that are assumed to be preferred (represented by large crosses in Figure 2.15) may be derived from the one vertex larger, most spherical *closo*-deltahedra, *by the removal of the highest connected vertex* with the single exception that the *nido*-12-vertex fragment with a VI-gonal open face that would be selected in this fashion has one remaining overly connected 6*k* vertex. No known polyboranes or carboranes adhere to this configuration, thus it is identified by the encircled T in Figure 2.15 which signifies that the configuration is unknown unless a transition element group is present to occupy the 6*k* site.

If the definition of the preferred fragments is simply changed to read "the *nido*-fragments that are assumed to be preferred (represented by large crosses in Figure 2.15) may be derived from the one vertex larger, most spherical *closo*-deltahedra *by the removal of that particular vertex that results in the reduction of the largest number of "highest connected vertices*" then a different "tentatively preferred" *nido*-12-vertex fragment with a V-gonal open face is selected and the lone exception disappears. This tentatively preferred fragment for the *nido*-12-vertex configuration is represented by a large cross in Figure 2.15, while the second choice is represented by a double cross.

Small species, related to *nido*-B_5H_9 have no choice but to have IV-gonal open faces due to limited size.

All carborane, carbocation, and polyborane derivatives of *nido*-B_6H_{10} (see Figure 2.6), have V-gonal open faces; although isoelectronic derivatives with IV-gonal open faces are found when nitrogen, sulfur, and transitional element groups are incorporated in the skeleton (see the encircled NST in Figure 2.15). Even the best carborane candidate, that is, the carbons-apart C,C'-dimethyl derivative of $nido\text{-}C_2B_4H_6^{2-}$ was found by Bausch, Prakash, Onak, and Williams to assume the pyramidal structure with the V-gonal open face [68].

Only one confirmed $nido$-7-vertex species has been reported [51], the carborane anion, $nido\text{-}C_2B_5H_8^-$, in Figure 2.11 with a V-gonal open face (Stx = 441); the $nido$-7-vertex configuration is not large enough to have a VI-gonal opening.

Derivatives of $nido\text{-}[B_8H_8^{4-}]$ assume structures [69] with (1) V-gonal open faces either favored by charge smoothing, that is, $S(CpCo)_2B_5H_7$ [50] or (2) VI-gonal open faces also favored either (i) by charge smoothing, that is, $R_4C_4B_4H_4$ [62], $R_4C_4B_3H_3(CpCo)$ [63], and $C_2B_6H_{10}$ [70, 71] or (ii) by allowing bridge hydrogens to be associated with sufficiently low coordinated borons (i.e., $\leqslant 66$-bridge hydrogens), for example B_8H_{12} [32] (for a definition of the $\leqslant 66$-bridge hydrogen numbering convention [72], see Section 2.4.1).

The structures of the unknown species, $B_8H_{11}^-$ or $B_8H_{10}^{2-}$ where there are neither heteroatoms nor compromised bridge hydrogens, should reveal which geometrical configuration is actually preferred.

In the case of carborane derivatives of $nido\text{-}[B_9H_9^{4-}]$, there appears to be no alternative to the V-gonal open faced structure.

Prior to 1987, no $nido\text{-}[B_{10}H_{10}^{4-}]$ configuration with a V-gonal open face had been suggested [33, 34], but numerous $nido$-decaborane analogs with VI-gonal open faces were known. As will be discussed in Section 2.6.2, charge smoothing almost certainly causes the $nido$-10-vertex configuration with a V-gonal open face to be either (1) the favored structure or at least (2) prevalent structures in rearrangements involving Koster's $C_4B_6H_{10}$ derivative [64] and Plesek's and Stibr's optically active $C_2B_8H_{11}^-$ [73] anion derived from $5,6\text{-}C_2B_8H_{12}$. All other compounds have $nido$-10-vertex configurations with VI-gonal open faces, either favored by charge smoothing, for example, $6,9\text{-}C_2B_8H_{10}^{2-}$ [61] or because bridge hydrogens must have access to a sufficient number of low coordinate borons to produce $\leqslant 66$-bridge hydrogens, for example, $B_{10}H_{14}$, $B_{10}H_{13}^-$, 5,7- [74], and $5,6\text{-}C_2B_8H_{12}$ [75] etcetera.

The structure of $B_{10}H_{12}^{2-}$ should be structurally independent of both heteroatom charge smoothing and bridge hydrogen constraints and thus reveal the geometrical preference. Shore et al. [76] are currently studying $B_{10}H_{12}^{2-}$ in the crystal and favor the ni-10⟨VI⟩ configuration resembling those illustrated in Figure 2.13.

The 11-vertex-$nido$-configuration with a V-gonal open face is apparently the only viable alternative. Many existing compounds should seemingly favor the $nido$-11-vertex configuration with a VI-gonal open face (not shown), if a satisfactory one existed; all opt for the V-gonal open faced skeletal configuration (resembling $C_4B_7H_{11}$ in Figure 2.12) even though the bonding (see Figure 2.8) differs dramatically, that is, Stx = 670 in $B_{11}H_{13}^{2-}$, 481 in $B_{11}H_{14}^-$ and probably 292 in $B_{11}H_{15}$ (see Figure 2.8, Section 2.6.6 and Figures 2.57 and 2.58).

Listed alongside of $[B_{12}H_{12}^{4-}]$ in Figure 2.15 are (1) a large cross, (2) a large double cross, (3) an encircled T, and (4) a small cross in quotes. The encircled T signifies the currently out of favor deltahedral fragment with a VI*-gonal open face produced from the most spherical 13-vertex deltahedron by the removal of one of the two highest connected $6k$ vertices. This leaves the other $6k$ vertex (its presence indicated by the asterisk) of the 13-vertex deltahedron intact; this structure has never been observed for any polyborane or carborane *unless a transition element group is present* to occupy the remaining $6k$ site. When one transition element group is present however, the "encircled T" isomer is always the favored *nido*-12-vertex deltahedral fragment.

The second aberrant fragment is identified by the small cross in quotes neighboring $[B_{12}H_{12}^{4-}]$ in Figure 2.15. It incorporates two adjacent IV-gonal open faces [66] which in one sense is equivalent to one V-gonal open face. Two nonaberrant *nido*-12-vertex deltahedral fragments remain; the one with a V-gonal open face is identified by a large cross while the other with a VI-gonal open face is identified by a large double cross. Both are condidates for the most stable $[B_{12}H_{12}^{4-}]$ structure.

Just as the encircled T beside *nido*-$[B_{12}H_{12}^{4-}]$ identifies an aberrant configuration (unknown unless an electropositive transition element group is present), there are outriding Ts in the row of alternative structures for *nido*-10-vertex compounds in Figure 2.15. These refer to an iridium compound [77, 78] with a IV-gonal open face and a ruthenium compound [79] with a VII-gonal open face reported by the Greenwood, Kennedy group. The presence of transition element groups apparently make possible these configurations. Other Ts in Figure 2.15 indicate the existence of aberrant ni-7⟨IV⟩ [80] and ni-9⟨IV⟩ [81] structures also due to the presence of transition element groups.

The examples discussed highlight the fact that the geometric systematics apply best to polyboranes, carboranes [38, 39] and carbocations [82]; when transition element groups are incorporated, nonconforming deltahedral fragment structures are frequently produced. In fact, there are no proven *nido*-carboranes or *nido*-polyboranes with 8-, 10- or 12-vertices which have V-gonal open faces unless there are heteroelements other than carbon in the structures (see Section 2.6.7).

Although the primary objective of this chapter is to systematize *nido*-compounds, it is illustrative to compare the *nido*-compounds with their nearest neighbors, the *closo*- and *arachno*-compounds (see Figure 2.16).

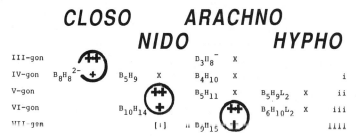

Figure 2.16. Correlation of aperture size and class, *closo*-, *nido*-, *arachno*- and *hypho*-.

2.3.6. Comparison of *Closo-*, *Nido-*, and *Arachno*-Configurations

All *closo*-species, for example, $B_nH_n^{2-}$ and $C_2B_nH_{n+2}$, have deltahedral structures (all III-gonal facets) in the crystal. *Closo*-$B_8H_8^{2-}$ in solution, however, exists in two alternative liquid phase conformers (see Figure 2.17) [83, 83b, 83c]. *Closo*-$B_{11}H_{11}^{2-}$ [84] probably also has different structures in solution as opposed to the crystal.

Arachno-compounds (see Figure 2.16) have a preference for VI-gonal and sometimes VII-gonal open faces if they have six or more vertices. The most stable isomer of *arachno*-B_9H_{15} [85] (perhaps the only isomer), has a VII-gonal open face to accommodate four bridge and two endo-hydrogens (see Figure 2.9). Conforming *arachno*-compounds have 6-, 8-, 9-, 10-, and 11-vertices while the *arachno*-compounds $B_3H_8^-$, B_4H_{10}, and B_5H_{11} are too small to have VI-gonal open faces.

In summary, it appears (Figure 2.16) that *closo*-compounds almost always prefer III-gonal faces but rarely accommodate IV-gonal open faces; *nido*-compounds usually V-gonal, but frequently, VI-gonal open faces; and *arachno*-compounds usually VI-gonal but frequently VII-gonal open faces. There are no examples of *hypho*-compounds large enough to allow us to generalize about the preferred open face size for *hypho*-compounds.

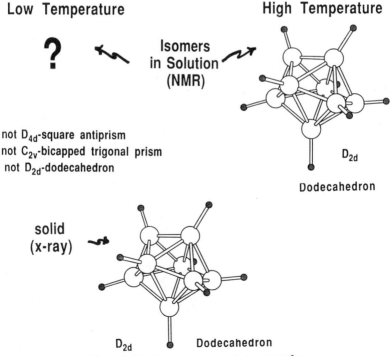

Figure 2.17. The structures of *closo*-$B_8H_8^{2-}$.

Before leaving the *closo*-systems and their built-in preference for III-gonal facets, the entire series of $B_n H_n^{2-}$ species (see Figure 2.18) should be considered.

At one extreme is $[B_5 H_5^{2-}]$. Only its isoelectronic carborane $1,5\text{-}C_2 B_3 H_5$ analog [7, 8, 38, 39] has been discovered, see Figure 2.4. At the other extreme, $[B_{13} H_{13}^{2-}]$ and $[B_{14} H_{14}^{2-}]$ have never been found. However, metallacarborane analogs of the 13- [66] and 14-vertex *closo*-species have been reported.

Two metallacarborane analogs of *closo*-$[B_{14} H_{14}^{2-}]$ have been isolated which have metastable structures with IV-gonal and V-gonal open faces [66]. Upon heating, both of these aberrant *closo*-structures rearrange into the expected most spherical *closo*-14-vertex deltahedral structures with III-gonal open faces only.

We advocate that even though the origins of the terms *closo*-, *nido*, *arachno*-, and *hypho*- evolved originally from considerations of morphology, that these same terms should now be assigned as a function of empirical formula and skeletal electron count and that it be recognized that a given empirical formula with a given skeletal electron count may have several configurations. To name such species based on configuration alone rather than electron count will lead to chaos.

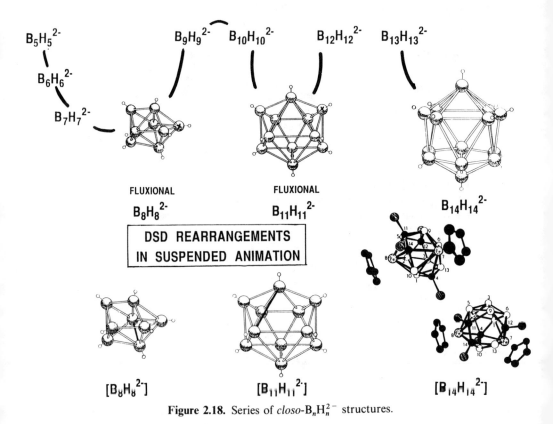

Figure 2.18. Series of *closo*-$B_n H_n^{2-}$ structures.

Closo-$B_{11}H_{11}^{2-}$ has five different kinds of borons in its known *closo*-deltahedral crystal structure (see top of Figure 2.18) but all of these appear as one kind of boron in solution (^{11}B-NMR). It was suggested [34] that *closo*-$B_{11}H_{11}^{2-}$ structures with aberrant IV-gonal open faces (see bottom middle of Figure 2.18), or aberrant V-gonal open faces probably exist in solution. Indeed, Nestor et al. [86] had already come to this conclusion and proved it with the crystal structure of a transition element containing derivative.

2.3.7. *Nido*-Configurations; Vertex Removal versus Connection Removal

The open faces of the *nido*-species were originally [38] ascribed as arising solely from the removal of the highest coordinated 4k, 5k, and 6k vertices available from the most spherical one-vertex-larger series of *closo*-deltahedra thus producing IV-gonal, V-gonal, and VI-gonal open faces, respectively (see Figure 2.9). Alternatively, it was frequently (but not always) possible to derive these same *nido*-configurations from *closo*-deltahedra with the same number of vertices by the removal of one to three adjacent connectivities. These "removable" connections are identified by blackened lines in Figure 2.19.

As an example, either (1) one belt connection ⟨i⟩ may be removed from the

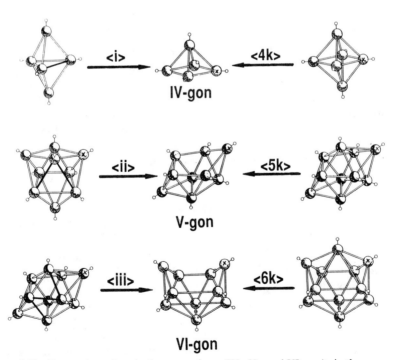

Figure 2.19. Generation of equivalent open faces (IV-, V-, and VI-gon) via the removal of connections (⟨i⟩, ⟨ii⟩, and ⟨iii⟩) or vertices (4k, 5k, and 6k).

surface of the *closo*-5-vertex-trigonal bipyramid which produces the *nido*-5-vertex skeleton with a IV-gonal open face characteristic of *nido*-B_5H_9 or (2) a $4k$ vertex may be removed from the *closo*-6-vertex octahedron. In similar fashion, to produce the V-gonal and VI-gonal open faces found in the *nido*-core-clusters $B_9H_9^{4-}$ and $B_{10}H_{10}^{4-}$, one may consider either (1) the removal of $5k$ or $6k$ vertices or (2) the removal of two connections ⟨ii⟩ or three connections ⟨iii⟩ from the deltahedral surfaces.

In conclusion, in electron deficient *nido*-compounds, it would seem that (1) pentagonal open faces are generally preferred, however, (2) overriding circumstances are frequently present resulting in hexagonal open faces due to:

1. Insufficient precursor deltahedron size, for example, B_5H_9 is too small to have a V-gonal open face.
2. Insufficient electron density or space for bridge hydrogens, for example, B_8H_{12} and $B_{10}H_{14}$.
3. Insufficient electron density for optimal heteroatom charge smoothing, for example, $C_4B_4H_8$.

Nothing can be done about (1) insufficient deltahedron size so (2) insufficient electron density and/or steric congestion involving bridge hydrogens will be considered first.

2.4. COORDINATION NUMBER PATTERN RECOGNITION (CNPR) CONCEPT OF CARBORANE STRUCTURES [39]

In Sections 2.2.2 and 2.3.4, it has been pointed out that carbons are found in low coordinated, carbons-apart sites within the carborane structures (see Figures 2.4, 2.5, 2.11, and 2.12), as the result of charge smoothing driving forces. Later, it was found that bridge hydrogens will only associate with sufficiently electron rich (low coordinated) boron vertices about the open face and that the bridge hydrogen's requirements are more important (see Figure 2.13) than carbon's requirements [39].

It was recognized that these patterns were based on coordination numbers and the approach [39] was termed "The Coordination Number Pattern Recognition (CNPR) Concept of Carborane Structures".

Bridge hydrogens differ in kind from the carbons in their quest for low coordination sites. Bridge hydrogens are like "monkeys" perched on boron-boron bonds (branches). If bridge hydrogens find their environments too electron poor, they simply "jump away" as protons or "brachiate" [28] along boron-boron branches as fluxional bridge-endo-hydrogens, in order to relocate in more electron-sufficient locations, on the same or on different molecules. A number of carborane and polyborane anions are derived in this fashion.

Carbons cannot migrate as easily and thus are found tolerating less desirable $4k$ and $5k$ sites.

2.4.1. "Hydrogencentric" Viewpoint [39]

Figure 2.20 tabulates the bridge hydrogens in the various *nido*-compounds and identifies them by the coordination numbers of the two neighboring borons with which each bridge hydrogen is affiliated.

In B_5H_9, four bridge hydrogens completely surround the IV-gonal open face (see Figure 2.2). Each of these bridge hydrogens is associated with *two* borons, each boron is coordinated with six neighboring atoms (i.e., 3 other borons, 2 bridge hydrogens and 1 terminal hydrogen). Each of the four bridge hydrogens in *nido*-B_5H_9 is therefore identified as a *66*-bridge hydrogen in Figure 2.21 reflecting the coordination numbers of each of the two borons neighboring the bridge hydrogens.

In *nido*-hexaborane, B_6H_{10}, there are two *66*-bridge hydrogens and two *65*-bridge hydrogens (see Figure 2.2). Bridge hydrogens are less stable (more labile) when the coordination numbers of the two associated borons are high, therefore, both compounds, B_5H_9 and B_6H_{10}, are listed in Figure 2.20 under the least stable "highest coordinated" bridge hydrogen in each molecule, that is, their *66*-bridge hydrogens. *Nido*-$B_{10}H_{14}$ incorporates four *66*-bridge hydrogens.

The bridge hydrogens in B_2H_6 are identified as *5'5'*-bridge hydrogens. Both of the borons neighboring the bridge hydrogens in B_2H_6 are associated with two terminal hydrogens, that is, they are BH_2 groups. The primes associated with each of the 5s indicate that the borons are coordinated with five neighboring atoms and two terminal hydrogens. A more complete description is outlined by Williams [39].

A few bridge hydrogens have been alleged to neighbor 7-coordinate borons in *nido*-carboranes and boranes reported to be stable at ambient conditions. Reinvestigation of such molecules will, in all probability, show such assignments are incorrect. It is predicted that *77*-, *76*-, and *75*-bridge hydrogens will not be found in stable polyboranes or carboranes at ambient conditions although *76*- and *75*-bridge hydrogens, along with endo-hydrogens, are probably present in selected compounds, stable only at low temperatures (see Section 2.6.6).

Nido-B_7H_{11} (Stx = 630) and *nido*-B_9H_{13} (Stx = 650), which were indicated [38] (Figure 2.9) to have structures with unfavorably puckered V-gonal open faces would, of necessity, incorporate extremely congested *76*- and *77*-bridge

77	76'	6'6'								B_9H_{13}	$[B_{11}H_{15}]$
76	75'	6'6						B_7H_{11}	B_8H_{12}		?
66	65'	5'5'	B_2H_6		B_5H_9	B_6H_{10}			B_8H_{12}	$B_{10}H_{14}$	
65	5'5		2	3	4	5	6	7	8	9	10	11

Figure 2.20. Bridge hydrogens identified by coordination numbers of neighboring boron atoms.

hydrogens, respectively, if all four "extra" hydrogens were bridge hydrogens; neither of these compounds has been discovered (see Section 2.6.6).

Anions and carboranes, isoelectronic and isostructural with both *nido*-B_7H_{11} and *nido*-B_9H_{13} (but with fewer bridge hydrogens and thus no 77-, 76- or 75-bridge hydrogens) have been discovered [39]; they are stable and have the structures [38] predicted in Figure 2.9.

The suggested [38] incorrect structure for B_8H_{12} (Stx = 640) (see top left in Figure 2.14), with a favorably puckered V-gonal open face, would have necessarily incorporated two 76-bridge hydrogens. In the actual *nido*-B_8H_{12} crystal structure [32], one additional edge-connection is broken, producing a VI-gonal open face, which allows 66-bridge hydrogens to be produced (see top right in Figure 2.14).

One problem remains, that is, all 66-bridge hydrogens are not equal! In the series of most stable *nido*-boranes, B_5H_9, B_6H_{10}, and $B_{10}H_{14}$, the 66-bridge hydrogens differ in degree, that is, they are more labile with increasing size; $B_5H_9 < B_6H_{10} < B_{10}H_{14}$. Simply stating that the highest coordinated bridge hydrogens in all three cases are of one kind, that is, 66-bridge hydrogens in

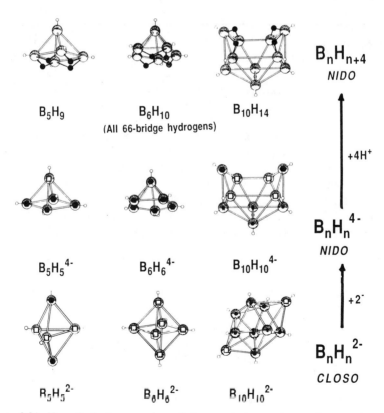

Figure 2.21. Hypothetical conversion of *closo*-core-clusters into *nido*-core-clusters and protonation into neutral polyboranes.

Figure 2.20, does not allow one to ascertain the degree to which they are known to be different.

Each of the neutral *nido*-boron hydrides discussed has, notionally, a $B_nH_n^{4-}$ core cluster (see Figures 2.9 and 2.15) upon which may be placed four protons (as bridge hydrogens) which would neutralize the core charge. It seemed reasonable that if one had two 4^- charged clusters, one of which was composed of five BH groups, $B_5H_5^{4-}$, while the other one was composed of ten BH groups, $B_{10}H_{10}^{4-}$, that the former would tend to attract and capture four added protons (as bridge hydrogens), about twice as strongly as would the latter since the 4^- would be spread over twice the number of BH-vertices in the latter case (see Figure 2.21).

This oversimplified view roughly explains the relative acidities of $B_{10}H_{14} > B_6H_{10} > B_5H_9$ [44] but it is cumbersome to (1) couple bridge hydrogen acidity to (2) the overall size of the molecules as well as to (3) the coordination numbers of the two adjacent borons and at the same time to (4) ignore the differences in aperture size which will be shown to be critically important [33, 34].

It seemed that there might be a more accurate way to characterize the bridge hydrogens on *nido*-skeletal clusters, other than by their electron richness (as extrapolated from their neighboring boron's coordination numbers, i.e., *66*- etc. in Figure 2.20), which would allow the rationalization of the differing acidities of similarly coordinated bridge hydrogens and which might subsequently allow the understanding of the influences of both bridge hydrogens, carbons, and other heteroatoms upon structural selection.

2.5. MODIFIED CNPR CONCEPT INCLUDES APERTURE DEPENDENT ELECTRON DISTRIBUTION

Simple first-order coordination number considerations have been found wanting. The highest coordinated bridge hydrogens on B_5H_9, B_6H_{10}, and $B_{10}H_{14}$ are labeled *66*-bridge hydrogens (Figure 2.20) but it is known that if they are more acidic, the larger is the cluster.

The carbons in the various *nido*-carboranes are found in the lowest, most electron-rich, coordination sites in their respective configurations but frequently configurations with differently sized open faces are seemingly available but there are no obvious trends signaling why any particular configuration prevails for a given composition of matter. An improved second-order, electron distribution estimating procedure is consequently needed.

2.5.1. Boroncentric Viewpoint

Nido-core-clusters with variously sized open faces are more complicated than *closo*-clusters; the latter vary in total size but, unless aberrant, all facets are III-gons. The *closo*-core is 2^- charged rather than 4^- but the relationship is constant. It was felt possible to (1) *easily* develop a general estimating procedure to rationalize the charge distribution in the *one* kind of uniformly "apertured" *closo*-deltahedra, (2) *easily* make the estimating procedure independent of charge (2^- versus 4^-), (3) *probably* extrapolate the estimating procedure from the *one* kind of uniformly-apertured *closo*-clusters to the *three* kinds of variably-apertured *nido*-core-clusters, (4) add four protons to the variably apertured *nido*-core-clusters and *possibly* generate numbers that reflect the "contentment" of the bridge hydrogens on each *nido*-core-cluster (better than the "first-order" 66-bridge hydrogen nomenclature) and (5) *possibly* these same second-order numbers would differentiate between the preferred sites for carbons and other heteroatoms as a function of electron distribution and charge smoothing on alternatively apertured configurations. It was possible to satisfy the first four counts and make a substantial contribution to the fifth.

At the bottom of Figure 2.21 are listed three examples of *closo*-deltahedra.

2.5.2. Charge Distribution In *Closo*-$B_nH_n^{2-}$ and *Nido*-$B_nH_n^{4-}$ Clusters

The three $B_nH_n^{2-}$ deltahedra at the bottom of Figure 2.21 incorporate "differently connected" vertices. The trigonal bipyramid ($B_5H_5^{2-}$) has two $3k$ vertices identified by solid hexagons and three $4k$ vertices denoted by open squares. All six vertices of the $B_6H_6^{2-}$ deltahedron are $4k$ vertices, identified by open squares. The $B_{10}H_{10}^{2-}$ deltahedron has two $4k$ vertices denoted by squares and eight $5k$ vertices identified by solid circles. *Nido*-clusters result when two additional electrons are hypothetically added to each of the *closo*-deltahedra and connections are removed as illustrated on the left-hand side of Figure 2.19.

In notionally transforming the *closo*-deltahedra, $B_nH_n^{2-}$ (bottom row of Figure 2.21) into the *nido*-deltahedral fragments (middle row of Figure 2.21), by the addition of two electrons, the $B_nH_n^{4-}$ *nido*-core-clusters become enriched in lower connected vertices. As will subsequently be explained, $3k$ vertices (identified by hexagons) have a greater negative charge than intermediately connected $4k$ vertices (identified by squares), which in turn, have greater negative charge than do the $5k$ vertices (identified by solid circles). The informal colloquial terms, "hot", "warm", and "cool", are used in the following discussions to reflect negative charge, thus strongly negatively charged $3k$ vertices are termed "hot", the intermediately charged $4k$ vertices "warm" and the weakly negatively charged $5k$ vertices are termed "cool".

The problem of just how asymmetrically the 4^- charge should be distributed remained to be solved. In looking at the resulting *nido*-polyboranes (top row of Figure 2.21), it is seen that the four bridge hydrogens are notionally "attracted" to the "formerly" $3k$, highest negatively charged, hot vertices of the *nido*-core clusters displayed in the middle row of Figure 2.21.

Williams reported a procedure [33, 34] for estimating the *nido*-fragment charge distributions by extrapolation from the easily estimated *closo*-electron charge distributions (see Figure 2.22). This strategy is an exercise in arithmetic, geometry, electron distribution, and charge smoothing; connectivities and vertices are discussed, not "bonds" or molecular orbitals.

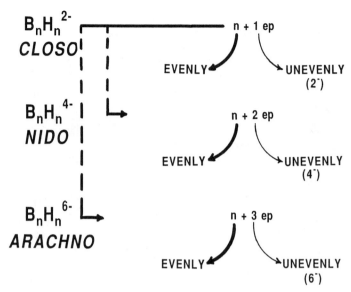

$B_nH_n^{2-}$
CLOSO

$n + 1$ ep

EVENLY ◄ ➤ UNEVENLY
(2^-)

$B_nH_n^{4-}$
NIDO

$n + 2$ ep

EVENLY ◄ ➤ UNEVENLY
(4^-)

$B_nH_n^{6-}$
ARACHNO

$n + 3$ ep

EVENLY ◄ ➤ UNEVENLY
(6^-)

Figure 2.22. Strategy for electron distribution.

Porterfield et al. [87] simultaneously reported what they termed a "vertex electron pair" scheme which does involve bonds and molecular orbitals. The two approaches, although starting from different bases, having different objectives and using totally different rationales to attain these objectives, gave results which are mutually reinforcing, complementary, and in conflict on minor points only. See Chapter 3.

The various $closo\text{-}B_nH_n^{2-}$ species incorporate $n + 1$ skeletal electron pairs and since trends were desired, as opposed to absolute values, the $n + 1$ electron pairs were distributed via two different modes. In Williams' approach [33, 34] (Figure 2.22), the n of the $n + 1$ electron pairs were distributed evenly (i.e., one electron pair was assigned to each BH vertex independent of its connectivity), however, the "extra" electron pair, or the 1 of the $n + 1$ electron pairs, was distributed unevenly, strictly as a function of vertex connectivity. More of the 2^- electron charge was assigned to the lower connected vertices in the order $3k > 4k > 5k > 6k$. Later, when the $nido\text{-}B_nH_n^{4-}$ compounds were considered, the n (of the $n + 2$) electrons pairs were distributed evenly (one electron pair per vertex), and the 2 "extra" electron pairs spread unevenly as a function of connectivity.

In order to achieve the goal of estimating electron distributions in the variably faceted $nido\text{-}B_nH_n^{4-}$ core clusters, a scheme was first developed for estimating such distributions unambiguously for the uniformly faceted $closo\text{-}B_nH_n^{2-}$ clusters [33, 34].

Three regular deltahedra, the $closo$-icosahedron, $B_{12}H_{12}^{2-}$, the $closo$-octahedron, $B_6H_6^{2-}$, and the hypothetical $closo$-tetrahedron, $[B_4H_4^{2-}]$, are displayed in Figure 2.23.

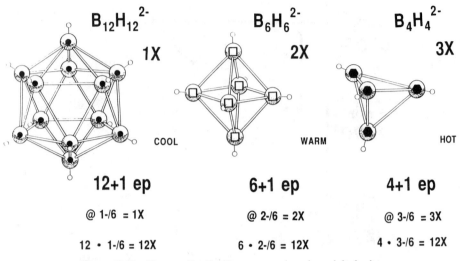

Figure 2.23. Charge distribution on regular *closo*-deltahedra.

In the completely symmetrical $B_{12}H_{12}^{2-}$, $B_6H_6^{2-}$, and hypothetical $[B_4H_4^{2-}]$ clusters, the 2^- is uniformly distributed over all of the BH vertices in all three cases. All of the vertices in a regular icosahedron are $5k$ vertices, in a regular octahedron, $4k$, and in a tetrahedron, $3k$. The $5k$ vertices are coded with solid circles, $4k$ with open squares and $3k$ with solid hexagons. These three regular deltahedra all incorporate $n + 1$ skeletal electron pairs, the icosahedron ($12 + 1$ electron pairs), the octahedron ($6 + 1$ electron pairs), and the tetrahedron ($4 + 1$ electron pairs).

Within these symmetrical deltahedra, all vertices within each deltahedron are identical; therefore, each of the BH groups of the 12 vertex, $B_{12}H_{12}^{2-}$ system, *must* have $1/6^-$ charge per each $5k$ vertex. It follows that there *must* be $2/6^-$ charge per $4k$ vertex in the octahedral $B_6H_6^{2-}$ and *must* be $3/6^-$ charge per $3k$ vertex in the hypothetical tetrahedral species $[B_4H_4^{2-}]$.

All other less symmetrical deltahedra incorporate mixtures of the same $3k$, $4k$, and $5k$ vertices listed above and the same charges may be assigned to each BH vertex, that is, $3k = 3/6^-$, $4k = 2/6^-$, and $5k = 1/6^-$ in all cases, exactly as in the three symmetrical deltahedral [33, 34]. The total charge on each deltahedron always adds up to $12/6^-$ or 2^- no matter what the shape or size of the deltahedron or whether it is most spherical or least spherical.

Since the ultimate objective is to extrapolate from the charge distributions characteristic of the *closo*-$B_nH_n^{2-}$ deltahedra of Figures 2.23 and 2.24 (where *each* deltahedron has a 2^- charge), to the *nido*-$B_nH_n^{4-}$ core clusters (where each deltahedral fragment has a 4^- charge), it was desirable to become independent of the cluster charge so that the electron distribution procedure would be the same no matter whether the cluster had a charge of 1^-, 6^-, or anything in between. This may be achieved as follows.

In Figure 2.23, the smallest aliquot of negative charge is a $1/6^-$ charge which, for convenience, is made equal to $1X$. If each BH group in the icosahedral, $B_{12}H_{12}^{2-}$ has a $1/6^-$ or $1X$ charge, then the charge per vertex in the octahedral $B_6H_6^{2-}$ is $2/6^-$ or $2X$ and the charge per BH group in the tetrahedral $B_4H_4^{2-}$ is $3/6^-$ or $3X$.

The immutable total value for each deltahedron of whatever size or shape, is $12X$; twelve $1X$ vertices in the case of the icosahedron; six $2X$ vertices in the case of the octahedron, and four $3X$ vertices in the case of the tetrahedron.

The three regular deltahedra of Figure 2.23 are the only deltahedra that have uniformly connected vertices; all other deltahedra (see Figure 2.24) are less uniform as far as vertex connectivity is concerned, but are simply related to the three regular deltahedra of Figure 2.23. In Figure 2.24 are shown all of the most spherical *closo*-deltahedra from 4 vertices to 14 vertices. The vertices are labeled hot ($3X$), warm ($2X$) or cool ($1X$) by the identification symbols, solid hexagon, open square, and solid circle, reflecting the relative charge values on each vertex; that is, $3/6^-$, $2/6^-$, and $1/6^-$, respectively, for *closo*-deltahedra.

The types and kinds of vertices are summarized below each deltahedron and the types of vertices lead to total negative charge values of 2^- and/or $12X$ for all *closo*-deltahedra.

Several trends are illustrated, that is, the smaller deltahedra incorporate more hot vertices than the larger deltahedra; these give way to warm vertices and to cool vertices as the deltahedra become progressively larger. The 11-, 13-, and 14-vertex deltahedra even contain vertices of skeletal connectivity 6 ($6k$) to which we assign a charge value of $0/6^-$ or $0X$.

What happens when one of these *closo*-deltahedra is converted into a *nido*-deltahedral fragment? Consider $B_8H_8^{2-}$ specifically in Figure 2.24; for comparison, it is reproduced in Figure 2.25.

The 8-vertex *closo*-deltahedron (all facets are III-gons) is reproduced on the left-hand side of Figure 2.25 and one connection is identified for subsequent removal by being "blackened" (Structure "III"). When the "blackened" connection is removed, two III-gonal facets "disappear" and one IV-gonal open face of a potential *nido*-structure is produced (Structure "IV") as is depicted to the right of the *closo*-structure in Figure 2.25. Following removal of the "blackened" connection, two cool $5k$ vertices in *closo*-$B_8H_8^{2-}$ become two warm $4k$ vertices as illustrated in the *nido*-$B_8H_8^{4-}$ structure with a IV-gonal open face to its right. The value of both previously cool vertices ($1X$) are increased to $2X$ with this *closo*- to *nido*-transformation.

It follows that the $12X$ value (the immutable total value characteristic of all nonaberrant *closo*-deltahedra) is inappropriate and must be changed into a total of $14X$ for this potential *nido*-8-vertex core-cluster with a IV-gonal open face. It will be shown below that the value of $14X$ is the immutable total value for all deltahedral fragments with one IV-gonal open face.

At the top of Figure 2.25, we are reminded that $12X$ equals 2^- in the case of the *closo*-compounds with III-gonal facets, and that $14X$ equals 4^- in the case of *nido*-deltahedral fragments with a IV-gonal open face. The value of X itself must

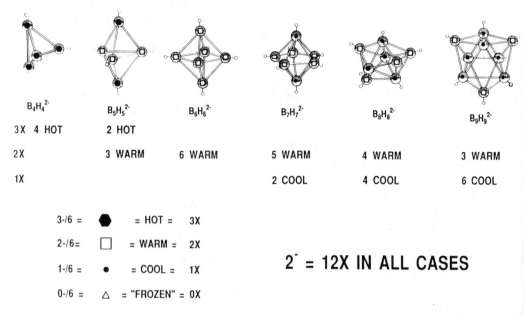

$B_4H_4{}^{2-}$ $B_5H_5{}^{2-}$ $B_6H_6{}^{2-}$ $B_7H_7{}^{2-}$ $B_8H_8{}^{2-}$ $B_9H_9{}^{2-}$

3X 4 HOT   2 HOT

2X   3 WARM   6 WARM   5 WARM   4 WARM   3 WARM

1X   2 COOL   4 COOL   6 COOL

3-/6 = ⬢ = HOT = 3X
2-/6 = ☐ = WARM = 2X
1-/6 = ● = COOL = 1X
0-/6 = △ = "FROZEN" = 0X

$2^- = $ **12X IN ALL CASES**

Figure 2.24. Charge distribution extrapolated

be different in these two cases. In other words, a 5*k* cool (1*X*) vertex in the *closo*-compound has a value of $1/6^-$ or 0.167^- charge whereas a 5*k*, cool (1*X*) vertex in the *nido*-deltahedral fragment with a IV-gonal open face is 0.29^-.

Removing another connection (also blackened), at the perimeter of the IV-gonal *nido*-deltahedral open face converts the IV-gonal open face (Structure "IV") into a V-gonal open face (Structure "V"). This is accompanied by the conversion of two warm vertices of value 2*X* into two hot vertices of value 3*X* (again, an increase of 2*X* in the total *nido*-cluster). It follows that deltahedral core-clusters with lone V-gonal open faces always have a total charge of 16*X*. The value of *X* becomes smaller upon conversion from a *nido*-deltahedral fragment with a IV-gonal open face (wherein the value of *X* is 0.29^-) into a *nido*-deltahedral fragment, with a V-gonal open face (i.e., the value of *X* reduces to 0.25^-).

Further edge connectivity breaking to produce a VI-gonal open face (Structure "VI") from a V-gonal open face (Structure "V") results in a total value of 18*X* for all deltahedral fragments with VI-gonal open faces and the value of *X* declines further to 0.22^-.

In summary, "one kind of" *closo*-deltahedra with III-gonal facets all have total values of 12*X* (which equal 2^-) and as connectivities of the original *closo*-deltahedra are broken, sequentially yielding "three different kinds" of *nido*-configurations with IV-gonal, V-gonal, and VI-gonal open faces, values of 14*X*, 16*X*, and 18*X* (all of which total 4^-) are produced.

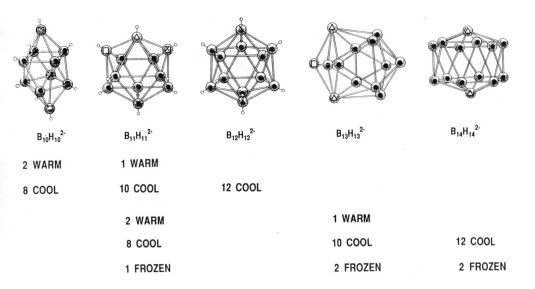

$B_{10}H_{10}^{2-}$	$B_{11}H_{11}^{2-}$	$B_{12}H_{12}^{2-}$	$B_{13}H_{13}^{2-}$	$B_{14}H_{14}^{2-}$
2 WARM	1 WARM			
8 COOL	10 COOL	12 COOL		
	2 WARM		1 WARM	
	8 COOL		10 COOL	12 COOL
	1 FROZEN		2 FROZEN	2 FROZEN

to all most spherical *closo*-deltahedra.

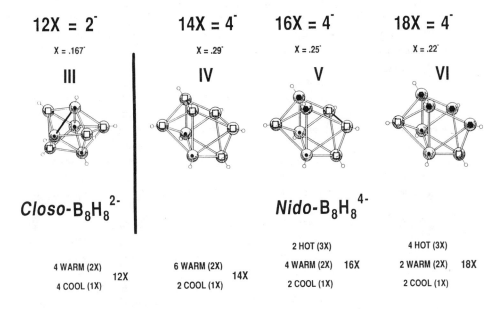

12X = 2⁻	14X = 4⁻	16X = 4⁻	18X = 4⁻
X = .167⁻	X = .29⁻	X = .25⁻	X = .22⁻
III	IV	V	VI

Closo-$B_8H_8^{2-}$		*Nido*-$B_8H_8^{4-}$	
		2 HOT (3X)	4 HOT (3X)
4 WARM (2X) 12X	6 WARM (2X) 14X	4 WARM (2X) 16X	2 WARM (2X) 18X
4 COOL (1X)	2 COOL (1X)	2 COOL (1X)	2 COOL (1X)

Figure 2.25. Hypothetical conversion of *closo*-$B_8H_8^{2-}$ into *nido*-$B_8H_8^{4-}$ core-clusters with IV-, V-, and VI-gonal open faces.

	PREVALENT	RARE	Aperture Size	Value of X
CLOSO 2⁻	12X		III	0.167⁻
		14X	IV	0.143⁻
		[16X]	V	0.125⁻
NIDO 4⁻		14X	IV	0.286⁻
	16X		V	0.250⁻
	18X		VI	0.222⁻
		20X	VII	0.200⁻
ARACHNO 6⁻		12X	III	0.500⁻
		14X	IV	0.429⁻
		16X	V	0.375⁻
	18X		VI	0.333⁻
	20X		VII	0.300⁻

Figure 2.26. Charge distribution in the various *closo*-, *nido*-, and *arachno*-core-clusters as a function of the size of the open face.

The two light-hand structures of Figure 2.25, with V-gonal and VI-gonal open faces, are the two structures actually observed in a number of 8-vertex *nido*-polyboranes, carboranes, thiaboranes, metallacarboranes, and Lewis base adducts (see Figures 2.11 and 2.14). This configuration with the larger open face embraces more hot, $3X$, vertices but the hot vertices are of lesser value (0.75^- declines to 0.67^-). These two compensatory changes in value reflect the all important quid pro quo that allows the evaluation of various alternative locations for bridge hydrogens, carbons, and other heteroatoms in the various *nido*-configurations as a function of the size of the open face (V-gonal versus VI-gonal).

In Figure 2.26, these observations are generalized to show that the favored structures have $12X$ charge in *closo*-clusters, $16X$ and $18X$ charge in *nido*-core clusters, and $18X$ and $20X$ charge in *arachno*-compounds, and how the charge on each vertex differs and is distributed as a function of aperture size and connectivity. These latter values constitute the spectrum of second-order "*nido*-numbers" that are used subsequently to differentiate between the various kinds of 66-bridge hydrogens and their influence on structure, and to couple carbon and heteroatom locations with charge smoothing in order to rationalize the final structures.

In Figure 2.27, the suggested [38] structure (Figure 2.9) for $B_8H_8^{4-}$ with its V-gonal open face is reproduced in the center flanked by the *nido*-core clusters for hexaborane and decaborane, namely $B_6H_6^{4-}$ (with a V-gonal open face) and $B_{10}H_{10}^{4-}$ (with a VI-gonal open face).

The value of X is 0.25^- for the two structures with V-gonal open faces but drops to 0.22^- when the open face is VI-gonal. The smaller *nido*-deltahedral fragments have higher populations of hot and warm vertices which give way to cool vertices as the *nido*-cage fragments become larger.

The consequences of adding four protons to the three *nido*-core-clusters of Figure 2.27 and converting them into neutral *nido*-polyboranes is considered next.

2.5.3. Bridge Hydrogen Index

As protons have a 1^+ total charge, the proton's charge will be split arbitrarily three ways $[3(0.33^+) = 1^+]$ as each of the protons notionally "settles upon" the targeted boron-boron bonds to become bridging hydrogens (equation 2.1).

$$H^+ + B\text{---}B \rightarrow 1/3^+B \overset{\displaystyle \overset{H\,1/3^+}{\big|}}{\bigwedge} B\,1/3^+ \tag{2.1}$$

In Figure 2.27, it is possible to place substantive values on each of the hot $(3X)$ and warm $(2X)$ borons bordering the open face prior to adding any protons since the values of X and how the value of X differs as a function of aperture size are known. Equation (2.1) also suggests that we can put "proportional" numbers on the edge boron atoms after adding the four protons (which become bridge hydrogens). It is not suggested that these proportional numbers (after adding the protons) are accurate values, only that these numbers parallel accurate values

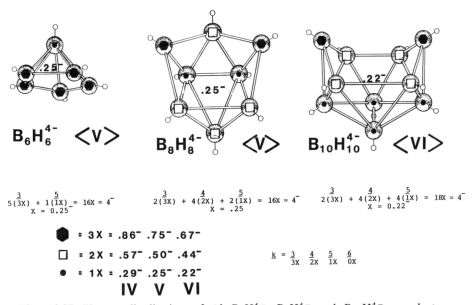

$$5(\tfrac{3}{3X}) + 1(\tfrac{5}{1X}) = 16X = 4^-$$
$$X = 0.25^-$$

$$2(\tfrac{3}{3X}) + 4(\tfrac{4}{2X}) + 2(\tfrac{5}{1X}) = 16X = 4^-$$
$$X = .25$$

$$2(\tfrac{3}{3X}) + 4(\tfrac{4}{2X}) + 4(\tfrac{5}{1X}) = 18X = 4^-$$
$$X = 0.22^-$$

⬢ ⫶ 3X ⫶ .86⁻ .75⁻ .67⁻

☐ ⫶ 2X ⫶ .57⁻ .50⁻ .44⁻

● ⫶ 1X ⫶ .29⁻ .25⁻ .22⁻

$$k = \tfrac{3}{3X} \quad \tfrac{4}{2X} \quad \tfrac{5}{1X} \quad \tfrac{6}{0X}$$

IV   V   VI

Figure 2.27. Electron distributions of *nido*-$B_6H_6^{4-}$, -$B_8H_8^{4-}$, and -$B_{10}H_{10}^{4-}$-core-clusters prior to protonation.

and have predictive value as far as bridge hydrogen placement and acidities are concerned.

As an example, in Figures 2.27 and 2.28, each of the perimeter borons in $B_6H_6^{4-}$ had $3X$ values, and since X equaled 0.25^-, then each edge boron prior to adding protons, has a value of 0.75^-.

Following the addition of four protons (splitting the 1^+ charges three ways), two basal borons in B_6H_{10} are produced which neighbor one bridging hydrogen (Figure 2.28) and have residual values of 0.42^- $(0.75^- + 0.33^+ = 0.42^-)$ and three borons are produced which are adjacent to two bridge hydrogens and have residual values of 0.09^- $(0.75^- + 2(0.33^+) = 0.09^-)$.

Having put values on all five edge borons following the addition of four protons, the "electron richness" of the "newly formed" bridge hydrogens in B_6H_{10} may be evaluated as a function of their two nearest boron neighbors (not as simple first-order functions of the coordination numbers of the two adjacent borons [39], that is, identifying them as 66- and 65-bridge hydrogens as in Figures 2.2 and 2.20), but as a function of the sums of the estimated, second-order residual charges on the two neighboring borons.

Tabulated at the bottom of Figure 2.28, are the estimated sums for each of the four bridge hydrogens in all three cases. It is seen that there are two bridge hydrogens in B_6H_{10} which have values of 0.51^- $(0.42^- + 0.09^- = 0.51^-)$ and two bridge hydrogens that have values of 0.18^- $(0.09^- + 0.09^- = 0.18^-)$. In these delocalized systems, it is not believed that these bridge hydrogens are very different from one another (i.e., it is assumed that the electron-rich share with the electron-poor) and that averaging the four bridge hydrogen numbers might

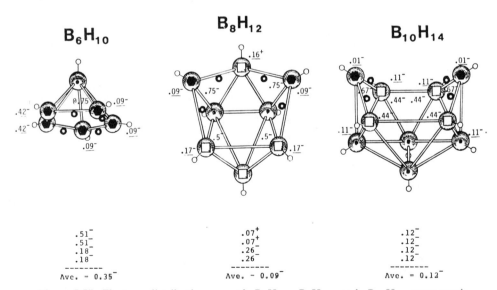

Figure 2.28. Electron distribution on *nido*-B_6H_{10}, -B_8H_{12}, and -$B_{10}H_{14}$ compounds following addition of protons.

yield a crude estimate of the electron density available to the four bridge hydrogens.

The average bridge hydrogen index (H-index) number for the four bridge hydrogens in nido-B_6H_{10} is 0.35^- which is not considered to be meaningful in its numerical size, but probably parallels (or is proportional to) as yet unknown real values. No estimate can be made at this time what the correlation factor might be.

Repeating the same procedure for nido-B_8H_{12} and nido-$B_{10}H_{14}$ (see Figure 2.28), yields average values of 0.09^- for B_8H_{12} and 0.12^- for $B_{10}H_{14}$. There is a seven coordinate boron in the nido-B_8H_{12} fragment (after the addition of two protons) and it has an estimated net charge of 0.16^+. In spite of the somewhat improbable seven coordinate boron, the previously discussed arithmetic was carried out.

In Figure 2.29, the three values for B_6H_{10}, B_8H_{12} (V-gonal open faces), and $B_{10}H_{14}$ (VI-gonal open face) as well as the values for B_5H_9 (IV-gonal open face) and the hypothetical species $[B_7H_{11}]$ (Stx = 630) and $[B_9H_{13}]$ (650) (with four bridge hydrogens) were plotted [33, 34]. (See also Section 2.6.6 where more likely 441 and 461 structures for B_7H_{11} and B_9H_{13} are discussed.) Surprisingly, a straight line through the three known structures for the Stx = 610, 620, and 660 compounds, that is, B_5H_9, B_6H_{10}, and $B_{10}H_{14}$ was obtained (Figure 2.29).

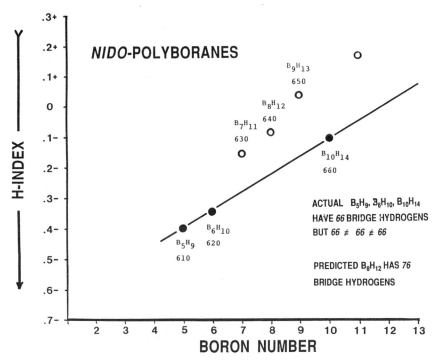

Figure 2.29. Bridge hydrogen index of nido-polyboranes versus size of core-clusters.

Recall that all of these *nido*-compounds, B_5H_9, B_6H_{10}, and $B_{10}H_{14}$, were previously characterized as having one kind of undifferentiated *66*-bridge hydrogens (see Figure 2.2) in accordance with the original electron distribution convention [39], but now they have values which reflect the degree to which their experimentally established acidities differ [44]. *Nido*-$B_{10}H_{14}$ is more acidic than B_6H_{10} in that it will donate a proton to $B_6H_9^-$ to produce B_6H_{10} and $B_{10}H_{13}^-$. In like manner, hexaborane, B_6H_{10}, is more acidic than B_5H_9 (see equation (2.2) and (2.3))

$$B_{10}H_{14} + B_6H_9^- \rightarrow B_{10}H_{13}^- + B_6H_{10} \tag{2.2}$$

$$B_6H_{10} + B_5H_8^- \rightarrow B_6H_9^- + B_5H_9 \tag{2.3}$$

Returning to Figure 2.29, the *H*-index value for *nido*-B_8H_{12} does not fall on the line defined by B_5H_9, B_6H_{10}, and $B_{10}H_{14}$. However, since the structure used for B_8H_{12} in Figure 2.28, is known to be incorrect in the crystal [32], it was suspected that the value of 0.09^- in "falling off-the-line" might reflect the "incorrectness" of this structure.

In Figure 2.30, two optional Stx = 640, structures for B_8H_{12} are reproduced with the predicted [38], but incorrect, structure of Figure 2.28 on the left and the actually observed crystal structure to the right.

Williams anticipated structure [38] has one 7-coordinate boron neighboring a bridge hydrogen which in retrospect is absent in all other confirmed structures for *nido*-polyboranes, *nido*-carboranes, and/or their derivatives. A recalculation

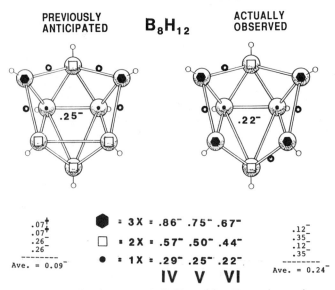

Figure 2.30. Electron distributions on *nido*-B_8H_{12} with a V-gonal open face versus *nido*-B_8H_{12} with a VI-gonal open face.

based on the actual B_8H_{12} crystal structure yields an average, H-index value for the four bridge hydrogens of 0.24^-. This is plotted in Figure 2.31 and to our pleased, but bemused amazement, the H-index value for the more open structure is colinear with those of B_5H_9, B_6H_{10}, and $B_{10}H_{14}$.

There are other structural reasons, for example, unfavorably puckered open faces, seven coordinate borons and endo-hydrogens, etcetera, that explain why the parent *nido*-species B_7H_{11} (630), B_9H_{13} (650), and $B_{11}H_{15}$ (670) do not exist as molecules that are stable at ambient conditions. Breaking additional edge connections does not help in these cases. We will consider these compounds in Sections 2.6.5 and 2.6.6.

At this point, a few caveats are in order. The results illustrated in Figures 2.29 and 2.31 look unexpectedly good but the validity of the straight line displayed in Figure 2.31 should be questioned. The last two assumptions in particular, that is, splitting the proton's charge three ways and averaging the four bridge hydrogen values, in route to generating the data in Figure 2.31 are particularly questionable. Consequently, in order to test the validity of our assumptions from another point of view, the data points shown in Figure 2.31 were replotted and values for the various anions and one cation of known structure were added (see Figure 2.32).

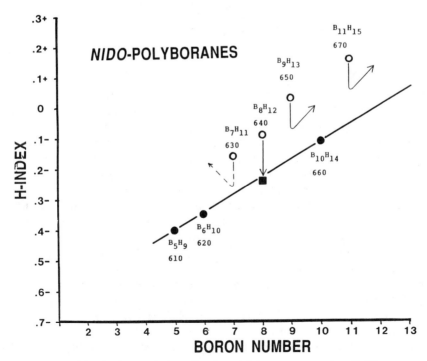

Figure 2.31. Bridge hydrogen index of *nido*-polyboranes versus size; B_8H_{12} structure corrected.

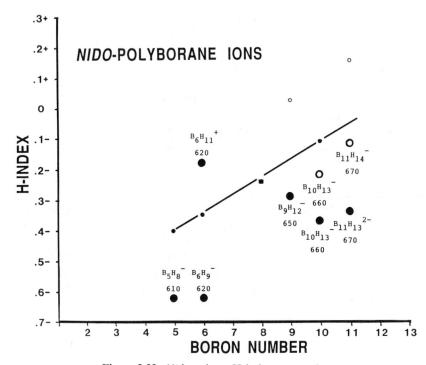

Figure 2.32. *Nido*-anions; H-index versus size.

The anions (produced by the removal of bridging protons) fall as expected below the line of Figures 2.31 and 2.32 which reflect the fact that the remaining, less acidic bridge hydrogens of the anions are in more hospitable (electron-rich) environments than were the four bridge hydrogens in their neutral precursors.

A proton may be added to B_6H_{10} without producing 76-bridge hydrogens; its value is found as expected above the line. $B_6H_{11}^+$ [29, 30] (Figure 2.6) is only stable at very low temperatures. Anions such as $B_9H_{12}^-$ [88] and $B_{11}H_{13}^{2-}$ [89, 90] are very stable while their neutral parents are understandably unknown at ambient conditions as sufficiently hospitable environments for four "extra" hydrogens are not available under any circumstances.

Similar treatment of the *arachno*-polyboranes roughly yields a straight line relationship also as will be shown in a future publication. The values for the *arachno*-polyborane anions also fall below the line of the parent *arachno*-polyboranes reflecting their greater stability.

In summary, it seems the second-order bridge hydrogen hypothesis can account for the stability of the bridge hydrogens in the various polyboranes better than did the first-order simple 66-notation (see Figure 2.20) of 1976 [39]. It apparently rationalizes the breaking of an additional connection to produce preferred environments for the four bridge hydrogens in B_8H_{12}.

It is no longer valid to state [38] (see Figure 2.9), that all *nido*-polyborane structures are simply derived from the one vertex larger *closo*-deltahedra by the removal of one highest coordination vertex in each case. While highest coordination vertex removal may reveal the preferred geometrical configurations in most cases (see dotted line in Figure 2.15), other *nido*-8-vertex and *nido*-12-vertex configurations wherein additional edge connections are broken are also observed (Figure 2.14).

This realization in the case of *nido*-B_8H_{12} caused the reevaluation [33, 34] of the original deltahedra-deltahedral fragment systematics [38].

2.6. MODIFIED DELTAHEDRA-DELTAHEDRAL FRAGMENT SYSTEMATICS

In introducing the modified approach to the generation of *nido*-deltahedral fragments from *closo*-deltahedra, it is necessary first to discard the highest coordination vertex removal as the sole method by which *nido*-structures may be generated. Not only the highest coordinated vertex, but also the lowest coordinated vertex is removed from the *closo*-9-vertex deltahedron (see Figure 2.33). Subsequently, the high coordinated edge connectivities are sequentially broken to see how many different *nido*-8-vertex fragments may reasonably be generated.

2.6.1. *Nido*-8-Vertex Compounds

In Figure 2.33, the *closo*-9-vertex deltahedron is illustrated and both the 5k vertex and the 4k vertex are removed to produce two fragments.

These 8-vertex *nido*-fragments are identified as ni-8⟨IV⟩ and ni-8⟨V⟩; the Roman numerals indicate the size of the open face which is equal to the connectivity of the vertices removed from their one vertex larger *closo*-precursor. Subsequently, adjacent edge connections ⟨i⟩ (backened) about the open face are removed. As seen in Figure 2.33, the removal of the 4k and 5k vertices plus subsequent edge connection removal yields five candidate structures. Two of the five candidates are redundant as indicated by two vertical arrows through their structures which point to the same structures. Thus, the number of candidate structures drops to three in Figure 2.33 which coincides with the three ni-8-configurations of Figure 2.25.

The structure with the IV-gonal open face is not anticipated to be prevalent in any carboranes or polyboranes. *Nido*-8-vertex compounds are too large to necessitate a IV-gonal open face unless a transition element group or selected heteroatom were present.

The vertices of the surviving structures, ni-8⟨V⟩ and ni-8⟨V + i⟩ (the latter is reidentified as ni-8⟨VI⟩) in Figure 2.33, are coded in Figure 2.34 as to their connectivity, that is, hot, warm, and cool. Adjacent to the coded structures are listed the compounds that apparently utilize the two alternative structures.

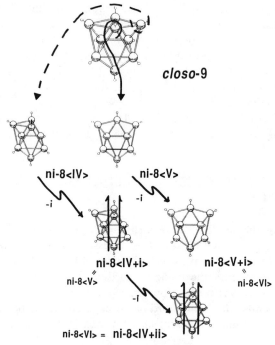

Figure 2.33. Dissection of the *closo*-9-vertex deltahedron into *nido*-8-vertex deltahedral fragments.

Under each compound listed are placed one or more solid dots and open circles which identify the atoms in each molecule that must have (solid dots) or desirably should have (open circles) lower coordination electron-rich sites. Bridge hydrogens are found neighboring, and carbons are found occupying, low coordination sites about the open face as are nitrogen, sulfur, etcetera. The bridge hydrogens outrank the carbons and each bridge hydrogen must be associated with two adjacent low coordination sites necessarily about the open face. On the other hand, carbons, less demanding than bridge hydrogens, should have access to only one low coordination site apiece and one of the carbons may occupy a low coordination site on the cage if available. Known structures are displayed in Figure 2.35.

In the following discussion, we assume that both ni-8$\langle$V$\rangle$ and ni-8$\langle$VI$\rangle$ configurations are equally available to *nido*-carboranes. This may not be the case, however (see Section 2.3.5 versus Section 2.6.7).

2.6.1.1. Nido-B_8H_{12} (*structure known*).

Compound B_8H_{12} has four problematic bridge hydrogens and only the ni 8$\langle$VI$\rangle$ configuration has enough hot and warm locations to avoid a 76-bridge hydrogen. The bridge hydrogens force the selection of the ni-8$\langle$VI$\rangle$ structure [32] as seen previously (Figures 2.14, 2.30, and 2.31).

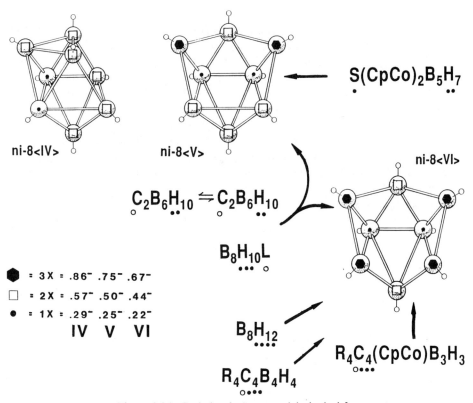

Figure 2.34. Coded *nido*-8-vertex deltahedral fragments.

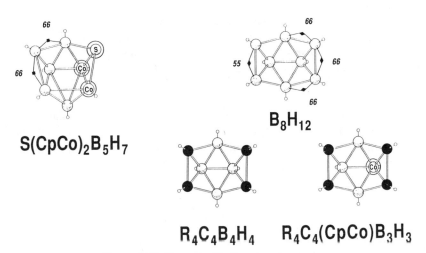

Figure 2.35. Known *nido*-8-vertex structures.

2.6.1.2. *Nido*-$C_4B_4H_8$ [62], *Nido*-$C_4(CpCo)B_3H_7$ [63] (*structures known*). The two 4-carbon compounds (Figure 2.35) contain three carbons that must have edge sites and one that could occupy a cage position in the ni-8⟨V⟩ configuration. Fehlner's $C_4B_4H_8$ [62] determines its ultimate structural choice based on the total negative charge notionally available to the four carbons in the $[B_8H_8^{4-}]$ precursor. The more open ni-8⟨VI⟩ structure, experimentally observed, is favored (driven by charge smoothing) because the four carbons have access to almost 7% greater negative charge as calculated below.

In such treatments, the four most promising (most negatively charged) BH-vertices in the hypothetical precursors $[B_8H_8^{4-}]$ for both ni-8⟨V⟩ and ni-8⟨VI⟩ configurations are listed, summed, and compared. Replacement of four negatively charged BH groups by four CH groups (each CH group adds 1.0^+) leads to four positive numbers that differ least from the remaining BH groups in the most favored configuration, that is, charge smoothing is maximized.

ni-8⟨V⟩		ni-8⟨VI⟩	
0.75^-		0.67^-	
0.75^-		0.67^-	
0.50^-		0.67^-	$<7\%$
0.50^-	$\xrightarrow{C_4B_4H_8}$	0.67^-	(exp. obs.)
2.50^-		2.68^-	

2.6.1.3. $S(CpCo)_2B_5H_7$ [50] (*structure known*). The two problematic bridge hydrogens in $S(CpCo)_2B_5H_7$ (Figures 2.11 and 2.35) utilize one hot and two neighboring warm edge sites becoming *66*-bridge hydrogens in the experimentally observed ni-8⟨V⟩ configuration while the sulfur (having offset the other half of the *nido*-$[B_8H_8^{4-}]$ negative charge) becomes far more electronegative than a comparable carbon atom and occupies the other hot (0.75^-) location. The two CpCo groups (isoelectronic with BH groups [39, 42]) are assumed to be electropositive (i.e., "electron generous" [91]) relative to BH groups and thus they cluster as close as possible to the electronegative (i.e., "electron hungry") sulfur atom. This indicates that "appeasing" the sulfur atom is more important than minor optimization [92] of the bridge hydrogen environments. If $S(CpCo)_2B_5H_7$ opened up to the ni-8⟨VI⟩ configuration, the value of the sulfur's hot location would worsen from 0.75^- to 0.67^- while the value of the bridge hydrogens would improve. In this case, charge smoothing involving sulfur seems to prevail.

	ni-8⟨V⟩		ni-8⟨VI⟩
$>11\%$	0.75^-	$\xrightarrow{S(CpCo)_2B_5H_7}$	0.67^-
(exp. obs.)			

2.6.1.4. *Nido*-$C_2B_6H_{10}$ [70, 71] *(structure uncertain)*.

A special case is $C_2B_6H_{10}$ which has posed a structural dilemma for over 15 years. Although compound $C_2B_6H_{10}$ had been known and identified (mass spectroscopically) even longer, it was resynthesized to solve the discrepancy between the X-ray structure [32] for B_8H_{12}, ni-8⟨VI⟩, and its alternative configuration, ni-8⟨V⟩, compared at the top of Figure 2.14.

The ^{11}B-NMR spectrum was not definitive and it was originally suggested [70] that either the "proposed-static" ni-8⟨VI⟩ or the fluxional "proposed-dynamic" ni-8⟨V⟩ structures (see Figure 2.36) which differed by the presence or absence of one edge connection, were correct.

Another ni-8⟨VI⟩ structure was rejected [70] in 1973 as it was assumed such a structure would be static and therefore could not be compatible with the ^{11}B-NMR spectrum. This second ni-8⟨VI⟩ configuration is now labeled as "possible-dynamic" in Figure 2.36. Two synchronized skeletal DSD rearrangements would be required to make the four borons neighboring the bridge hydrogens equivalent as required by the ^{11}B-NMR spectrum.

Support for the latter rearrangement has accumulated since 1980, for example, two sets of two skeletal carbons [67] and several sets of borons in $R_4C_4B_8H_8$ coalesce into "one kind" of carbon and fewer kinds of boron at almost ambient conditions (see Section 2.6.3.1 and Figure 2.50). Thus, the existence of fluxional intermediates, more complicated than those shown for the possible-dynamic, ni-8⟨VI⟩ structures of Figure 2.36, must be invoked.

The possible-dynamic ni-8⟨VI⟩ structure offers the greatest amount of negative charge (27%) to the carbon sites (offsetting the positive charge of the carbons) which would lead to the greatest amount of charge smoothing for carbon. However, minor optimization of the bridge hydrogen environments

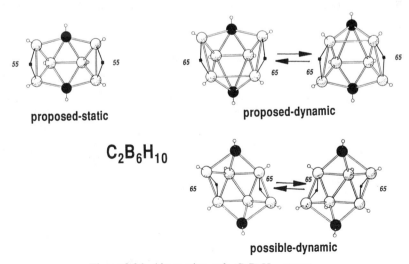

proposed-static   **proposed-dynamic**

$C_2B_6H_{10}$

possible-dynamic

Figure 2.36. Alternative *nido*-$C_2B_6H_{10}$ structures.

[72, 92] is suggested to be more important than carbon locations and thus the "proposed-static" configuration should be favored. An X-ray structural determination is needed [see 92b].

			ni-8⟨VI⟩
			(proposed-static)
13%	ni-8⟨V⟩	⟵	0.44⁻
	(proposed-dynamic)		0.44⁻
			———————
	0.50⁻		0.88⁻
	0.50⁻	$C_2B_6H_{10}$	
	———————		ni-8⟨VI⟩
	1.00⁻		(possible-dynamic) >27%
		⟶	0.67⁻
			0.67⁻
			———————
			1.34⁻

2.6.1.5. *Nido*-$B_8H_{10}L$ [69] (*structure uncertain*).

$Nido$-$B_8H_{10}L$ has been isolated [69] and a structure proposed (see Figure 2.37). The "proposed" structure incorporates an unacceptable 77-bridge hydrogen and two undesirable 75-bridge hydrogens. By simply relocating the Lewis base to another 4k boron and the three bridge hydrogens, as illustrated in the "possible" configuration, the unacceptable 77- and 75-bridge hydrogens are converted into desirable 66- and 65-bridge hydrogens (a major optimization [92]). In a Lewis base-on-boron moiety, the boron becomes quasi-isoelectronic with a carbon; it is treated here as a "surrogate" carbon.

Should the "possible" ni-8⟨V⟩ cluster hypothetically assume the more open "probable" ni-8⟨VI⟩ configuration (Figure 2.37), the Lewis base-on-boron moiety (surrogate carbon) would lose about 13% of its electron density as the

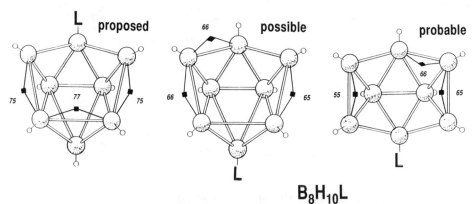

Figure 2.37. Alternative $B_8H_{10}L$ structures.

value 0.50^- would drop to 0.44^-. However, further minor optimization [92] of the bridge hydrogen [72] environments and other evidence (Section 2.6.7) suggests that all *nido*-8-vertex carboranes and polyboranes (including the isoelectronic $B_8H_{11}^-$) will be restricted to the heavily favored ni-8$\langle$VI$\rangle$ configuration excepting species incorporating strongly electronegative heteroatoms such as $S(CpCo)_2B_5H_7$ [50] (see Figure 2.35).

$$\begin{array}{cccc} & \text{ni-8}\langle\text{V}\rangle & & \text{ni-8}\langle\text{VI}\rangle \\ <13\% & 0.50^- & \xleftarrow{B_8H_{10}L} & 0.44^- \end{array}$$

2.6.2. *Nido*-10-Vertex Compounds

Three *nido*-10-vertex fragments may be generated from the *closo*-11-vertex deltahedron by the removal of 6*k*, 5*k*, and 4*k* vertices (see Figure 2.38). Of the various 5*k* vertices, only that specific kind of 5*k* vertex (which simultaneously removes the "offending" 6*k* (7-coordinate) vertex) was selected for removal; *nido*-

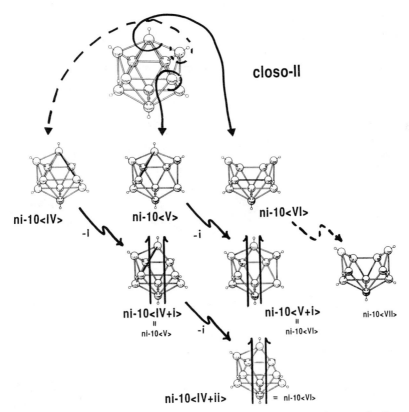

Figure 2.38. Dissection of the *closo*-11-vertex deltahedron into *nido*-10-vertex deltahedral fragments.

fragments retaining the "overly coordinated" 6*k* vertex are not expected, unless a transition element group, that could accommodate a 6*k* environment, is present.

Of the three 10-vertex *nido*-deltahedral fragments, the one with a IV-gonal open face (ni-10⟨IV⟩) is not expected to be represented within the carboranes as the *nido*-molecules are too large to opt for such a configuration in the absence of transition element groups. As if to illustrate the point, one Iridium derivative has been reported by Greenwood et al. with exactly the unanticipated ni-10⟨IV⟩ structure [77, 78].

Selected edge connections (blackened in Figure 2.38) are then removed and four more candidate fragments are generated. Close examination of these seven candidates reveals that three are redundant. Representative molecules with V-gonal (ni-10⟨V⟩) and VI-gonal (ni-10⟨VI⟩) open faces are anticipated.

One example of a rhodium compound, ni-10⟨VI + i⟩, with an aberrant VII-gonal open face has also been identified by Brown et al. [79] (see dotted arrow at the far right).

In Figure 2.39, are coded the hot, warm, and cool vertices of three 10-vertex *nido*-structures.

Representative structures are reproduced in Figure 2.40.

2.6.2.1. Nido-$B_{10}H_{14}$, $B_{10}H_{13}^-$, 5,6-$C_2B_8H_{12}$, and 5,7-$C_2B_8H_{12}$ (*structures known*). The ni-10⟨VI⟩ structure is preferred by most *nido*-10-vertex compounds as overriding considerations are almost always present. The ni-10⟨V⟩ configuration can never accommodate more than two bridge hydro-

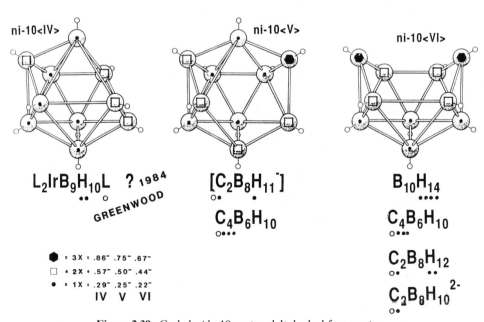

Figure 2.39. Coded *nido*-10-vertex deltahedral fragments.

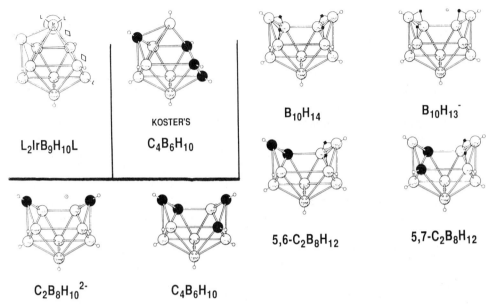

Figure 2.40. Known *nido*-10-vertex structures.

gens about the open face and if two bridge hydrogens are present, then only one carbon can be accommodated easily on the remaining 4*k* cage location; if two carbons are present, only one bridge hydrogen can be accommodated. It follows that $B_{10}H_{14}$, $B_{10}H_{13}^-$ [93,94] and the 5,6- [75] and 5,7- [74] isomers of $C_2B_8H_{12}$ must have ni-10⟨VI⟩ structures, as is experimentally observed.

2.6.2.2. Nido-6,9-C₂B₈H₁₀²⁻ [61] (structure known).

The carbons in $C_2B_8H_{10}^{2-}$ could be accommodated by either the ni-10⟨V⟩ or ni-10⟨VI⟩ configurations but the two carbons prefer the experimentally observed ni-10⟨VI⟩ configuration (see Figure 2.40), aided by charge smoothing, as its two hot locations are more negatively charged by about 7%.

ni-10⟨V⟩		ni-10⟨VI⟩	
0.75^-		0.67^-	
0.50^-	$\xrightarrow{C_2B_8H_{10}^{2-}}$	0.67^-	$<7\%$
1.25^-		1.34^-	(exp. obs.)

2.6.2.3. Koster's peralkyl derivative of nido-C₄B₆H₁₀ [64] (structure uncertain).

The structure of Koster's $C_4B_6H_{10}$ isomer, in contrast to the known structure of 5,6,8,9-$C_4B_6H_{10}$ [59], is of greatest interest. In Figure 2.41, the synthesis of Koster's peralkyl derivative of $C_4B_6H_{10}$ is shown. In both the "adamantane-like" parent and the *closo*-carborane "grandparent" [64], the carbons are in lowest possible coordination sites and are separated as far apart as possible. Reagglomeration of the carbons into three highest coordinated vertices and into a triangle of adjacent carbons in any carboranes by further heating are in direct opposition to charge smoothing forces and without precedent. The only carborane compounds where triangles of carbons are known to be incorporated with certainty are in the all carbon alkyl derivatives of $C_5H_5^+$ [54] (Figure 2.12) and $C_6H_6^{2+}$ [31] (Figure 2.6).

Nevertheless, if one combines the elegant two-dimensional (2D) NMR data of Koster and Wrackmeyer with a *static-traditional* structure of decaborane, ni-10⟨VI⟩, then the "published" structure at the right in Figure 2.41 has to be correct.

As Koster's peralkyl derivative of $C_4B_6H_{10}$ was fluxional above room temperature [64], therefore fluxional during synthesis and probably fluxional at and below room temperature, it seemed inconceivable that the carbons would not have had access, through rearrangement processes, to relocate into preferred low coordination sites.

For these reasons, the published structure of Koster's $C_4B_6H_{10}$ illustrated on the right in Figure 2.41 is highly unlikely. There are potential answers to this structure problem. The "published" structure (top right in Figure 2.42) is compared with three other structures, labeled "improbable", "candidate" and "possible", that could also be compatible with the 2D NMR data.

The "improbable" structure is shown as static and symmetrical. To properly visualize the "candidate" and "possible" configurations on the other hand, one must think of each of them as fluxional *dl* pairs rapidly interconverting between the enantiomers shown in Figure 2.42 and their mirror images (not shown) to make them compatible with the 2D NMR spectra.

The "improbable", static ni-10⟨IV⟩ structure incorporates four $4k$ sites, (optimal for carbon) and six $5k$ sites (ideal for boron) and appears satisfactory except that if the number of skeletal atoms is large enough, nature apparently prefers the more open "candidate" ni-10⟨V⟩ or "possible" ni-10⟨VI⟩ structures of Figure 2.42. A second reservation is that there would be no lower-coordinate $3k$ or $4k$ boron atoms in the "improbable" ni-10⟨IV⟩ structure to account for the lone ^{11}B-NMR resonance observed at low field representing one of the six borons; this argument was suggested by Hermanek [95a]. The "improbable" ni-10⟨VI⟩ isomer may be converted into the fluxional "candidate" ni-10⟨V⟩ configuration by the removal of one edge connection.

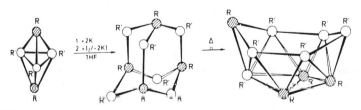

Figure 2.41. Synthesis of Koster's peralkyl derivative of *nido*-$C_4B_6H_{10}$.

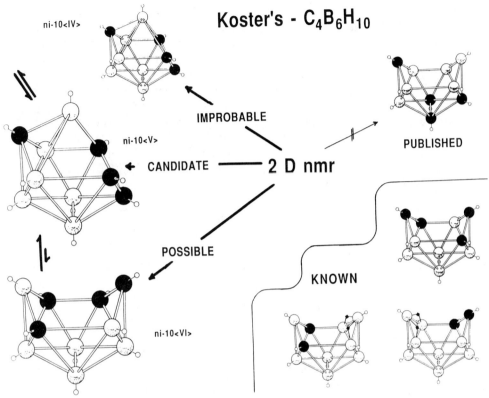

Figure 2.42. Alternative structures for Koster's *nido*-$C_4B_6H_{10}$.

The reversible relocation of the unique edge connection from the topmost $4k$ boron in the "candidate" ni-10⟨V⟩ configurations (Figure 2.42) to the two otherwise equivalent carbons, in a minimum motion, "wind shield wiper" type action, would make this tautomeric pair of "candidate" structures (with V-gonal open faces) totally compatible with the 2D NMR spectrum. Moreover, the topmost boron, $5k$ in the "improbable" ni-10⟨IV⟩ isomer becomes $4k$ in the "candidate" ni-10⟨V⟩ configuration which would give rise to the low field resonance representing one boron in the ^{11}B-NMR spectrum [95a]. The resonances representing the other five borons, in $5k$ positions, are appropriately clustered at higher field.

At the lower left in Figure 2.42, the traditional decaborane-like structure, ni-10⟨VI⟩, is labeled as "possible". The two halves of this "possible" ni-10⟨VI⟩ structure correspond to the carbon containing halves of two known compounds, $5,6$-$C_2B_8H_{12}$ [75] and $5,7$-$C_2B_8H_{12}$ [74] (shown at lower right), and therefore the "possible" isomer should be stable when the two halves are joined.

If this "possible" pair of tautomeric isomers turns out to be the most stable (see Section 2.6.7), then a convoluted rearrangement, almost certainly involving the "candidate" isomers as intermediates, would be required to take place at room temperature in order for compatibility with the 2D NMR spectra to be realized. The "possible" configuration also incorporates one $3k$ boron [95a].

The competitive "candidate" (broken arrows) and "possible" (solid arrows) alternative rearrangements are illustrated in Figure 2.43.

Charge smoothing analysis of the carbon locations favors the "candidate" ni-10⟨V⟩ configuration over the "possible" ni-10⟨VI⟩ configuration by 12% and the "possible" and "candidate" structures over the "published" structure by 51% and 40%, respectively. The "candidate", ni-10⟨V⟩ and "improbable", ni-10⟨V⟩, structures differ by little more thsn 1% so that even if a IV-gonal open faced configuration were acceptable, it would offer no charge smoothing advantage over the ni-10⟨V⟩ configuration.

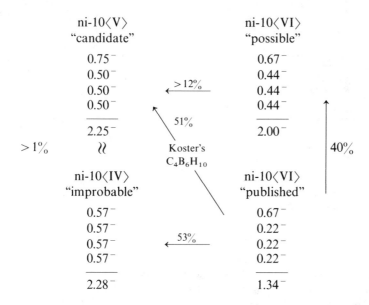

Initially the "candidate wind shield wiper" pair of tautomeric, ni-10⟨V⟩ structures for Koster's peralkyl derivative of $C_4B_6H_{10}$ [64] (Figure 2.43) were preferred [33] by geometrical systematics considerations as the major contributing isomers for both 2D NMR and charge smoothing arguments but in the context of present discussions, the dynamic "possible" isomers are favored (Figure 2.43, see also Section 2.6.7); an X-ray crystal structure is needed.

2.6.2.4. Hermanek et al.'s Nido-$C_4B_6H_{10}$ [59] (structure known).

A charge smoothing comparison of the isomers of $C_4B_6H_{10}$ is illuminating. In Figure 2.44, are shown the alternative ni-10⟨V⟩ and ni-10⟨VI⟩ structures in both cases. Recall that $C_2B_8H_{10}^2$ [61] (Figure 2.40) and $C_4B_4H_8$ [62] (Figure 2.14) could have assumed structures with V-gonal open faces, but instead chose carbon-preferred configurations, with more open VI-gonal open faces, assisted

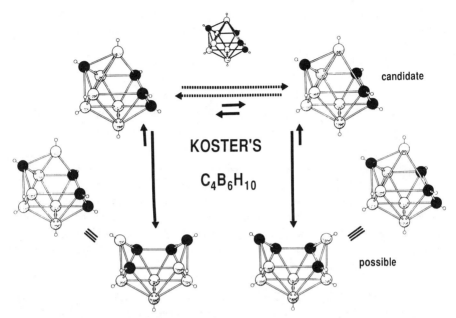

Figure 2.43. "Candidate" structures versus "possible" structures.

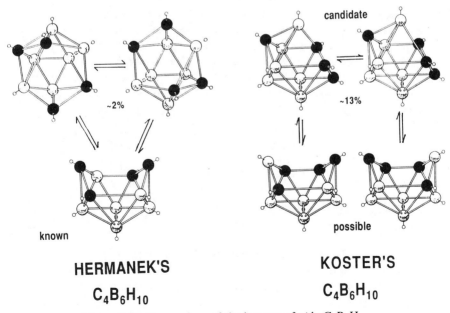

Figure 2.44. Comparison of the isomers of *nido*-$C_4B_6H_{10}$.

by charge smoothing in both cases. By analogy, then the same could be true for both Hermanek's [59] and Koster's [64] isomers of $C_4B_6H_{10}$.

	ni-10⟨V⟩		ni-10⟨VI⟩	
	0.75^-		0.67^-	
>1%	0.50^-		0.67^-	
	0.50^-	Hermanek's $C_4B_6H_{10}$	0.44^-	experimentally
	0.50^-	⟷	0.44^-	observed
	2.25^-		2.22^-	

The heretofore unexpected ni-10⟨V⟩ structure is favored over the ni-10⟨VI⟩ alternative by values of over 12% for Koster's compound, but by less than 2% (negligible), for Hermanek's compound. Perhaps nature selected the traditional ni-10⟨VI⟩ structures for the $C_4B_6H_{10}$ due to symmetry considerations or due to crystal packing forces. Alternatively, the ni-10⟨V⟩ structures might exist or coexist undetected in solution.

2.6.2.5. Nido-[$C_2B_8H_{11}^-$] [73] (structure unknown). Derived From 5,6-$C_2B_8H_{12}$ [75].

It is known that chiral, *carbons-adjacent* 5,6-$C_2B_8H_{12}$ (ni-10⟨VI⟩, structure known, Figure 2.40), can be resolved into one of its *d* or *l* isomers by conversion into its chiral monoanion of empirical formula $C_2B_8H_{11}^-$ (structure unknown) and that further deprotonation produces the nonchiral, *carbons-apart*, dianion 6,9-$C_2B_8H_{10}^{2-}$ [61] (ni-10⟨VI⟩ structure known, Figure 2.40). A tightly sterically controlled mechanism for the migrating carbons must be involved. We suggest that the monoanion $C_2B_8H_{11}^-$ (necessarily chiral) is central to both of these processes and that it, as a stable entity or at least as an intermediate, almost certainly has a ni-10⟨V⟩ structure (helped by a <12% charge smoothing advantage on carbon and possibly by an even greater influence of the lone bridge hydrogen) which in turn could control the carbon migrations during rearrangement (see Figure 2.45); an X-ray crystal structure of $C_2B_8H_{11}^-$ is needed.

	ni-10⟨V⟩		ni-10⟨VI⟩
<12%	0.75^-		0.67^-
expected	0.50^-	$C_2B_8H_{11}^-$	0.44^-
	1.25^-		1.11^-

2.6.2.6. Nido-$B_{10}H_{12}^{2-}$ (structure probably known).

$Nido$-$B_{10}H_{12}^{2-}$ could accommodate either the ni-10⟨V⟩ or the ni-10⟨VI⟩ configurations (Figure 2.39) with ≤66-bridge hydrogens. Preliminary X-ray data from Shore et al. [76] favor the latter ni-10⟨VI⟩ configuration for $nido$-$B_{10}H_{12}^{2-}$ in the crystal.

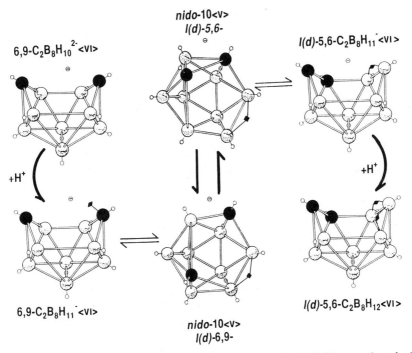

$6,9\text{-}C_2B_8H_{10}^{2-}$ <VI>

$nido\text{-}10$<v>
$I(d)\text{-}5,6\text{-}$

$I(d)\text{-}5,6\text{-}C_2B_8H_{11}^-$ <VI>

+H⁺

+H⁺

$6,9\text{-}C_2B_8H_{11}^-$ <VI>

$nido\text{-}10$<v>
$I(d)\text{-}6,9\text{-}$

$I(d)\text{-}5,6\text{-}C_2B_8H_{12}$<VI>

Figure 2.45. Interconversions of $nido\text{-}5,6\text{-}C_2B_8H_{12}$, $nido\text{-}C_2B_8H_{11}^-$, and $nido\text{-}6,9\text{-}C_2B_8H_{10}^{2-}$.

2.6.3. *Nido*-12-Vertex Compounds

The top half of Figure 2.46 shows the thirteen vertex deltahedron and the three primary fragments that result from the removal of the lone $4k$ vertex, a selected $5k$ vertex, and one of the two $6k$ vertices, respectively. These fragments are labeled ni-12⟨IV⟩, ni-12⟨V⟩ (Type 7) and ni-12⟨VI*⟩ (Type 5) in the top row. In the ni-12⟨IV⟩ and ni-12⟨V⟩ (Type 7) cases in Figure 2.46, both of the overly coordinated $6k$ vertices in the parent *closo*-13-vertex deltahedron are eliminated. We will return to the "Type" labels below [65–67].

The removal of one of the two $6k$ vertices to produce the *nido*-fragment ni-12⟨VI*⟩ (Type 5) leaves the other $6k$ vertex intact. The asterisk in ni-12⟨VI*⟩ indicates the presence of the remaining $6k$ vertex and distinguishes this VI*-gonal isomer from another ni-12⟨VI⟩ (Type 4/2) isomer with no $6k$ vertices and therefore no asterisk (see below).

Configuration ni-12⟨VI*⟩ (Type 5) has only been observed when there is an appropriately electropositive (i.e., "electron generous") transition element group available to occupy the remaining $6k$ vertex on the ni-12⟨VI*⟩ configuration (see Figure 2.15). It is assumed that a BH group in a $6k$ site would necessarily be involved in three skeletal 3c2e bonds to six neighboring cage atoms and that an

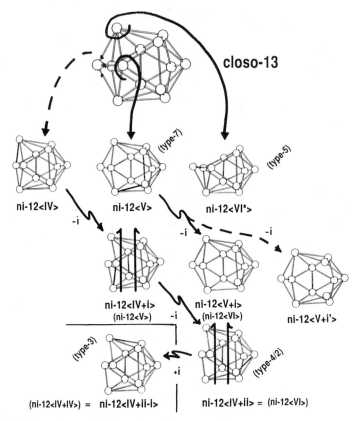

Figure 2.46. Dissection of the *closo*-13-vertex deltahedron into *nido*-12-vertex deltahedral fragments [33, 34].

electropositive transition element group would be more amenable to occupying such a $6k$ site than would a BH group. It is not expected that either the ni-12⟨IV⟩ or ni-12⟨VI*⟩ (Type 5) configurations will be observed in "pure" polyborane or carborane structures in the absence of skeletal transition element groups or appropriate heteroatoms.

Various edge connections may be broken (blackened in Figure 2.46) generating four additional structures in Figure 2.46 (the boxed "Type 3" structure is not counted among the four). The one at the far right, ni-12⟨V + i′⟩ with alternating hot and cool sites about the open VI-gonal face, is not expected to be observed because of its inability to accommodate any bridge hydrogens; however, it is included in Figure 2.46 for completeness. Two of the four additional structures are redundant and elemenated as is indicated by the double vertical sets of arrows through their structures in Figure 2.46. If the improbable outriding structures (ni-12⟨IV⟩ and ni-12⟨V + i′⟩) are excluded, there are then three remaining fragments with "expected" configurations (unboxed) in the center two columns of Figure 2.46. These three are clockwise from top left, ni-12⟨V⟩ (Type 7), top right, ni-12⟨VI*⟩ (Type 5) and middle right, ni-12⟨IV + ii⟩ (the latter is re-identified as ni-12⟨VI⟩) (Type 4/2).

This latter species has a VI-gonal open face and if, in an unorthodox and unexpected way, a bond is reintroduced symmetrically across the VI-gonal open face, a pair of IV-gonal open faces are produced. This fourth unexpected configuration is shown in the box at lower left as the ni-12⟨IV + ii − i⟩ structure and is re-identified as the ni-12⟨IV + IV⟩ (Type 3) configuration.

These four remaining structures, after re-identification, are clockwise from top left, ni-12⟨V⟩ (Type 7), top right, ni-12⟨VI*⟩ (Type 5), middle right, ni-12⟨VI⟩ (Type 4/2) and lower left, ni-12⟨IV + IV⟩ (Type 3) in the center two columns of Figure 2.46. These four structures coincide with four of Grimes' structures in Figure 2.47.

Grimes [66] listed no less than seven types of *nido*-12-vertex configurations of which four are illustrated in Figure 2.47. This was surprising as all other *nido* families had no more than two candidates (e.g., ni-8- and ni-10-compounds) and most only one allowed *nido*-configuration (e.g., ni-5-, -6-, -7-, -9-, and -11-compounds) (see Figure 2.15).

In Figure 2.47, the earlier Grimes/Sinn figure [65, 66] was rearranged in order to show four of their seven types in the same relationship to each other as these same configurations were illustrated in Figure 2.46. They later found their Type 2 compound (structure unknown at the time) to be isostructural with Type 4. Our main point is that the deltahedron dissection systematics (Figure 2.46) produced three reasonable structures [33, 34] [ni-12⟨V⟩ (Type 7), ni-12⟨VI*⟩ (Type 5), and ni-12⟨VI⟩ (Types 2 and 4] that match three of the four structures that had previously been reported by Grimes et al. (Figure 2.47) and that by suitable manipulation, the fourth ni-12⟨IV + IV⟩ (Type 3), could be generated (see Figure 2.46).

Types 1 and 6 are missing and are considered to be outside of the present discussion; they are viewed arbitrarily as related to *nido*- and *arachno*-11-vertex

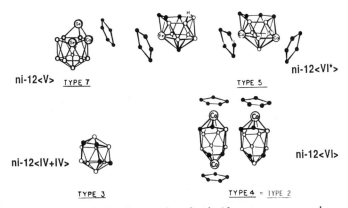

ni-12<V> TYPE 7   TYPE 5   ni-12<VI*>

ni-12<IV+IV>   ni-12<VI>

TYPE 3   TYPE 4 = TYPE 2

Figure 2.47. Grimes' examples of *nido*-12-vertex compounds.

compounds with electron precise external scaffolding which will not be included in this analysis of cluster structures. In the present systematics, all electron precise "external scaffolding" and terminal groups are "notionally sheared off" and hypothetically replaced with terminal hydrogens (see Figure 2.48). The carbons in external scaffolding are ignored just as the carbons in terminal methyl or ethyl groups are not counted as skeletal atoms. Others [42, 66, 67] do not adhere to this practice.

Following the preparation of this chapter, Hawthorne [95b] reported the structure of the unstable C,C′-dialkyl isomer of $C_2B_{10}H_{13}^-$ (to be compared to the stable isomer, Type -1 at the left of Figure 2.48). Hawthorne's new structure [95b] (not shown) is related to the ni-12⟨VI*⟩ (Type 5) structure of Figure 2.49 but differs in that one $6k$ to $4k$ connection is removed to eliminate the residual $6k$ vertex (undesirable for boron). A metastable ni-12⟨VI + IV⟩ configuration is produced.

If the $3k$ skeletal carbon is considered to be electron precise external scaffolding, then the RC≡ group may notionally be replaced by three endo hydrogens to produce an *arachno*-11⟨VII⟩ configuration (Stx = 383) similar to that illustrated at the right of Figure 2.48 but with one less edge connection leading to the VII-gonal open face. Both VI-gonal and VII-gonal open faces are favored among *arachno* compounds (see Figures 2.16 and 2.26).

One other caveat is necessary, Grimes' unique Type 3 compound, ni-12⟨IV + IV⟩ in Figure 2.47, reflects a configuration which would never have been contemplated [33, 34] unless deliberately attempting to make the deltahedral fragment systematics, displayed in Figure 2.46, correlate with Grimes' four types in Figure 2.47.

Nido-$R_4C_4B_8H_8$, with the Type 3 or ni-12⟨IV + IV⟩ structure in Figure 2.47 has one localized 2c2e bond between the central carbons. If this isolated electron precise 2c2e bond was hypothetically removed and replaced with two endo terminal hydrogens, an *arachno*-$R_4C_4B_8H_{10}$ would result (not shown in Figure 2.47). Such an *arachno*-compound, ara-12⟨VI⟩ (Stx = 582), would be "almost isoelectronic" with its *nido*-$R_4C_4B_8H_8$ precursor, ni-12⟨IV + IV⟩. Certainly if

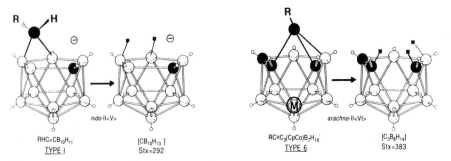

Figure 2.48. External scaffolding removal converts pseudo-12-vertex *nido*-compounds to 11-vertex *nido*- and *arachno*-compounds.

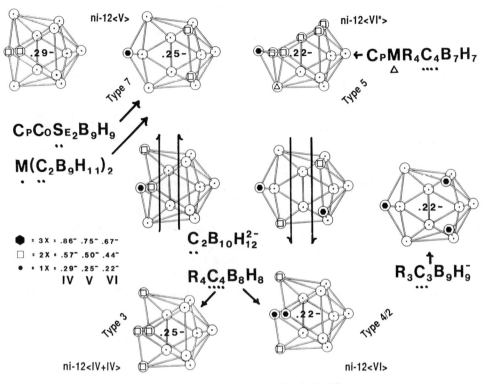

Figure 2.49. Coded *nido*-12-vertex deltahedral fragments.

the isolated 2c2e bond were notionally replaced by an endo-methylene ($-CH_2-$) or ethylene group ($-CH_2CH_2-$), such groups would be excised as electron-precise scaffolding and replaced with two endo terminal hydrogens. Why not a similar treatment for an isolated "endo-2c2e bond" ($-(CH_2)_o-$)?

The pseudo *arachno*-structure so generated is probably representative of nature's preferred (but as yet unknown) *arachno*-12-vertex deltahedral fragment as it would be the only *arachno*-12-vertex fragment that could be produced by removing two adjacent 5k vertices from the most spherical *closo*-14-vertex deltahedron that simultaneously would remove both 6k vertices (see Figure 2.24).

Two examples are shown for Grimes' Type 4 compounds, ni-12⟨VI⟩, in Figure 2.47. They incorporate two transition element groups in one case and one transition element group in the other. Why not a Type 4 structure with no transition element groups? In fact, Grimes' Type 2 structure was later found [67] to have the same Type 4 structure with no transition element group; thus Type 2 and Type 4 compounds are grouped together in Figure 2.47.

In Figure 2.49, the vertices in all of the *nido*-12-vertex structures were coded and the four acceptable Grimes' configurations of Figure 2.47 were double-labeled according to both Grimes' and Williams' conventions. The outriding structures on the far left and far right will be ignored in the following discussion. The structures with double arrows in the middle of Figure 2.49 are redundant; the double arrows pointing out the identical configurations.

Structure ni-12⟨V⟩ (Type 7 in Figure 2.49) incorporates only one hot and two warm locations. There are no hot or warm positions on the cages in any of the four acceptable (double-labeled) ni-12-vertex-structures thus, any bridge hydrogens and heteroelements, C, N, S, etcetera, must compete for hot and warm sites about the various open faces.

2.6.3.1. The $R_4C_4B_8H_8$ Compounds [66] (structures known).

At the bottom of Figure 2.49 are the two structures, ni-12⟨IV + IV⟩ (Type 3) and ni-12⟨VI⟩ (Type 2/4). The empirical formula $R_4C_4B_8H_8$ is allocated to both structures. Each of these structures have four sites about the open face for the four carbons, that is, four warm $4k$ locations on the ni-12⟨IV + IV⟩ (Type 3) configuration and two hot, $3k$, and two warm, $4k$, locations on the ni-12⟨VI⟩ (Type 4/2) configuration. The $R_4C_4B_8H_8$ structure wherein R = Me, Et, and Pr, use all four hot and warm locations in both configurations simultaneously when in solution (Figure 2.50)!

ni-12⟨IV + IV⟩ (ara-12⟨VI⟩) (Type 3)	ni-12⟨V⟩ (Type 7)		ni-12⟨VI⟩ (Type 4/2)	
	0.75^-			
	0.50^-			
	0.50^-			
	0.25^-			
	——		0.67^-	
	2.00^-		0.67^-	
0.50^-		Grimes'	0.44^-	$<10\%$
0.50^-		$\underrightarrow{R_4C_4B_8H_8}$	0.44^-	
0.50^-			——	
0.50^-			2.22^-	
——				
2.00^-				

When R = methyl in $R_4C_4B_8H_8$, only the ni-12⟨IV + IV⟩ (Type 3) structure is observed in the crystal. In contrast, when R = ethyl in $R_4C_4B_8H_8$ only the alternative ni-12⟨VI⟩ (Type 4/2) structure is observed in the crystal. When crystals of either $Me_4C_4B_8H_8$ or $Et_4C_4B_8H_8$ are dissolved, ^{11}B-NMR spectra reveal that both the ni-12⟨IV + IV⟩ and ni-12⟨VI⟩ structures are fluxional and coexist in solution in equilibrium with each other [67]. Equilibrium is attained slowly enough so that when dissolved, the crystalline configurations predominate at first, but as time passes, the other configuration "grows" into the NMR spectrum at the expense of the structure seen in the crystal.

The ni-12⟨VI*⟩ (Type 5) structure has one hot and five warm locations about

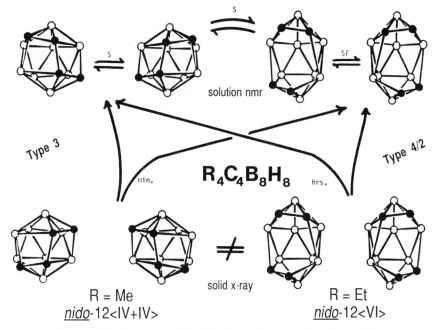

Figure 2.50. Structures of $R_4C_4B_8H_8$ in the crystal and in solution.

the open face (Figure 2.49) and seemingly could also accommodate at least four carbons with ease; this structure, incorrectly considered Williams' "first choice" [39] in 1976, is seen only when a transition element group is present to occupy the remaining $6k$ vertex [66] (identified by the open triangle).

Returning to a comparison of the four competitive structural types illustrated in Figures 2.47 and 2.49, it is obvious that if there is a need to find warm or hot positions for four carbon atoms, the ni-12⟨V⟩ configuration cannot optimally accommodate four carbons as it has only three hot/warm locations.

Certain molecules which contain a transition element group can exist in both ni-12⟨VI⟩ (Type 4/2) and ni-12⟨VI*⟩ (Type 5) configurations in Figures 2.47 and 2.49. Compounds have been prepared, initially having a ni-12⟨VI⟩ (Type 4/2) structure, which incorporate one or two transition element groups. Such compounds, upon heating, rearrange to produce ni-12⟨VI*⟩ (Type 5) structures [66] so the ni-12⟨VI*⟩ (Type 5) structure is more stable when one transition element group is present.

2.6.3.2. The Nido-Dianions Produced From Closo-1,2-, 1,7-, and 1,12-$C_2B_{10}H_{12}$ [96–102] (structures unknown).

It has been known for over 20 years that the icosahedral *closo*-carboranes, 1,2-, 1,7-, and 1,12-$C_2B_{10}H_{12}$ can be converted into two *nido*-12-vertex dianions by the addition of two electrons (the same dianion, $C_2B_{10}H_{12}^{2-}$ (1,2; 1,7) is produced from both 1,2- and 1,7-$C_2B_{10}H_{12}$. Such dianions have only two carbons that should require hot or warm locations; the NMR spectra of these two dianions are indistinct and have not been elucidated. The ni-12⟨V⟩ (Type 7) structure (see Figure 2.49) would make only 1.25⁻ charge available to the two carbons. The $C_2B_{10}H_{12}^{2-}$ (1,2; 1,7) dianion could also be accommodated by the ni-12⟨IV + IV⟩ (Type 3) and ni-12⟨VI⟩ (Type 4/2) structures at the bottom of Figure 2.49 and reproduced in Figure 2.51. The latter, ni-12⟨VI⟩ (Type 4/2) configuration, would make 1.34⁻ charge available to the two carbons which is 29% greater than in ni-12⟨IV + IV⟩ and > 14% greater than in ni-12⟨V⟩. The ni-12⟨VI⟩ configuration should be favored.

ni-12⟨IV + IV⟩ (ara-12⟨VI⟩) (Type 3)	ni-12⟨V⟩ (Type 7)		ni-12⟨VI⟩ (Type 4/2)
0.50⁻			
0.50⁻			
———		⟶	29%
1.00⁻			(if comparison is justified)
		$C_2B_{10}H_{12}^{2-}$ (1,2; 1,7)	0.67⁻
			0.67⁻
	0.75⁻		1.34⁻
	0.50⁻	⟶	> 14%
	———		
	1.25⁻		

The marginal equivalence of two or three acceptable *nido*-12-vertex configurations for a compound such as $C_2B_{10}H_{12}^{2-}$ and the probable fluxional behavior involving one or two, if not all three structures, might account for the indistinct NMR spectra.

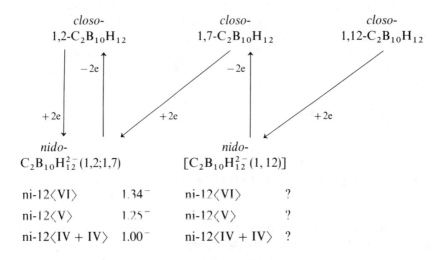

ni-12⟨VI⟩	1.34⁻	ni-12⟨VI⟩	?
ni-12⟨V⟩	1.25⁻	ni-12⟨V⟩	?
ni-12⟨IV + IV⟩	1.00⁻	ni-12⟨IV + IV⟩	?

The question of why the *nido*-$C_2B_{10}H_{12}^{2-}$ (1,2; 1,7) dianion (lower left above), formed from the two electron reduction of *closo*-1,7-$C_2B_{10}H_{12}$ (with nonadjacent carbons), upon reoxidation to a neutral species, produces 1,2-$C_2B_{10}H_{12}$ (with adjacent carbons) is buried in the two or three structures listed below *nido*-$C_2B_{10}H_{12}^{2-}$ (1,2; 1,7). The conversion of *closo*-1,7-$C_2B_{10}H_{12}$ into the thermodynamically less stable *closo*-1,2-$C_2B_{10}H_{12}$ must be the result of kinetic control. The two carbons probably opt for the two hot locations in the anionic ni-12⟨VI⟩ (Type 4/2) isomer (bottom left in Figure 2.51) and thus are preconditioned to form a single 2c2e bond between the two carbons (forming the anionic ni-12⟨IV + IV⟩ (Type 3) isomer (top left in Figure 2.51). This manner is the same as was demonstrated by Grimes [66] to take place in the isoelectronic analogs which contain four carbons, $R_4C_4B_8H_8$, (ni-12⟨VI⟩ ↔ ni-12⟨IV + IV⟩) (compare Figures 2.50 and 2.51). This process could then go to completion as the dianion is oxidized to the neutral species; that is, two IV-gons collapse to four III-gons as illustrated at the far left in Figure 2.51.

2.6.3.3. *"Slipped" Compounds* [103, 104] *(structures known)*. One
of the two preferred configurations, ni-12⟨V⟩ (Type 7), in Figures 2.47 and 2.49, incorporates only one hot and two warm locations. Such a structure can only

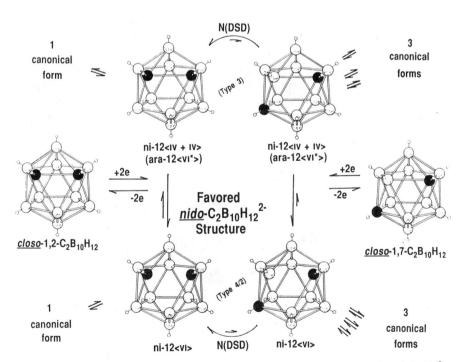

Figure 2.51. Alternative ni-12-⟨VI⟩ and ni-12⟨IV+IV⟩ structures for $C_2B_{10}H_{12}^{2-}$ (1,2;1,7) derived from both *closo*-1,2 and 1,7-$C_2B_{10}H_{12}$.

accommodate one bridge hydrogen and no more than one carbon "easily" if there is one bridge hydrogen and never more than three carbons.

There are a number of compounds that would require three or less hot and warm locations. One such compound is the diselenium molecule illustrated in Figures 2.47 and 2.52 which was discovered by Todd et al. [105] and listed as Type 7 by Grimes [66]. The two seleniums need hot or warm locations and indeed (if the ni-12⟨V⟩ configuration is preferred), the two electronegative seleniums occupy the hot location, 0.75^- and one of the warm locations, 0.50^- for a total of 1.25^-. Another important feature to note is the fact that the electropositive (relative to BH groups) and therefore "electron generous" CpCo group assumes the highest coordinated location [91] that is available, adjacent to both electronegative selenium atoms. If the seleniums were hypothetically relocated to the two hot, 0.67^-, locations (total 1.34^-) of the ni-12⟨VI⟩ (Type 4/2) configuration, they would gain 7% towards charge smoothing but would lose the support of the "electron generous" CpCo group to one of the seleniums atoms.

As the presence of electronegative sulfur seems to be associated with several smaller apertured configurations in *nido*-6-, -8- and -9-vertex compounds; selenium probably plays a similar role (see Section 2.6.7).

Compounds long misidentified as distorted *closo*-compounds were discovered by Hawthorne in the late 1960s. Such a "slipped" compound [103, 104] is reproduced at the left in Figure 2.25. The connectivities from the metal to three borons in each of the two cages were added to emphasize the symmetry; the metal is not connected to the carbons.

These compounds have *nido*-electron counts and *nido*-empirical formulae but have generally been considered to simply have slightly distorted or "slipped" *closo*-iscosahedral structures. In actuality, these slipped configurations sub-

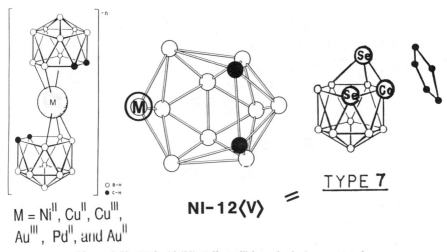

$M = Ni^{II}, Cu^{II}, Cu^{III},$
$Au^{III}, Pd^{II},$ and Au^{II}

NI-12⟨V⟩ $=$ TYPE 7

Figure 2.52. *Nido*-12⟨V⟩ "slipped" icosahedral compounds.

scribe to the expected ni-12⟨V⟩ (Type 7) deltahedral fragment produced from the 13 vertex deltahedron as shown in Figures 2.49 and 2.52 [39].

The metal atoms occupy the hot locations of two interlocking *nido*-12-vertex fragments with the carbons occupying, as expected, the warm locations after the metal occupation of the hot position(s). The metal atoms between the two cage moieties, while occupying a hot location with respect to each *nido*-12-vertex deltahedral fragment, is actually in a "very highly connected" location when the entire bicage system is considered. The electropositive metal atom, in "generously" satisfying the electronic appetite of the common "very highly connected" vertex environment, is compensated by being allowed access to the "two" 0.75⁻ "hottest" locations as far as each *nido*-12-vertex cage is considered. The two "hot" locations (0.75⁻) in the two conjoined ni-12⟨V⟩ (Type 7) configurations are of greater value to the metal atom than two "hot" locations (0.67⁻) in the alternative ni-12⟨VI⟩ (Type 4/2) configurations.

2.6.3.4. *Nido*-[CB₁₁H₁₂³⁻], *Nido*-CB₁₁H₁₃²⁻ (*structures unknown*).
Closo-CB₁₁H₁₂⁻ is a known compound [106] (see Figure 2.53) and should be capable of adding two electrons to open into *nido*-[CB₁₁H₁₂³⁻]. The ni-12⟨V⟩ (Type 7) structure would be favored by charge smoothing with the lone carbon in the 0.75⁻ site; partial neutralization with one proton could place a 66-bridge hydrogen between the two 0.50⁻ borons in the ni-12⟨V⟩ (Type 7) configuration producing *nido*-[CB₁₁H₁₃²⁻] (see Figure 2.53) if electron-precise –CH₂– extrusion does not take place.

If ni-12⟨VI⟩ (Type 4/2) structures are favored for these boron-carbon skeletons (see Section 2.6.7) then the "blackened" connections in the *nido*-configurations in Figure 2.53 should be removed. The carbons in both structures would lose electron density (0.67⁻ decreasing to 0.50⁻) but the resulting 65-

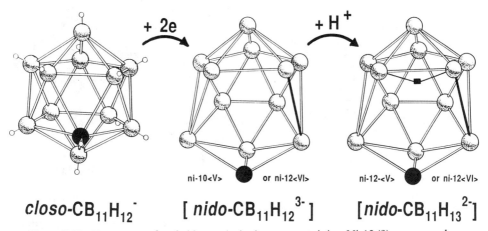

Figure 2.53. Heteroatom free bridge-endo hydrogen containing Ni-12⟨?⟩ compounds.

bridge hydrogen would gain (minor optimization [92], 1.00^- to 1.11^-). Both species would be expected to be fluxional.

The two foregoing possibilities should be compared to the partial neutralization of $C_2B_{10}H_{12}^{2-}$ (1,2) (illustrated in Figure 2.51) to form $C_2B_{10}H_{13}^-$ [107, 108] following which one $-CH_2-$ group is "extruded" and a derivative of the *nido*-11-vertex compound $CH_2CB_{10}H_{11}^-$ is formed (Stx = 292) (see Figure 2.48).

2.6.4. *Nido*-6-Vertex Compounds [39] (*structures known*)

The even-numbered 12-, 10-, and 8-vertex classes of *nido*-compounds have been discussed as well as many *nido*-6-vertex compounds (see Figure 2.6) related to B_6H_{10}. However, possible alternative *nido*-6-vertex structures have not been investigated.

If $4k$ and $5k$ vertices are removed from the *closo*-7-vertex deltahedron (see Figure 2.54), both the pyramidal configuration, ni-6⟨V⟩, is produced as well as an "envelope" shaped configuration with a IV-gonal open face, ni-6⟨IV⟩ (see Figures 2.15 and 2.54).

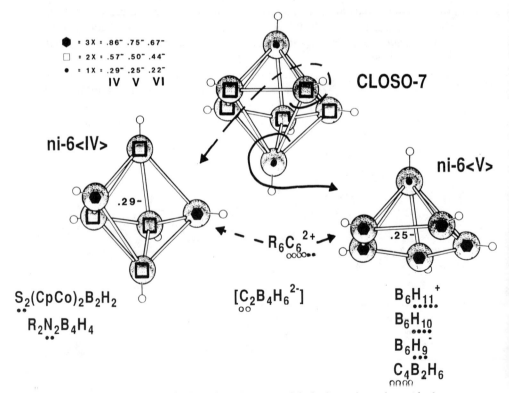

Figure 2.54. Dissection of the *closo*-7-vertex deltahedron into its *nido*-6-vertex deltahedral fragments.

The small IV-gonal open face of pentaborane (Figure 2.2) was rationalized because its precursor, the *closo*-octahedron (Figure 2.3), had only $4k$ vertices available for removal. The probabilities that larger *nido*-boron/carbon skeletons will have IV-gonal open faces were de-emphasized in the absence of transition element groups, because all are large enough to have V-gonal and frequently VI-gonal open faces (see Figure 2.15). While this reluctance to accept *nido*-compounds with IV-gonal open faces seems justified, the *closo*-deltahedron in Figure 2.54 is only one vertex larger than the *closo*-octahedron. Consequently, it was felt that $4k$ vertex removal should be considered here at the boundary between (1) deltahedra that are too small to generate deltahedral fragments with V-gonal open faces and (2) deltahedra that are just large enough to have V-gonal open faces.

In Figure 2.54 are displayed eight different empirical formulae. Solid dots underline bridge hydrogens that *must* have edge positions about the open face. The carbons are underlain by circles to indicate that they should have access to hot or warm locations while the sulfurs and nitrogens have solid dots reflecting the empirical observation that the more electronegative sulfur and nitrogen atoms always occupy hot edge sites in any *nido*-derivatives.

Both the envelope (ni-$6\langle IV \rangle$) and pentagonal pyramid (ni-$6\langle V \rangle$) structures are richly endowed with hot and warm vertices. It is apparent that the ni-$6\langle IV \rangle$ (envelope) configuration could support no more than two bridge hydrogens about the open face and thus compounds such as $B_6H_9^-$, B_6H_{10} [29], and $B_6H_{11}^+$ [30] with three or more bridge hydrogens (see Figure 2.6) must subscribe to the pentagonal pyramidal configuration.

It seemed similarly, if other things were equal, that the carbons in the $R_6C_6^{2+}$ cation [31] should have preferred the "envelope" configuration as all the vertices of the envelope configuration are hot and warm locations (see ni-$6\langle IV \rangle$; Figure 2.54) in contrast to the pentagonal pyramidal alternative (see ni-$6\langle V \rangle$; Figure 2.54) which incorporates a single electron poor, cool location which should not be attractive for carbon. The pyramidal ni-$6\langle V \rangle$ structure for $R_6C_6^{2+}$ prevails [31], however, in spite of the cool apex position, so all other things are not equal and the envelope ni-$6\langle IV \rangle$ configuration must be less favored.

Ignoring the cool apex position, the envelope "alternative" has two very hot locations of 0.86^- value in contrast to the hottest locations of 0.75^- value in the pentagonal pyramid. Hence, even though the pyramidal configuration may be geometrically favored, it was wondered [33, 34] if charge smoothing (i.e., carbons gaining access to greater electron density) could possibly cause the ni-$6\langle IV \rangle$ configuration to be favored over the ni-$6\langle V \rangle$ structure in selected cases.

What carborane structure could optimally utilize the two hot, 0.86^-, locations in the "envelope" alternative and possibly override the geometrical preference for the pentagonal pyramid? Williams suggested [33, 34] the dianion, $C_2B_4H_6^{2-}$ with nonadjacent carbons and no bridge hydrogens as containing two carbons that could ideally utilize the two hot 0.86^- hot locations in the "envelope" configuration.

Hosmane's related carbons-adjacent isomer [109] of $C_2B_4H_6^{2-}$ has the pyramidal structure, but this is expected as the two carbons are adjacent in the necessarily pyramidal precursor, $C_2B_4H_8$, and the carbons in Hosmane's $C_2B_4H_6^{2-}$-isomer could not occupy the envelope's two 0.86^- hot vertices without rearrangement.

The envelope structure at the left of Figure 2.54 should be ideal for $C_2B_4H_6^{2-}$ (with nonadjacent carbons) if charge smoothing was of overriding importance. It is favored by $<14\%$ but in spite of the charge smoothing advantage, the ni-$6\langle V\rangle$, pyramidal structure has recently been experimentally observed [68].

	ni-6⟨IV⟩		ni-6⟨V⟩	
	0.86^-		0.75^-	(experimentally
$<14\%$	0.86^-	$\xrightarrow{C_2B_4H_6^{2-}}$	0.75^-	observed)
	$\overline{1.72^-}$		$\overline{1.50^-}$	

Some time before the structure [68] of 2,4-$(CH_3)_2C_2B_4H_4^{2-}$ was determined, Schleyer was queried about the possibility of the carbons-nonadjacent $C_2B_4H_6^{2-}$-dianion having the envelope structure. Schleyer's ab initio calculations [110] favored the pentagonal pyramid structure for $C_2B_4H_6^{2-}$ as illustrated in Figure 2.55.

Paetzold [111] reported an isoelectronic and isostructural nitrogen analog, $R_2N_2B_4R_4$, that had exactly the envelope structure that Williams had anticipated [33, 34] for $C_2B_4H_6^{2-}$. Sneddon and Kang [112] subsequently reported the compound $S_2(CpCo)_2B_2H_2$ and proposed that it had the envelope structure wherein the electronegative sulfurs occupied the hot 0.86^- vertices. Thus, it seems that a reasonable rationale was put forth and that the envelope structure was indeed a viable candidate structure for some *nido*-6-vertex compounds, but the two carbons in $C_2B_4H_6^{2-}$, having offset only half of the negative charge of the *nido*-core-cluster $[B_6H_6^{4-}]$, are not nearly as electronegative as the two nitrogens or sulfurs that offset all of the 4^- charge. The charge smoothing forces in $C_2B_4H_6^{2-}$ are apparently not strong enough to override the geometrical preference for the ni-$6\langle V\rangle$, pentagonal pyramid.

In $S_2(CpCo)_2B_2H_2$, the two electropositive CpCo groups are located at those two specific $4k$ vertices in the ni-$6\langle IV\rangle$ configuration that allow them to contribute electrons to both sulfur atoms simultaneously. If the electronegative sulfurs cause the envelope structure to be preferred, to satisfy their electron appetite for the 0.86^- vertices, then the electropositive (electron generous) CpCo moieties [91] probably tend to favor the same configuration.

2.6.5. *Nido*-11-,9-, and 7-Vertex Configurations Have Only V-gonal Open Faces

In Figure 2.15, circles are found opposite both $B_9H_9^{4-}$ and $B_{11}H_{11}^{4-}$ in the VI-gonally open faced column. These circles suggest that there are no satisfactory *nido*-9-vertex and *nido*-11-vertex configurations with VI-gonal open faces. *Nido*-7-vertex compounds are simply too small to have ni-$7\langle VI\rangle$ structures.

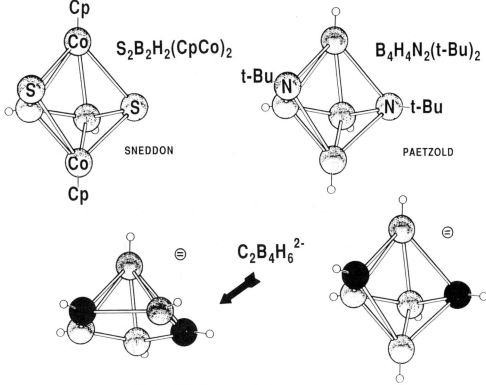

Figure 2.55. Known *nido*-6-vertex structures.

2.6.5.1. *Nido-11-Vertex Compounds.*

Consider removing one edge connection ("blackened" in Figure 2.56) between adjacent carbons about the V-gonal open face of the traditional ni-11⟨V⟩ configurations to produce ni-11⟨VI⟩ configurations which resemble the ara-11⟨VI⟩ configuration at far right in Figure 2.46. Two 0.67^- and two 0.44^- vertices would notionally be produced from four previously 0.50^- vertices. Such 0.67^- vertices could harbor two carbons in Hermanek's and Sneddon's "probable" $C_4B_7H_{11}$ ni-11⟨VI⟩ compounds (see Figure 2.56).

The carbons would have over a 10% advantage in charge smoothing in such ni-11⟨VI⟩ configurations but no examples of ni-11⟨VI⟩ structures have been observed. At least three isomers of $C_4B_7H_{11}$ [58, 113] have been found, but all have the ni-11⟨V⟩ configuration. This reasoning indicates that ni-11⟨VI⟩ configurations are unsatisfactory, hence the circle in Figure 2.15.

There is also another line of evidence; $B_{11}H_{13}^{2-}$ has the expected [89, 90] (Stx = 670, ni-11⟨V⟩), structure and while $B_{11}H_{14}^-$ also has the ni-11⟨V⟩ configuration [114], one of its "extra three" hydrogens becomes an endo hydrogen (Stx = 481) to avoid the incorporation of two 76-bridge hydrogens in a Stx = 670 configuration (Figure 2.56). *Nido*-11-vertex compounds, of either

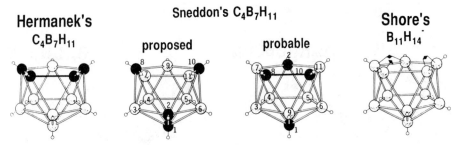

Hermanek's
$C_4B_7H_{11}$

Sneddon's $C_4B_7H_{11}$

proposed

probable

Shore's
$B_{11}H_{14}^-$

Figure 2.56. *Nido*-11-vertex compounds have ni-11⟨V⟩ structures despite charge smoothing advantages of a larger aperture.

Stx = 670 or 481, apparently have only one candidate skeletal configuration, ni-11⟨V⟩.

The alternative known structures of $B_{11}H_{14}^-$, and Hermanek's et al.'s $C_4B_7H_{11}$ isomer [58] and the "proposed" and "probable" structures for Sneddon's high temperature stable $C_4B_7H_{11}$ isomer [113] are displayed in Figure 2.56.

The five 4k edge vertices of $C_4B_7H_{11}$'s hypothetical ni-11⟨V⟩ precursor, $[B_{11}H_{11}^{4-}]$, would have negative values of 0.5^- prior to carbon substitution. Following substitution of four of the five edge borons with carbons, Hermanek's isomer [58] has four 0.5^+ edge vertices occupied by carbons in close proximity and only one 0.5^- edge vertex occupied by boron.

If charge smoothing is truly a major influence, then perhaps an even more stable $C_4B_7H_{11}$ isomer would have one carbon relocated on the cage (see "probable" isomer) substituting for a $5k, 0.25^-$ boron and notionally producing a 0.75^+ carbon vertex. One carbon would "lose" in the charge smoothing "competition" but the other three carbons and the entire system would gain in overall charge smoothing. The vertices in Sneddon's "proposed" $C_4B_7H_{11}$ structure are numbered correctly in accordance with Sneddon's report [113] while the suggested "probable" configuration based on the present treatment is numbered arbitrarily in order that the "proposed" [113] and "probable" $C_4B_7H_{11}$ structures may be compared as follows.

In NMR spectroscopy, the borons and carbons that are connected to the greatest number of adjacent borons and carbons (within a given molecule) are almost always found at higher field than are lower connected borons or carbons. The fact that the 1-^{13}C resonance is found at highest field [113] (as befits 5k carbons) favors both the "proposed and probable" structures of Sneddon's high temperature isomer but the fact that the 8-, 10-, and 2-^{13}Cs (all 4k in the "probable" structure) are found almost superpositioned at lower field supports the "probable" structure over the "proposed" structure wherein the 2-^{13}C is a 5k carbon and should have been found at higher field. Secondly, that the resonance for the 8- and 10-^{13}Cs are equivalent quartets, that is, each ^{13}C is coupled with *one* ^{11}B (the 8-^{13}C with the 3-^{11}B and the 10-^{13}C with the 6-^{11}B) and the 2-^{13}C is similarly coupled with *two* ^{11}Bs (in 64% of the cases) yielding a not quite resolved septet strongly favors the "probable" configuration. Thirdly, the incorporation of three rather than two carbons in the low coordinated, $4k$, edge sites, favored by charge smoothing, also supports the "probable" configuration over the "proposed structure" which only incorporates two carbons in low connected, $4k$, edge positions.

Hermanek's isomer of $C_4B_7H_{11}$ [58] might be expected to rearrange into Sneddon's "probable" high temperature isomer upon heating.

2.6.5.2. Nido-9-Vertex Compounds.

The symmetrical edge connection of the ni-9⟨V⟩ configuration, characteristic of *nido*-compounds such as $B_9H_{12}^-$ [88] or C,C-$Me_2C_2B_7H_9$ [49] in Figure 2.11 was notionally broken to see if a satisfactory more open ni-9⟨VI⟩ configuration (not shown) could be produced which might accommodate the parent *nido*-B_9H_{13} without producing a *7X*-bridge hydrogen. The resulting ni-9⟨VI⟩ configuration incorporates alternating hot and warm vertices about the open face and there is no way to place four bridge hydrogens into such a *nido*-B_9H_{13} structure without producing a *75*-bridge hydrogen.

Secondly, in the more open ni-9⟨VI⟩ configuration, no way was found to rationally distribute the requisite closed BBB 3c2e bonds and 2c2e bonds about the boron skeleton in either the Stx = 650 or 461 configurations without violating Lipscomb's rules [32]. It is concluded that the ni-9⟨VI⟩ configuration probably will not be found and that conclusion is indicated by the open circle in Figure 2.15.

2.6.6. Shore Compounds; Endo- and 76-Bridge Hydrogens at Very Low Temperatures

There are several *nido*-compounds with surprising arrangements of bridge and endo-hydrogens about the open faces. Examples are $B_{11}H_{14}^-$ [114] and $B_{11}H_{15}$ [114] both of which have been prepared by Shore et al. They are also studying the potential synthesis of B_9H_{13} and a number of other elusive compounds at very low temperature.

In the present discussion, the presumed "nonexistence" of *nido*-carboranes and boranes with 76-bridge hydrogens at ambient conditions has been emphasized. At low temperatures, however, species suspected of containing 76-bridge hydrogens are probably capable of existence although repeated attempts to prepare crystals of, for example, $B_{11}H_{15}$ for X-ray structural determinations have failed [114]. Surrogate "76"-bridge hydrogens have been found by Greenwood et al. [115] in stable ni-11⟨V⟩ compounds wherein the seven coordinate BH group has been replaced by a sufficiently electropositive ("electron generous") transition element grouping [91] (see also Chapter 6).

For the sake of comparison, Olah, Prakash et al. have unambiguously identified numerous nonclassical electron deficient carbocations in superacids at low temperatures [82, 116] (see Chapter 14). The low temperatures are necessary to sufficiently suppress the fluxional behavior of the cations for structural identification via ^{1}H- and ^{13}C-NMR and/or to prevent decomposition of the carbocations. In a similar fashion, Shore's methodology allows the stabilization of fluxional polyboranes (that probably incorporate 76-bridge hydrogens) at low temperature.

If, as suggested above, all species related to B_7H_{11}, B_9H_{13}, and $B_{11}H_{15}$ are restricted to ni-7⟨V⟩, ni-9⟨V⟩, and ni-11⟨V⟩ configurations (see Figure 2.15), how could the four extra hydrogens in each case possibly be arranged so that 77-bridge hydrogens could be avoided and the number of 76-bridge hydrogens minimized in the parent compounds? In addition, the ni-7⟨V⟩ and ni-9⟨V⟩ structures (see Figure 2.9) are uniquely unfavorably puckered to allow the accommodation of four bridge-endo hydrogens.

2.6.6.1. The Probable Structure of Nido-B₁₁H₁₅.

Starting with the series of known *nido*-compounds $B_{11}H_{13}^{2-}$ (known structure, Stx = 670) [89, 90], $B_{11}H_{14}^-$ (known structure, 481) [114], and $B_{11}H_{15}$ (predicted structure, Stx = 292), why would such compounds have such similar boron skeletons (see Figure 2.57) but such different electronic structures as reflected by the Stx-numbers that count numbers and locations of various 3c2e and 2c2e bonds? Do precedents exist?

At the top of Figure 2.57, the preferred Stx = 670 configuration for $B_{11}H_{13}^-$ is displayed; it is the very stable structure experimentally observed [89, 90] with two 66-bridge hydrogens. In the center row are two possible $B_{11}H_{14}^-$ structures;

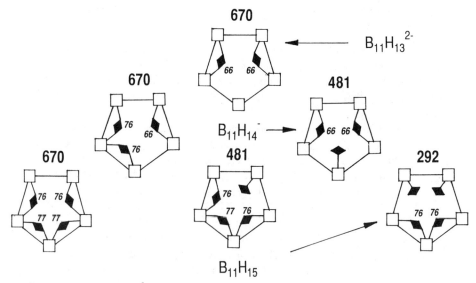

Figure 2.57. *Nido*-$B_{11}H_{13}^{2-}$, $B_{11}H_{14}^-$, and [$B_{11}H_{15}$] compounds favor Stx equals 670, 481 and 292, respectively.

the 670 configuration with two *76*-bridge hydrogens, as compared with Shore's experimentally observed alternative 481 configuration [114] which has two *66*-bridge hydrogens and one endo hydrogen.

On the bottom row are the 670, 481, and 292 (see also Figure 2.8) alternatives for Shore's $B_{11}H_{15}$ [114]. The 670 and 481 alternatives have *77*-bridge hydrogens and should be ruled out in favor of the 292 option with two *76*-bridge hydrogens, which make it very unstable, but no *77*-bridge hydrogens are incorporated. Of greater significance, the 292 alternative has a structural precedent (Sections 2.6.5.1 versus 2.6.6.1).

Closo-1,2-$C_2B_{10}H_{12}$ adds two electrons to produce *nido*-$C_2B_{10}H_{12}^{2-}$ (1,2) of unconfirmed structure [96–102] (see Figure 2.51). Subsequent partial neutralization with one proton yields two isomers of *nido*-$C_2B_{10}H_{13}^-$, one of which has the known "Type 1" structure [107, 108] illustrated in Figure 2.48. Following "scaffolding replacement" with two endohydrogens, the hypothetical *nido*-$CB_{10}H_{13}^-$ ion (Stx = 292) is produced (see Figure 2.48) which would be isoelectronic with that suggested for $B_{11}H_{15}$, Stx = 292 as compared in Figures 2.57 and 2.58.

2.6.6.2. The Structure of Nido-[B₉H₁₃].

2.6.6.2. The Structure of Nido-[B_9H_{13}]. Shore is investigating the possibility of producing *nido*-B_9H_{13}. In Figure 2.59, the V-gonal open face available to both $B_9H_{12}^-$ and the parent B_9H_{13} is displayed along with the known 650 configuration for $B_9H_{12}^-$ [88].

The 650 description for B_9H_{13} is unacceptable due to the presence of a *77*-bridge hydrogen. The best candidate would be an extremely congested unstable 461, ni-9⟨V⟩ fluxional configuration for B_9H_{13} which would incorporate two undesirable *76*-bridge hydrogens.

2.6.6.3. The Structure of Nido-[B_7H_{11}]. If B_7H_{11} can be produced at very low temperatures, the extremely congested 441, ni-7⟨V⟩ structure with one

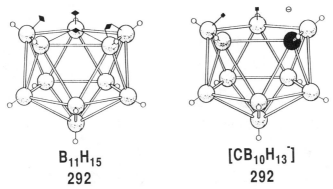

$B_{11}H_{15}$
292

$[CB_{10}H_{13}^-]$
292

Figure 2.58. [*Nido*-$B_{11}H_{15}$] (Stx = 292) has a precedent in [*nido*-$CB_{10}H_{13}^-$] (Stx = 292).

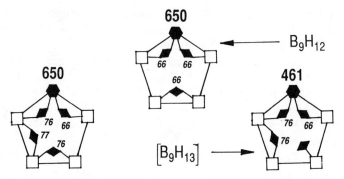

Figure 2.59. *Nido*-B₉H₁₂⁻ and [B₉H₁₃] compounds favor Stx equals 650 and 461, respectively.

endo hydrogen seems most promising as the 630 description would include *76*- and *75*-bridge hydrogens (see Figure 2.60).

Superficially, it would seem the anion $B_7H_{10}^-$ should have the 630 configuration unless congestion forced the 441 conformer. There is uncertainty in *nido*-7-vertex compounds as *nido*-$C_2B_5H_8^-$ has the 441 conformer [51] in the crystal (see Figure 2.11) when it would seem that the 630 conformer should be preferred as there is no congestion.

2.6.7. *Nido*-Configurations, V-Gonal versus VI-Gonal Open Faces, Differentiated Evaluation

Figures 2.9 and 2.15, summarize attempts to cross-correlate the "high coordination vertex removal pattern" [38] with the actual structures of all relevant *nido*-compounds that have been reported to date irrespective of elemental composition. Figure 2.15 identifies (by large crosses), to so-called

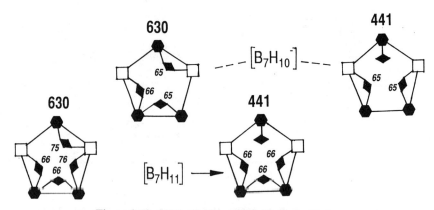

Figure 2.60. [*Nido*-B₇H₁₀⁻ and B₇H₁₁] structures.

normal "preferred" configurations produced in each case by the removal of the highest coordinated 4k, 5k, and 6k vertices. The alternative configurations are identified by double crosses in Figure 2.15 and these are presumed to be almost as favorable as the configurations derived by high coordination vertex removal. The underlying expectation has been that a unified set of geometrical systematics applicable to polyboranes, carboranes, and carbocations would yield a fundamental set of model clusters against which all other electron deficient families of clusters could and should be compared [33, 34].

Throughout this chapter, it has been emphasized that *nido*-6-, -7-, -8-, -9-, -11-, and -12-vertex compounds with V-gonal open faces are well established and that Koster's [64] *nido*-10-vertex $C_4B_6H_{10}$ may also have a V-gonal open face. Candidate compounds needing structural confirmation that they have V-gonal open faces were listed as "uncertain" structures along the right-hand margin of Figure 2.15.

During the preparation of this chapter, a review of the extensive literature led to the emergence of one additional pattern: there are no proven structures for 12-, 10-, and 8-vertex-*nido*-carboranes or *nido*-polyboranes with V-gonal open faces whose skeletons are composed solely of boron and carbon. Only carborane and polyborane derivatives incorporating heteroatoms, in addition to boron and carbon, have thus far had their structures unambiguously assigned ni-8- and ni-12-configurations with V-gonal open faces. No examples, with or without heteroatoms, having V-gonal apertures for ni-10-configurations have been confirmed. Moreover, ni-5-, -6-, and -7-configurations are too small to have VI-gonal open faces.

In the case of *nido*-8-vertex compounds, for example, $S(CpCo)_2B_5H_7$ was confirmed as having the ni-8⟨V⟩ configuration [50] (see Figures 2.9 and 2.11) while $B_8H_{10}L$ was "proposed" [69] to have the ni-8⟨V⟩ configuration. In the latter case, Sneddon stated, "final structural confirmation will have to await detailed x-ray analysis" [69]. In any event, when $B_8H_{10}L$ was "proposed" [69] to have the same skeletal ni-8⟨V⟩ structure ("proposed" versus "possible" in Figure 2.37) as the known $S(CpCo)_2B_5H_7$ [50] structure, we accepted this suggestion (except for bridge hydrogen locations) [92] as proven. An X-ray structure determination of either $B_8H_{10}L$ or $B_8H_{11}^-$ is sorely needed.

As to *nido*-10-vertex compounds, it may be that Koster's isomer of $C_4B_6H_{10}$ [64] has the ni-10⟨V⟩ configuration; we have argued that it is a good "candidate" structure (Figures 2.42 and 2.43). The "possible" alternative structure conforms to the ni-10⟨VI⟩ configuration but in this case, the "candidate" ni-10⟨V⟩ configurations would necessarily be involved as fluxional intermediates (Figure 2.43).

A *nido*-11-vertex compound with a IV-gonal open face and two transition element groups has currently been reported [117a,b] (see also Chapter 6).

In the case of *nido*-12-vertex systems, "slipped" compounds incorporating metal atoms [103, 104] and the selenium derivatives [105] have ni-12⟨V⟩ configurations with certainty (Figures 2.47 and 2.52), however, there are no examples of 12-vertex *nido*-carboranes or *nido*-polyboranes with V-gonal open faces unless skeletal atoms other than boron and carbon are present (although candidates are suggested along the right-hand border of Figures 2.15 and 2.61).

In Figure 2.61, the compounds with boron and carbon (only) skeletons were differentiated from those incorporating heteroatoms and we designate those boron-carbon skeletal compounds by the large symbol CB while showing the presence of nitrogen, sulfur, and selenium with the small symbols N, S, Se and signifying the presence of transition elements with a small T.

Looking at the data, recast in the form of Figure 2.61, it would appear that (1) *nido*-carborane, -polyborane and, -carbocation compounds probably adhere to one systematic structural pattern that applies to boron-carbon skeletons and that (2) when more electronegative sulfur, selenium, and nitrogen atoms are present, configurations are favored which have smaller open faces (if available) in order that the sulfur, selenium, and nitrogen atoms gain access to the greater electron density available around smaller apertures. In other words, the polyborane, carborane, carbocation continuum of geometrical cluster systematics may not model all (or most) other electron deficient families of compounds exactly, but it may be hoped that the various families will differ systematically according to regular patterns.

NIDO-CONFIGURATIONS

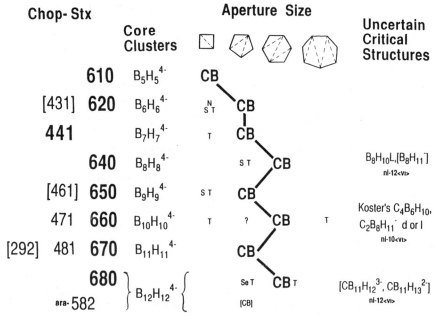

Figure 2.61. Cross-correlation between *nido*-skeleton size and aperture size (compare with Figure 2.15).

Along the right-hand column of Figure 2.61, are shown the same "uncertain" structures as were listed in Figure 2.15, but in Figure 2.61, it is questioned whether they have VI-gonal open faces rather than the V-gonal open faces anticipated in Figure 2.15. If the "uncertain" structures listed in both Figures 2.15 and 2.61 have VI-gonal open faces then the pattern illustrated in Figure 2.61 is to be preferred over that in Figure 2.15.

These considerations of Figure 2.61 (i.e., involving only C, B, and H) do not deter us from the conviction that charge smoothing forces strongly influence configuration selection whether the pattern illustrated in Figure 2.15 or the pattern illustrated in Figure 2.61 more closely approximates the real situation.

Figure 2.61 suggests that when more electronegative atoms such as nitrogens, sulfurs or seleniums replace the carbons, configurations with smaller open faces, if available, are preferred. The advantage of smaller open faces is to produce hot vertices of greater electron density even though the total number of hot vertices decreases. The compound $S_2(CpCo)_2B_2H_2$ favors the ni-6$\langle IV \rangle$ configuration [112] because the two hot vertices in the hypothetical core-cluster, $B_6H_6^{4-}$, had values of 0.86^- rather than 0.75^- in the pentagonal pyramid. The sulfurs in the *nido*-8-vertex and *nido*-9-vertex compounds, $S(CpCo)_2B_5H_7$ [50] and $S_2(CpCo)_2B_5H_5$ probably favor the ni-8$\langle V \rangle$ and ni-9$\langle IV \rangle$ configurations for the same reason. It may be desirable to compare only separated classes of compounds as a function of the heteroatoms incorporated.

If the systematics of Figure 2.61 are found to be applicable to most or all boron-carbon compounds and both Koster's compound [64], $C_4B_6H_{10}$, and the $C_2B_8H_{11}^-$ ion [73] are found not to have ni-10$\langle V \rangle$ configurations (see the question mark in Figure 2.61), then those ni-10$\langle V \rangle$ configurations certainly must be intermediates during the rearrangements involving Koster's $C_4B_6H_{10}$ (see Figure 2.43) and $C_2B_8H_{11}$ when it functions as an intermediate in the rearrangement of chiral 5,6-$C_2B_8H_{12}$ into nonchiral 6,9-$C_2B_8H_{10}^{2-}$ and vice versa (see Figure 2.45).

A sulfur containing *nido*-anion such as $SB_9H_{10}^-$, with only one bridge hydrogen, should be a very strong candidate to assume a ni-10$\langle V \rangle$ structure as sulfur atom charge smoothing is probably more influential than minor optimization [92] of the bridge hydrogen. The anion, $SB_9H_{10}^-$, should easily be synthesized from the known compound, SB_9H_{11}, which necessarily has a ni-10$\langle VI \rangle$ structure [118] to accommodate major optimization [92] of the two bridge hydrogens. An X-ray structure of $[SB_9H_{10}^-]$ is needed.

If one set of *nido*-configurations is optimal for boron-carbon skeletons and a somewhat different set with smaller open faces, is optimal when more electronegative heteroatoms are present, then possibly systems may be found wherein *nido*-compounds (by electron count) incorporating numerous sufficiently electronegative skeletal heteroatoms would approach deltahedral structures resembling configurations usually reserved for *closo*-compounds. The *nido*-compounds, reviewed by Halet et al. [119] and Richmond and Kochi [120] that have "*closo*-structures" might be examples of such compounds.

2.7. CONCLUSION

Geometrical systematics based on modified, aperture dependent, electron distribution, charge smoothing considerations, elaborated in this chapter, apply to all *nido*-species. Only slight modifications to accommodate the similarities of bridge and endo hydrogens (and endo hydrogens on carbon) are required to apply these same systematics to *arachno*-polyboranes, -carboranes, and -carbocations which will be discussed in future publications.

The originally suggested identification of bridge hydrogens as *66*-bridge hydrogens, *65*-bridge hydrogens, etcetera remains a useful procedure. It identifies those bridge hydrogens which will or will not accept, at ambient conditions, locations neighboring borons that are more than 6-coordinate.

In the order of decreasing influence on final configuration are apparently: (1) major optimization of bridge hydrogens [92], (2) heteroatoms such as sulfur and nitrogen, (3) minor optimization of bridge hydrogens [92], and (4) finally carbons. All exert their influences through charge smoothing driving forces.

When more complex, modified evaluations of bridge hydrogens are required, that is, to determine the relative acidities of otherwise equivalent *66*-bridge hydrogens, then the aperture dependent procedure which assigns charge distributions to the boron vertices about the open face (rather than coordination numbers) is much more revealing. The aperture dependent, electron distribution procedure is also amenable to evaluating the "hospitality" of the various environments for carbons and other heteroatoms in these fluxional species and influencing the final structure selection.

For the purposes of Figure 2.15, the assumption was made that the optional structures for all "undifferentiated" *nido*-compounds, irrespective of heteroatom identity, had either V-gonal or VI-gonal open faces and that in the case of *nido*-carboranes, the optional structures were selected to afford the carbons and other electronegative heteroelements access to the most electron rich vertices driven by charge smoothing driving forces. If it turns out that the pattern illustrated in Figure 2.61 is correct, it merely means that there is a single set of most favored configurations for the boron and carbon skeletons of the *nido*-carboranes and *nido*-polyboranes and that the most favored configurations for the 8-, 10-, and 12-vertex boron-carbon compounds have VI-gonal open faces while most favored configurations for the 6-, 7-, 9-, and 11-vertex compounds have V-gonal open faces.

Over fifteen known compounds are identified whose structures, if they can be determined, should remove most of the ambiguities.

Apparently, alternative geometrical configurations with V-gonal and VI-gonal open faces within the *nido*-boranes, are energetically quite similar. Bridge hydrogens (primarily) and carbons (secondly) influence the selection of VI-gonal versus V-gonal open face isomers. Sulfurs and nitrogens are clearly more influential on final configuration selection than carbon and are apparently more important than minor optimization [92] of bridge hydrogens. The underlying driving force in all cases is the quest for charge smoothing in these fluxional clusters.

ACKNOWLEDGMENT

Support over the years by the United States Navy and Air Force, the U.S. Office of Naval Research and the Loker Hydrocarbon Research Institute is gratefully acknowledged. Professors George Olah, Stan Hermanek, G. K. Prakash, Robert Bau, Thomas Onak, Sheldon Shore, and Ken Wade are thanked for sharing their scientific insights and discussions. Ms. Cheri Gilmore helped with the graphics and Mrs. Carolyn Stone typed the manuscript.

REFERENCES

1. (a) Mielcarek, J. J. and Keller, P. C., *J. Chem. Soc., Chem. Commun.*, 1090 (1972); (b) Mielcarek, J. J. and Keller, P. C., *J. Am. Chem. Soc.*, **96**, 7143 (1974).
2. Stock, A., *Hydrides of Boron and Silicon*, Cornell University Press, Ithaca, NY, 1933.
3. Dulmage, W. J. and Lipscomb, W. N., *Acta Crystallogr.*, **5** 260 (1952).
4. Kasper, J. S., Lucht, C. M. and Harker, D., *Acta Crystallogr.*, **3**, 436 (1950).
5. Eberhardt, W. H., Crawford, B. L. and Lipscomb, W. N., *J. Chem. Phys.*, **22**, 989 (1954).
6. Longuet-Higgins, H. C. and Roberts, M. De V., *Proc. R. Soc.*, **A230**, 110 (1955).
7. Williams, R. E., Good, C. D. and Shapiro, I., 140th Meeting, *J. Am. Chem. Soc.*, 14 N, p. 36, Chicago, IL (1961).
8. Good, C. D. and Williams, R. E., U.S. Patent No. 3030289 (1959); *Chem. Abstr.*, **57**, 12534b (1962).
9. Onak, T. P., Drake, R. P. and Dunks, G. B., *Inorg. Chem.*, **3**, 1686 (1964).
10. Onak, T. P., Gerhart, F. J. and Williams, R. E., *J. Am. Chem. Soc.*, **85**, 3378 (1963).
11. Beaudet, R. A. and Poynter, R. L., *J. Am. Chem. Soc.*, **86**, 1258 (1964).
12. Onak, T. P., Dunks, G. B., Beaudet, R. A. and Poynter, R. L., *J. Am. Chem. Soc.*, **88**, 4622 (1965).
13. Reitz, R. R. and Schaeffer, R., *J. Am. Chem. Soc.*, **93**, 1263 (1971); **95**, 6254 (1973).
14. Beck, J. S. Kahn, A. P. and Sneddon, L. G., *Organometallics*, **5**, 2552 (1986).
15. Williams, R. E. and Gerhart, F. J., *J. Am. Chem. Soc.*, **87**, 3513 (1965).
16. Bobinsky, J. J., *J. Chem. Ed.*, **41**, 500 (1964).
17. Heying, T. L., Ager, J. W., Clark, S. L., Mangold, D. J., Goldstein, H. L., Hillman, M., Polak, R. J. and Szymanski, J. W., *Inorg. Chem.*, **2**, 1089 (1963).
18. Potenza, J. A. and Lipscomb, W. N., *J. Am. Chem. Soc.*, **86**, 1874 (1964).
19. Potenza, J. A. and Lipscomb, W. N., *Inorg. Chem.*, **3**, 1673 (1964).
20. Schroeder, H. and Vickers, G. D., *Inorg. Chem.*, **2**, 1317 (1963).
21. Grafstein, D. and Dvorak, J., *Inorg. Chem.*, **2**, 1128 (1963).
22. Pepetti, S. and Heying, T. L., *J. Am. Chem. Soc.*, **86**, 2295 (1964).
23. Berry, T. E., Tebbe, F. N. and Hawthorne, M. F., *Tetrahedron Lett.*, **12**, 715 (1965).
24. Tebbe, F. N., Garrett, P. M. and Hawthorne, M. F., *J. Am. Chem. Soc.*, **86**, 4222 (1964).
25. Tsai, C. and Strieb, W. E., *J. Am. Chem. Soc.*, **88**, 4513 (1966).

26. Tebbe, F. N., Garrett, P. M., Young, D. C. and Hawthorne, M. F., *J. Am. Chem. Soc.*, **88**, 609 (1966).

27. Williams, R. E., *J. Inorg. Nucl. Chem.*, **20**, 198 (1961).

28. (a) Williams, R. E., *Inorg. Chem.*, **1**, 971 (1962); (b) Rickborn, B. and Wood, S. E., *J. Am. Chem. Soc.*, **93**, 3940 (1971).

29. Johnson, H. D., Brice, V. T., Brubaker, G. L. and Shore, S. G., *J. Am. Chem. Soc.*, **94**, 6711 (1972).

30. Johnson, H. D., Geanangel, R. A. and Shore, S. G., *Inorg. Chem.*, **9**, 908 (1970).

31. Hogeveen, H. and Kwant, P. W., *Tetrahedron Lett.*, **19**, 1665 (1973).

32. Lipscomb, W. N., *Boron Hybrides*, Benjamin, New York, 1963.

33. Williams, R. E., IMEBORON-VI, Bechyne Czechoslovakia, June 1987.

34. Williams, R. E., The Loker Symposium on Electron Deficient Compounds at the University of Southern California, Los Angeles, CA, January 1989.

35. Onak, T. P., Williams, R. E. and Weiss, H. G., *J. Am. Chem. Soc.*, **84**, 2830 (1962).

36. Williams, R. E. and Gerhart, F. J., *J. Am. Chem. Soc.*, **87**, 3513 (1965).

37. Williams, R. E., in: *Progress in Boron Chemistry* (R. J. Brotherton and H. Steinberg, Eds.), Vol. 2, Chap. 2, p. 37, Pergamon, Oxford, 1970.

38. Williams, R. E., *Inorg. Chem.*, **10**, 210 (1971).

39. Williams, R. E., *Adv. Inorg. Chem. Radiochem.*, **18**, 67 (1976).

40. Kameda, M. and Kodama, G., *Polyhedron*, **2**, 413 (1983).

41. Wade, K., *J. Chem. Soc., Chem. Commun.*, 792 (1971).

42. Wade, K., *Adv. Inorg. Chem. Radiochem.*, **18**, 1 (1976).

43. Rudolph, R. W. and Pretzer, W. R., *Inorg. Chem.*, **11**, 1974 (1972).

44. Parry, R. W. and Edwards, L. J., *J. Am. Chem. Soc.*, **81**, 3554 (1959).

45. (a) Stone, A. J., *Inorg. Chem.*, **20**, 563 (1981); *Polyhedron*, **3**, 1299 (1984). (b) Wales, D. J. and Stone, A. J., *Inorg. Chem.*, **26**, 3845 (1987).

46. (a) Mingos, D. M. P. and Wales, D. J., *Introduction to Cluster Chemistry*, Prentice Hall, Englewood Cliffs, NJ, 1989. (b) Mingos, D. M. P., *Acc. Chem. Res.*, **17**, 311 (1984). (c) Mason, R., Thomas, K. M. and Mingos, D. M. P., *J. Am. Chem. Soc.*, **95**, 3802 (1973).

47. Gimarc, B. M. and Ott, J. J., *Inorg. Chem.*, **25**, 2708 (1986).

48. King, R. B., *Inorg. Chem. Acta*, **49**, 237 (1981).

49. Huffman, J. C. and Streib, W. E., *Chem. Commun.*, 662 (1972).

50. Zimmerman, G. J. and Sneddon, L. G., *J. Am. Chem. Soc.*, **103**, 1102 (1981).

51. Beck, J. S., Quintana, W. and Sneddon, L. G., *Organometallics*, **7**, 1015 (1988).

52. See reference 38, footnote no. 7.

53. Stohrer, W.-D. and Hoffman, R., *J. Am. Chem. Soc.*, **94**, 1661 (1972).

54. Masamune, S., Sakai, M., Ona, H. and Jones, A. J., *J. Am. Chem. Soc.*, **94**, 8956 (1972).

55. Binger, P., *Tetrahedron Lett.*, **24**, 2675 (1966).

56. Onak, T. P. and Wong, G. T. F., *J. Am. Chem. Soc.*, **92**, 5226 (1970).

57. Pasinski, J. P. and Beaudet, R. A., *Chem. Commun.*, 928 (1973).

58. Stibr, B., Plesek, J., Jelinek, T., Base, K., Janousek, Z. and Hermanek, S., in: *Boron Chemistry* (S. Hermanek, Ed.) World Sci. Publ. Co. Inc., Singapore, 1987.

59. Stibr, B., Jelinek, T., Plesek, J., Drdakova, E., Plzak, Z. and Hermanek, S., *J. Chem. Soc., Chem. Commun.* (1987) in press.

60. Stibr, B., Plesek, J. and Hermanek, S., *Collect. Czech. Chem. Commun.*, **37**, 2696 (1972).

61. Stibr, B., Janousek, Z., Base, K., Hermanek, S., Plesek, J. and Zakharova, I. A., *Collect. Czech. Chem. Commun.*, **49**, 1891 (1984).

62. (a) Fehlner, T. P., *J. Am. Chem. Soc.*, **99**, 8355 (1977); (b) Fehlner, T. P., *J. Am. Chem. Soc.*, **102**, 3424 (1980); (c) Siebert, W. and El-Essawi, M. E. M., *Chem. Ber.*, **112**, 1480 (1979).

63. Zimmerman, G. J. and Sneddon, L. G., *Inorg. Chem.*, **19**, 3650 (1980).

64. Koster, R., Seidel, G. and Wrackmeyer, B., *Angew. Chem.*, **97**, 317 (1985).

65. Maynard, R. B., Sinn, E. and Grimes, R. N., *Inorg. Chem.*, **20**, 1201 (1981).

66. Grimes, R. N., *Adv. Inorg. Radiochem.*, **26**, 55 (1983).

67. Venable, T. L., Maynard, R. B. and Grimes, R. N., *J. Am. Chem. Soc.*, **106**, 6187 (1984).

68. Williams, R. E., Bausch, J. W., Prakash, G. K. S. and Onak, T. P., to be published.

69. Briguglio, J. J., Carroll, P. J., Corcoran, E. W. and Sneddon, L. G., *Inorg. Chem.*, **25**, 4621 (1986).

70. Gotcher, A. J., Ditter, J. F. and Williams, R. E., *J. Am. Chem. Soc.*, **95**, 7514 (1973).

71. Reilly, T. and Burg, A. B., *Inorg. Chem.*, **13**, 1250 (1974).

72. Reference 39, pp. 68, 69.

73. Stibr, B., Plesek, J. and Zobacova, A., *Polyhedron*, **1**, 824 (1982).

74. Garrett, P. M., Ditta, G. S. and Hawthorne, M. F., *J. Am. Chem. Soc.*, **93**, 1265 (1971).

75. Rietz, R. R. and Schaeffer, R., *J. Am. Chem. Soc.*, **93**, 1263 (1971).

76. Shore, S., Personal communication to R. E. Williams, August 1989.

77. Greenwood, N. N., *Chem. Soc. Rev.*, **13**, 353 (1984).

78. Bould, J., Greenwood, N. N., Kennedy, J. D. and McDonald, W. S., see Footnote 13, Reference 77, *J. Chem. Soc., Chem. Commun.*, 465 (1982).

79. Brown, M., Fontaine, X. L. R., Greenwood, N. N., Kennedy, J. D. and MacKinnon, P., *J. Chem. Soc., Chem. Commun.*, 817 (1987).

80. Venable, T. L. and Grimes, R. N., *Inorg. Chem.*, **21**, 887 (1982).

81. Kennedy, J. D., *Prog. Inorg. Chem.*, **32**, 819 (1984); **34**, 211 (1986).

82. Olah, G. A., Prakash, G. K. S., Williams, R. E., Field, L. D. and Wade, K., *Hypercarbon Chemistry*, Wiley-Interscience, New York, 1987.

83. (a) Muetterties, E. L., Wiersema, R. J. and Hawthorne, M. F., *J. Am. Chem. Soc.*, **95**, 7521 (1973). (b) The high temperature isomer of *closo*-$B_8H_8^{2-}$ in solution [see Reference 83c] has the same D_{2d} structure as the crystal structure (Figure 2.17), while the structure of the low temperature isomer of *closo*-$B_8H_8^{2-}$ in solution has neither the D_{4d} nor C_{2v} structures previously suggested [83a]. (c) Williams, R. E., Prakash, G. K. S. and Bausch, J. W., BUSA-2, Durham, NC, June 1990, to be published. Selected theoretically calculated (*Ab Initio*-IGLO) ^{11}B NMR spectra were found to match experimentally observed ^{11}B NMR spectra. In this fashion, the correct structures for a number of polyboranes and carboranes have been selected including the high temperature isomer of *closo*-$B_8H_8^{2-}$ (in solution) (Figure 2.17),

closo-$C_2B_6H_8$ (fluid) (Figure 2.4) and *nido*-$C_2B_6H_{10}$ (fluid) (Figure 2.36). See Chapter 4 for *Ab Initio*-IGLO background.

84. Klanberg, F. and Muetterties, E. L., *Inorg. Chem.*, **5**, 1955 (1966).

85. Schaeffer, R. and Walter, E., *Inorg. Chem.*, **12**, 2209 (1973).

86. Nestor, K., Fontaine, X. L. R., Greenwood, N. N., Kennedy, J. D., Plesek, J., Stibr, B. and Thornton-Pett, M., *Inorg. Chem.*, **28**, 2219 (1989).

87. (a) Porterfield, W. W., Stephenson, I. R. and Wade, K., in: *Boron Chemistry* (S. Hermanek, Ed.), World Scientific Publ. Co. Inc., Singapore, 1987; (b) IMEBORON-VI, Bechyne, Czechoslovakia, 1987.

88. Siedle, A. R., Ph.D. Thesis, p. 61, Indiana University, Bloomington, 1973; $B_9H_{12}^{-}$ data to be published with Garber, A. R., Bodner, G. M. and Todd, L. J.

89. Aftandilian, V. D., Miller, H. C., Parshall, G. W. and Muetterties, E. L., *Inorg. Chem.*, **1**, 734 (1962).

90. Fritchie, C. J., *Inorg. Chem.*, **6**, 1199 (1967).

91. Other transition element groups, in contrast, such as CpNi groups (isoelectronic with CH groups) are assumed to be electronegative relative to CH groups and thus, like CH groups, tend to be found in low coordinate sites and even cause smaller apertured configurations to be selected as do sulfur and nitrogen atoms; see Stibr, B., Plesek, J. and Hermanek, S., in: *Advances in Boron and the Boranes* (J. F. Liebman, A. Greenberg and R. E. Williams, Eds.), Chap. 3, p. 35, VCH Publishers, Inc., New York, 1988.

92. (a) Major optimization means changing *77*-, *76*-, and *75*-bridge hydrogens in *nido*-compounds into *66*- or "better" bridge hydrogens; minor optimization refers to improving *66*-bridge hydrogens to *65*- or *55*-bridge hydrogens. (b) The proposed-static structure for *nido*-$C_2B_6H_{10}$ (Figure 2.36) has been confirmed [Reference 83c].

93. Sneddon, L. G., Huffman, J. C., Schaeffer, R. O. and Streib, W. E., *Chem. Commun.*, 474 (1972).

94. Siedle, A. R., Bodner, G. M. and Todd, L. J., *J. Inorg. Nucl. Chem.*, **33**, 3671 (1971).

95. (a) Hermanek, S., personal communication to R. E. Williams, IMEBORON-VI, Bechyne, Czechoslovakia, June 1987; (b) Getman, T. D., Knobler, C. B. and Hawthorne, M. F., *Inorg. Chem.*, **29**, 158 (1990).

96. Fein, M. M., Bobinski, J., Mays, N., Schwartz, N. and Cohen, M. S., *Inorg. Chem.*, **2**, 1111 (1963).

97. Grafstein, D. and Dvorak, J., *Inorg. Chem.*, **2**, 1128 (1963).

98. Zakharkin, L. and Kalinin, V., *Izv. Akad. Nauk SSSR, Ser. Khim.*, 194 (1967).

99. Zakharkin, L. and Kalinin, V., *Izv. Akad. Nauk SSSR, Ser. Khim.*, **10**, 2310 (1967).

100. Zakharkin, L. I., Kalinin, V. N., Kvasov, B. A. and Synakin, A. P., *Zh. Obshch. Khim.*, **41**, 1726 (1971).

101. Zakharkin, L. I., Kalinin, V. N. and Podvistotskaya, L., *Izv. Akad. Nauk SSSR, Ser. Khim.*, 2310 (1967).

102. Dunks, G. B., Wiersema, R. J. and Hawthorne, M. F., *J. Am. Chem. Soc.*, **95**, 3174 (1973).

103. Warren, L. F. and Hawthorne, M. F., *J. Am. Chem. Soc.*, **90**, 4823 (1968).

104. Wing, R. M., *J. Am. Chem. Soc.*, **89**, 5599 (1967).

105. Fricsen, G. D., Barriola, A., Daluga, P., Ragatz, P., Huffman, J. C. and Todd, L. J., *Inorg. Chem.*, **19**, 458 (1980).

106. Knoth, W. H., *J. Am. Chem. Soc.*, **89**, 1274 (1967).

107. Dunks, G. B., Wiersema, R. J. and Hawthorne, M. F., *J. Am. Chem. Soc.*, **95**, 3174 (1973).

108. Tolpin, E. I. and Lipscomb, W. N., *Chem. Commun.*, 257 (1973).

109. Hosmane, N. S., de Meester, P., Siriwardane, V., Islam, M. S. and Chu, S. S. C., *J. Chem. Soc., Chem. Commun.*, 1421 (1986).

110. Schleyer, P. v. R., Personal communication to R. E. Williams, IMEBORON-VI, Bechyne, Czechoslovakia, June 1987.

111. Paetzold, P., in: *Boron Chemistry* (S. Hermanek, Ed.), World Sci. Publish. Co. Inc., Singapore, 1987; personal communication to R. E. Williams, IMEBORON-VI Bechyne, Czechoslovakia, June 1987.

112. Kang, S. O. and Sneddon, L. G., *Inorg. Chem.*, **27**, 3769 (1988).

113. Astheimer, R. J. and Sneddon, L. G., *Inorg. Chem.*, **22**, 1928 (1983).

114. (a) Getman, T. D., Krause, J. A. and Shore, S. G., *Inorg. Chem.*, **27**, 2398 (1988); (b) Footnote II, Reference 114a.

115. Brown, M., Fontaine, X. L. R., Greenwood, N. N., Kennedy, J. D. and Thornton-Pett, M., *J. Chem. Soc., Dalton Trans.*, 1169 (1987).

116. Olah, G. A., Prakash, G. K. S. and Sommer, J., *Superacid Chemistry*, Wiley-Interscience, New York, 1985.

117. (a) Nestor, K., Fontaine, X. L. R., Greenwood, N. N., Kennedy, J. D. and Thornton-Pett, M., *J. Chem. Soc., Chem. Commun.*, 455 (1989); (b) If a single DSD rearrangement is imposed upon an icosahedron (twelve 5k vertices), a deltahedron is produced with two 4k vertices, two 6k vertices, and ten 5k vertices. The subsequent removal of one of the two 4k vertices removes both offending 6k vertices simultaneously and produces the aberrant ni-11⟨IV⟩ configuration observed.

118. Hertler, W. R., Klanberg, F. and Muetterties, E. L., *Inorg. Chem.*, **6**, 1696 (1967).

119. Halet, J.-F., Hoffmann, R. and Saillard, J.-Y., *Inorg. Chem.*, **24**, 1695 (1985).

120. Richmond, M. G. and Kochi, J. K., *Inorg. Chem.*, **26**, 541 (1987).

CHAPTER 3

Electron Distribution in Boranes and Carboranes

K. WADE

3.1. INTRODUCTION

This chapter looks at some simple ways of predicting or exploring how uniformly the skeletal electrons in cluster compounds are distributed. Borane and carborane clusters are used as examples, but the arguments employed are applicable to cluster compounds in general. Though the emphasis is primarily on simple bonding models, the use of ^{11}B-NMR chemical shifts as a guide to electron distribution in icosahedral carborane derivatives is touched on briefly in a concluding section.

It seemed appropriate, in this volume dedicated to Professor Bill Lipscomb, to pay tribute to his enormous contribution to cluster chemistry by drawing attention in this chapter to some of the virtues of the 2- and 3-center bond schemes he devised and developed for treating the bonding in neutral boranes [1–3]. Though still the preferred treatment for *nido* and *arachno* boranes, they have been used less for *closo* clusters than they deserve. The first part of this chapter is therefore devoted to an outline of the assumptions on which such bond schemes are based, and an analysis of what they can tell us about the electron distribution in *closo* boranes and related clusters.

In cluster chemistry, treatment of the skeletal bonding in terms of 2- and 3-center bonds assigns skeletal electron pairs, respectively, to the edges and faces

Electron Deficient Boron and Carbon Clusters, Edited by George A. Olah,
Kenneth Wade, and Robert E. Williams.
ISBN 0-471-52795-5 © 1991 John Wiley & Sons, Inc.

of the skeletal polyhedra. From such a localized electron pair scheme, it is a small conceptual step to assign the skeletal pairs instead to polyhedral *vertices*, particularly since in borane-type clusters the commonest cluster-forming units contribute two electrons apiece for skeletal bonding. The insight provided by a vertex electron pair scheme for the skeletal bonding in borane-type clusters is therefore explored in a later section of this chapter.

Such schemes are not intended to be rigorous, of course, but rather to convey some feeling for how relatively thinly the electrons are distributed, and where they tend to accumulate in cluster polyhedra. Where precision is required, molecular orbital calculations are needed. The value of spectroscopic probes should not be overlooked, however, and to illustrate this, a final section of this chapter shows how the chemical shifts of the various types of boron atom in some icosahedral carboranes respond to the influence of various substituents.

3.2. TWO- AND THREE-CENTER BOND SCHEMES

The standard method of treating the bonding in *nido* and *arachno* boranes (B_2H_6, B_4H_{10}, B_5H_9, B_5H_{11}, B_6H_{10}, $B_{10}H_{14}$ etc.), though less often used for *closo* borane anions $B_nH_n^{2-}$ and related carboranes $CB_{n-1}H_n^-$, $C_2B_{n-2}H_n$ etcetera, employs the 2- and 3-center bond schemes developed by Lipscomb [1–3]. It was largely thanks to Lipscomb's painstakingly meticulous X-ray crystallographic studies [3] that the structural pattern [3–9] to which borane clusters conform was originally elucidated, and it was he who showed how the bonding between their intricate networks of atoms could be understood not only by carrying out rigorous molecular orbital calculations [2, 3, 11–14], but also, very simply and elegantly, by assigning the valence shell electrons to suitable combinations of 2-center 2-electron (2c2e) BH and BB bonds and 3-center 2-electron (3c2e) BHB and BBB bonds [1–3, 15]. As such localized bond schemes can be more informative than is commonly recognized, and can moreover be quite helpfully applied to *closo* systems, it is worth outlining here the assumptions on which they are based and the implications for electron distribution.

Lipscomb recognized and perceptively exploited the fact that, in the neutral boranes, B_nH_{n+m}, the atoms can be regarded as lying on two concentric spheres. The *n* boron atoms lie on the inner spherical surface, each with a single *exo* hydrogen atom attached to it by a 2c2e B–H bond radiating outwards, away from the common center of the two spheres. The remaining *m endo* hydrogen atoms, whether in terminal BH sites, or bridging BHB sites, lie on the same inner spherical surface as the *n* boron atoms. If one allocates a pair of electrons (and one boron atomic orbital) to each *exo* BH bond, then the bonding problem reduces to that of explaining how *n* BH units (each supplying two electrons and three AOs) and *m endo* hydrogen atoms (each supplying one electron and one AO) may be held together. Lipscomb assigned the available $(2n + m)$ skeletal bonding electrons—now known to indicate the polyhedron on which the structure is based [6–10, 16] (it will have $n + m/2\text{-}1$ vertices)—to s BHB bonds, t BBB bonds, y BB bonds and x BH_{endo} bonds (additional to the original n BH_{exo} bonds). Equating (1) the numbers of bond pairs to the numbers of bonds, (2) the numbers of boron atoms to the numbers of 3c2e bonds (because each boron atom provides three AOs for skeletal bonding but only two electrons), and (3) the numbers of *endo* hydrogen atoms to the numbers of BHB bonds and additional BH bonds, led to the following equations of balance:

1. $2n + m = 2(s + t + y + x)$.
2. $n = s + t$.
3. $m = s + x$ (whence $2y = s \quad x$).

With only three independent equations involving four unknown parameters (s, t, y and x), these equations do not have unique solutions for typical boranes $B_n H_{n+m}$, though the sets of styx numbers found to be compatible with a particular molecular formula invariably include that appropriate for the actual structure (where that is known) and one or more other bond networks through which scrambling of the *endo* hydrogen atoms may occur.

Applied to *closo* borane anions $B_n H_n^{2-}$, which have no *endo* hydrogen atoms, Lipscomb's equations of balance reduce to $t = n - 2$ and $y = 3$ [8]. The $(n + 1)$ pairs of skeletal bonding electrons in such systems, if assigned to three 2c2e BB (polyhedral edge) bonds and $(n - 2)$ 3c2e BBB (polyhedral face) bonds, ensure that each boron atom will be involved in three skeletal bonds. The bonding problem that these *closo* clusters pose is thus not what set of 2- and 3-center bonds to use, but how to deploy the three 2c2e and $(n - 2)$ 3c2e bonds most effectively over the surface of n-vertex polyhedra with $(3n - 6)$ edges and $(2n - 4)$ faces (Figure 3.1).

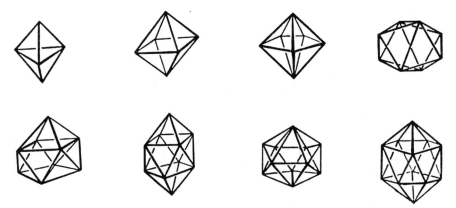

Figure 3.1. The deltahedral shapes of *closo*-borane anions $B_n H_n^{2-}$ ($n = 6 \rightarrow 12$) and carboranes $C_2 B_{n-2} H_n$ ($n = 5 \rightarrow 12$).

Figure 3.2. Localized 2c2e (edge) and 3c2e (face) bond arrangements by which individual polyhedron edges can be accounted for.

This problem is simplified by making the reasonable assumptions that a satisfactory bond network must account for all of the $(3n - 6)$ polyhedron edges by one or other of the bond arrangements shown in Figure 3.2, that is, by a 2c2e bond along that edge, or by a 3c2e bond in one or both of the faces adjacent to that edge [15, 17]. Allocation of a 2c2e bond to the edge of a face containing a 3c2e bond is excluded on the grounds that this would entail unrealistically close crowding of the skeletal bonds.

These constraints limit significantly the number of ways the 2- and 3-center bonds can be assigned to the edges and faces of the familiar *closo* polyhedra (Figure 3.1), though the relatively high symmetry of many of them leads to scope for extensive resonance between equivalent canonical forms. Examples of possible bond networks are shown in Figure 3.3. Note that for the trigonal

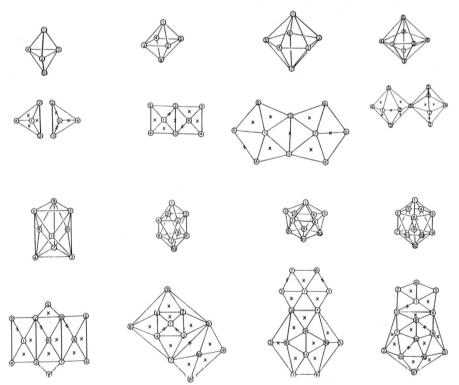

Figure 3.3. Skeletal 2c2e and 3c2e bond networks for *closo* species.

bipyramidal $C_2B_3H_5$ ($B_5H_5^{2-}$, expected to have a similar structure, is unknown), the three 2c2e and three 3c2e bonds needed have to be grouped on opposite sides of the equatorial plane. Allowing resonance between the two permitted forms, one infers that the bond orders of the equatorial-equatorial and axial-equatorial edges are 0.33′ and 0.83′, respectively (since a 2c2e bond confers a bond order of 1.0 on the edge it occupies, while a 3c2e bond contributes one-third of an electron pair to each edge of the face it occupies). Similar arguments can be used to deduce the bond orders, and so electron distribution, of the various types of edge in the higher *closo* borane polyhedra.

The electron distribution in these clusters, and the manner in which it reflects the skeletal coordination numbers of the skeletal atoms, can be inferred in another way if one considers the contributions atoms need to make to 2c2e and 3c2e bonds, respectively. A 2c2e bond between like atoms requires each to contribute one electron. A 3c2e bond between like atoms requires each to contribute 0.66′e. A neutral BH unit, having two electrons and three AOs available for skeletal bonding, has enough electrons to bond to its neighbors using three 3c2e bonds, but would need to acquire a further electron if it were to form three 2c2e bonds. The negative charge on a boron atom thus increases progressively with the number of 2c2e bonds it uses (Table 3.1).

Figure 3.4 shows how the skeletal coordination number of an atom, constrained to form only three bonds, influences the types of bond it can use. If it is to bond to as many as six neighbors, it must do so using exclusively 3c2e bonds. Use of two 3c2e bonds and one 2c2e bond becomes an alternative possibility for an atom with five neighbors. Four neighbors permit two 2c2e bonds, three neighbors permit three 2c2e bonds. Allowing the alternative possible bonding arrangements equal weight for a particular skeletal coordination number, one can see (Table 3.2) that the negative charge on a BH unit is estimated to increase by increments of 0.16′e as the skeletal coordination number of the boron atom decreases from six (zero charge) to three (charge 0.5e), thus underlining the association of negative charge with low coordination number of the skeletal atoms in such *closo* clusters. The implied proportions of the total charge carried by the different types of BH units in *closo* anions $B_nH_n^{2-}$ are listed in Table 3.3.

For *closo* carboranes $C_2B_{n-2}H_n$, such arguments allow one to understand why the thermodynamically preferred isomers are those with the carbon atoms in sites of lowest skeletal coordination number, because such sites provide the regions of greater electron density to which the carbon nuclei, more highly positively charged than boron nuclei, will be attracted. Indeed, one can deduce

TABLE 3.1. Variation of BH charge with skeletal bond type.

Number of 2c2e bonds	3	2	1	0
Number of 3c2e bonds	0	1	2	3
Charge on BH unit	-1	$-\frac{2}{3}$	$-\frac{1}{3}$	0

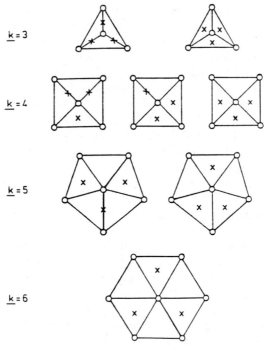

Figure 3.4. The bonding networks by which atoms of particular skeletal coordination numbers, k, can bond to their skeletal neighbors.

TABLE 3.2. Variation of BH charge with skeletal coordination number k.

Skeletal coordination number k	3	3	4	4	4	5	5	6
Number of 2c2e bonds	3	0	2	1	0	1	0	0
Number of 3c2e bonds	0	3	1	2	3	2	3	3
Charge on BH unit	-1	0	$-\frac{2}{3}$	$-\frac{1}{3}$	0	$-\frac{1}{3}$	0	0
Mean charge[a]	$-\frac{3}{6}$		$-\frac{2}{6}$			$-\frac{1}{6}$		0
k	3		4			5		6

[a]Giving equal weight to each of the possible bond networks for $k = 3$, 4 or 5.

TABLE 3.3. Charge distribution in *closo* borane anions $B_nH_n^{2-}$.

	$B_5H_5^{2-}$		$B_6H_6^{2-}$	$B_7H_7^{2-}$		$B_8H_8^{2-}$		$B_9H_9^{2-}$		$B_{10}H_{10}^{2-}$		$B_{11}H_{11}^{2-}$			$B_{12}H_{12}^{2-}$
k	3	4	4	4	5	4	5	4	5	4	5	4	5	6	5
$6 \times$ charge c	-3	-2	-2	-2	-1	-2	-1	-2	-1	-2	-1	-2	-1	0	-1
Number	2	3	6	5	2	4	4	3	6	2	8	2	8	1	12
Σc	-1	-1	-2	$-\frac{5}{3}$	$-\frac{1}{3}$	$-\frac{4}{3}$	$-\frac{2}{3}$	-1	-1	$-\frac{2}{3}$	$-\frac{4}{3}$	$-\frac{2}{3}$	$-\frac{4}{3}$	0	-2

that the *positive* charges on the CH units in carboranes will increase with increasing coordination number; a CH unit forming three 2c2e bonds to similar units is expected to be neutral, though if these bonds are in resonance with three 3c2e bonds, the charge on the CH unit will be $+0.5$. In $C_2B_3H_5$, for example, with the CH units in axial sites and the BH units in equatorial sites, one would infer that each CH unit has a positive charge of 0.5 electronic units, each BH unit a negative charge of 0.33 units. Yet higher positive charges, $+0.66$ units for skeletal connectivity 4, $+0.83'$ for $k = 5$, are implied for the CH units of higher carboranes. Clearly, these are overestimates of the actual charges; the skeletal electron distribution will become distorted from that implicit in Table 3.3, the carbon nuclei at the low connectivity sites attracting yet more electron density towards them than already assumed. In effect, the 2c2e bonds will become localized in the neighborhood of the carbon atoms. At best, therefore, these 2c2e and 3c2e bond schemes lead to limiting values for the positive charges on the CH units and negative charges on the BH units of carboranes; actual charges will be numerically smaller than those listed above [2, 3, 18, 19].

Before leaving this discussion of the value of 2c2e and 3c2e bond schemes for indicating how the skeletal electrons are distributed in cluster molecules, it should be acknowledged that there are two categories of cluster compound known for which skeletal electron pair numbers match the numbers of polyhedron edges or faces. The first category consists of tetrahedral clusters that formally contain six skeletal bond pairs (P_4, tetrahedrane, $Ru_4(CO)_{12}H_4$) or four skeletal bond pairs (B_4Cl_4, $(LiMe)_4$, $Re_4(CO)_{12}H_4$) that may be allocated to the tetrahedral edges or faces respectively [20]. The second category includes octahedral clusters containing eight or twelve skeletal bond pairs, assignable to the faces or edges, respectively, and typified by the early transition metal halide clusters $Nb_6Cl_{12}^{2+}$ and $Mo_6Cl_8^{4+}$. These latter clusters differ from those that are our main concern here in that each skeletal atom uses four atomic orbitals for cluster bonding. For both of these categories of cluster, however, localized 3c2e face bonds or 2c2e edge bonds clearly provide helpful and realistic indications of how the skeletal electrons are distributed [8, 9].

3.3. VERTEX ELECTRON PAIR SCHEMES

The localized bond schemes just discussed for *closo* borane anions $B_nH_n^{2-}$ and related species were complicated in many cases by the numbers of resonance canonical forms that had to be considered, and the difficulties of visualizing the three-dimensional networks involved. An alternative localized electron pair scheme that avoids resonance complications and conveys some feeling for the electron distribution in *closo*, *nido*, and *arachno* clusters is worth brief mention here. Instead of assigning electron pairs to polyhedron edges or faces, it assigns them primarily to polyhedron *vertices* [21].

The approach exploits the fundamental feature of *closo* borane anions $B_nH_n^{2-}$ that each of the n BH units contributes two electrons for skeletal bonding. Since in the cluster these electrons are shared with similar BH units, if these are in equivalent positions then the overall distribution of electron density over the surface of the polyhedron will leave each boron atom's share as two electrons, centered on the boron nucleus and so on the polyhedron vertex. (Actually, as outlined above, for clusters with more than one type of vertex, the atoms at vertices of lower connectivity end up with a slightly greater share of the skeletal electrons than those at vertices of higher connectivity, because there are fewer adjacent nuclei competing for their electrons.) Compared to an isolated BH unit, with its available skeletal electron pair cylindrically symmetrically disposed about its B–H axis, the skeletal pair associated with a BH unit in a cluster will become polarized under the attractive influence of the neighboring boron nuclei (Figure 3.5).

In addition to retaining effective control over the two skeletal electrons it contributes, each BH unit in a *closo* cluster anion $B_nH_n^{2-}$ will also acquire a supplementary electronic charge of $-2/n$, that being its share of the $(n + 1)$th skeletal electron pair.

This is illustrated schematically in Figure 3.6 for $B_7H_7^{2-}$. Figure 3.6a shows a vertical cross-section of this pentagonal bipyramidal cluster in exploded form, showing its axial BH units, and two of its equatorial BH units, each contributing a pair of electrons for skeletal bonding, surrounding the extra $(n + 1)$th pair of electrons. When these and the remaining three BH units (not shown) are brought together to form $B_7H_7^{2-}$, there are two major consequences for the skeletal electron pairs. Those associated with the BH units acquire the C_{5v} (axial) or C_{2v} (equatorial) distortion already referred to (Figure 3.5), as they are

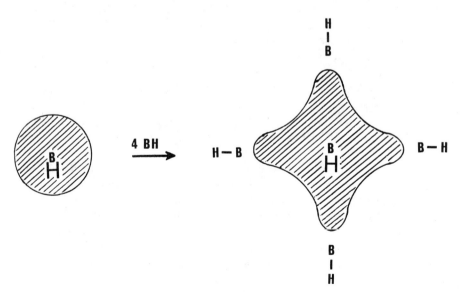

Figure 3.5. The polarization of a BH unit's skeletal electron pair caused by neighboring boron nuclei.

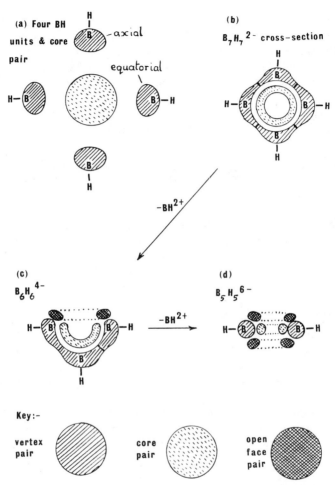

Figure 3.6. Schematic representation, in cross-section, of the distribution of skeletal electron pairs in the *closo* anion $B_7H_7^{2-}$ and in the *nido* and *arachno* anions $B_6H_6^{4-}$ and $B_5H_5^{6-}$ derived from $B_7H_7^{2-}$ by successive removal of BH^{2+} units from the axial vertices.

attracted towards adjacent nuclei, while the $(n+1)$th electron pair, which initially may be considered to lie at the cluster center, spreads out to form a pentagonal bipyramidally distorted spherical shell of electronic charge just inside the surface on which the boron nuclei lie.

This distribution of the $(n+1)$ skeletal electron pairs, with n pairs centered on the polyhedron vertices, the other pair delocalized within the polyhedron, corresponds well with the electron distribution implicit in MO treatments of *closo* clusters [2, 3, 14–19, 22–26]. Invariably, there is a unique, strongly bonding MO of A symmetry (S^σ-type in Stone's nomenclature [22–24]) resulting from a fully in phase combination of inward-pointing sp hybrid AOs. This MO concentrates electronic charge in a roughly spherical shell just inside the surface on which the boron atoms lie. It is this orbital that is represented by the unique $(n+1)$th delocalized electron pair in the vertex electron pair scheme, centered on the middle of the cluster but concentrated near to the boron nuclei. The n remaining bonding MOs all arise primarily from interactions of the $2n$ tangentially oriented AOs that concentrate electron density on the same surface as the boron nuclei.

The vertex electron pair scheme has been extended to estimate the polyhedral edge bond orders of borane anions $B_nH_n^{2-}$ [21], and can also be used to explore the relationship between *closo*, *nido*, and *arachno* species, as illustrated in Figure 3.6b–d. Since *nido* and *arachno* clusters contain the same numbers of electrons as their *closo* parent, though fewer skeletal atoms, their generation from the *closo* species may be represented by successive removal of BH^{2+} units from a parent *closo* anion $B_nH_n^{2-}$. The consequence for the skeletal electrons of removing an axial BH^{2+} unit from $B_7H_7^{2-}$ is shown in vertical cross-section in Figure 3.6c. The electrons originally associated with the vertex from which the BH^{2+} unit has been removed, losing their electrostatic attraction to that vertex, spread out towards the five neighboring equatorial nuclei. The core-centered delocalized skeletal pair also moves away from the vacant vertex towards the remaining nuclei. The effect of these shifts in the electron distribution is to build up electron density around the open face, with only a marginal increase in electron density elsewhere. This is nicely consistent with the picture conveyed by MO calculations. Typically the HOMO of *nido* clusters like $B_6H_6^{4-}$ concentrates electronic charge around the open face [3, 9, 13, 22–24, 27–30] making this region suitable for protonation in generating neutral species B_nH_{n+4}. The high connectivity of the vertex vacated (superficially puzzling because it requires more B...B links to be severed than would be the case if a low connectivity vertex had been vacated) is intelligible in that the residual ring of electronic charge will be stabilized increasingly with the number of nuclei around the open face.

Similar arguments apply to *arachno* clusters B_nH_{n+6}, whose skeletal anions $B_nH_n^{6-}$ have shapes formally derivable from *closo* parents $[B_{n+2}H_{n+2}]^{2-}$ by removal of two BH^{2+} units. Figure 3.6d illustrates this for the removal of a second axial BH^{2+} unit, that is, from a vertex opposite the one first vacated. Again, the electron pair originally assigned to the vertex vacated spreads out to form a ring of charge adjacent to the neighboring nuclei, and the core-centered, delocalized skeletal pair likewise relaxes towards those nuclei.

The *arachno* $B_5H_5^{6-}$ shown in cross-section in Figure 3.6d as the end product of these successive removals of two BH^{2+} units from $B_7H_7^{2-}$ is a pentagonal species isoelectronic with the cyclopentadienide anion $C_5H_5^-$. The rings of electronic charge above and below the B_5 pentagon correspond directly to the electron distribution in the doubly degenerate HOMO of an aromatic ring system such as $C_5H_5^-$.

It should be acknowledged that the argument just developed suggests that the most stable shape predicted for the *arachno* $B_5H_5^{6-}$ residue would be a pentagon, whereas the shape of *arachno* B_5H_{11} is that requiring the second BH^{2+} unit lost to come from an equatorial site instead of the second axial site shown in Figure 3.6. Pentaborane (11), B_5H_{11}, and the cyclopentadienide anion, $C_5H_5^-$, thus illustrate the two skeletal shapes possible for an *arachno* cluster derived from a pentagonal bipyramidal *closo* parent. The question arises as to why the boron species B_5H_{11}, indeed *arachno* boranes B_nH_{n+6} in general, have skeletal shapes that entail removal of one high connectivity skeletal atom and one *neighboring* skeletal atom from the *closo* parent, instead of two nonadjacent high connectivity atoms. The explanation for this can be found by considering the bonding requirements of the six *endo* hydrogen atoms of *arachno* boranes. The tent-shaped arrangement of the five boron atoms of B_5H_{11} can accommodate the six *endo* hydrogen atoms more effectively than a pentagonal shape could. The cyclopentadienide anion, with no *endo* hydrogen atoms to accommodate, has a regular pentagonal shape which suffers little distortion when it takes up a single proton in forming neutral cyclopentadiene, C_5H_6 [21, 31].

3.4. ELECTRON DISTRIBUTION IN SOME ICOSAHEDRAL CARBORANE DERIVATIVES

So far in this chapter we have been concerned with simple localized electron pair schemes by which one might get some feeling for the way the skeletal electrons are distributed in borane-type clusters. In their different ways these schemes have provided reminders of how significantly the numbers of skeletal bonding contacts in *closo* clusters $(3n-6)$ exceed the numbers of skeletal bond pairs $(n+1)$, making these bonding contacts have low formal bond orders (mean $(n+1)/(3n-6)$, i.e., 7/12 for octahedral species and 13/30 for icosahedral species). The use of 2c2e and 3c2e bond schemes allows one to deduce that the atoms of lower connectivity will bear a greater negative charge than their neighbors of higher connectivity, though overestimating the charge separation. The use of vertex electron pairs enables one to understand why, in *nido* clusters, high connectivity sites are left vacant, and provides a bridge between cluster chemistry and aromatic ring chemistry. However, if one wants to estimate more precisely the electron distribution in the less symmetrical clusters, or the effects of substituents on the electron distribution in relatively symmetrical clusters, NMR studies backed up by molecular orbital calculations offer considerable promise, as illustrated here for some derivatives of the icosahedral carborane $1,2\text{-}C_2B_{10}H_{12}$ [31].

The sensitivity of the electron distribution in carboranes to the presence of substituents has long been apparent [32]. For icosahedral carborane derivatives $RC_2B_{10}H_{11}$, ^{11}B-NMR studies have shown that the chemical shifts of the cage boron atoms, particularly the boron atom opposite to the point of attachment of the substituent (the antipodal atom) vary with the substituent R [33, 34], though difficulties of assignment have limited the scope of such studies. With the advent

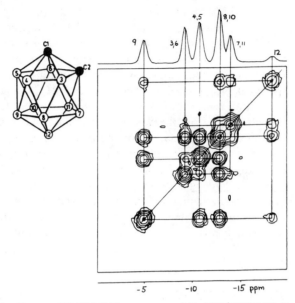

Figure 3.7. ^{11}B-^{11}B-2D-COSY-NMR spectrum of Et$_3$NH$^+$[O(Ph)C$_2$B$_{10}$H$_{10}$]$^-$.

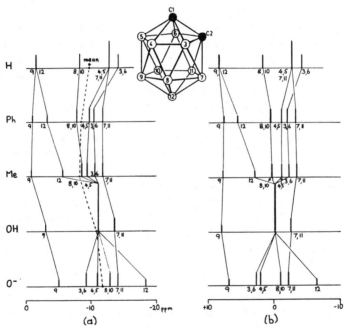

Figure 3.8. Peak assignments in the ^{11}B-NMR spectra of the *ortho*-carborane 1,2-C$_2$B$_{10}$H$_{12}$ and its derivatives 1-R-1,2-C$_2$B$_{10}$H$_{11}$ (R = Ph, Me, OH, O$^-$) (a) raw spectra; (b) spectra with mean resonances vertically aligned.

of 2D-COSY techniques [34–36] for establishing which resonances arise from directly linked boron atoms, uncertainties of assignment of peaks have been considerably reduced if not invariably eliminated. We have exploited this in studies of the ^{11}B-NMR spectra of a series of carborane derivatives 1,2-$R'R''C_2B_{10}H_{10}$ using such substituent groups R as Me, Ph, OH and O^-.

Figure 3.7 shows a typical spectrum, that of the anion [1,2-$O(Ph)C_2B_{10}H_{10}$]$^-$ recorded as its triethylammonium salt, with the assignments deduced from the off-diagonal resonances. Figure 3.8a shows how the resonances attributable to the different types of boron atom in compounds 1,2-$RC_2B_{10}H_{11}$ change in chemical shift for the sequence R = H, Ph, Me, OH, O^-. All of the resonances show some changes. The marked antipodal effect, on the resonance attributable to the boron atom in position 12, is clearly apparent, and the mean chemical shift, weighted to allow for the different numbers of boron atoms in the different sites, shows a small downfield shift in the sequence R = H, Ph, Me, but a more marked upfield shift in the sequence R = H, OH, O^-. This latter upfield shift can be explained in terms of π-delocalization from the substituent OH or O^- ligands into the C_2B_{10}-icosahedron, leading to an overall increase in the collective shielding experienced by the boron nuclei of a type familiar in the NMR spectra of 3-coordinate boron compounds, for example, compare BMe_3, $\delta(^{11}B)$ 86.6 ppm with $B(OH)_3$, $\delta(^{11}B)$ 19.6 ppm [37].

If one allows for this collective shielding by vertical alignment of the mean resonances (Figure 3.8b), it is clear that the boron resonances show two distinct types of behavior. Those due to atoms in sites 8, 9, 10 and (particularly) 12 show shifts to higher field in the sequence R = H, OH, O^-, whereas those due to the other boron atoms show downfield shifts in the same sequence. The effect on the boron atoms remote from the skeletal carbon atoms, particularly on the antipodal atom (site 12), is to become progressively more shielded, whereas the boron atoms near to the point of substitution experience decreased shielding, as the C-attached substituent R changes from H to OH to O^-.

Calculations on these compounds by the molecular orbital bond index (MOBI) [38] method have shown that, although these chemical shift changes do

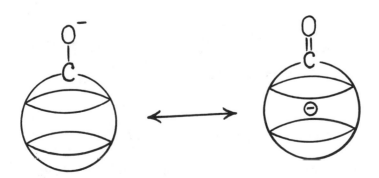

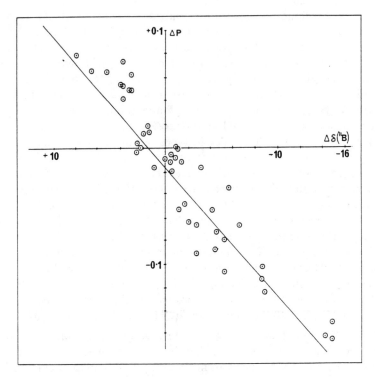

Figure 3.9. Correlation between chemical shift changes, $\Delta\delta$, and changes in cage polarity, Δp (calculated by the MOBI method [38]) for orthocarborane derivatives $RC_2B_{10}H_{11}$.

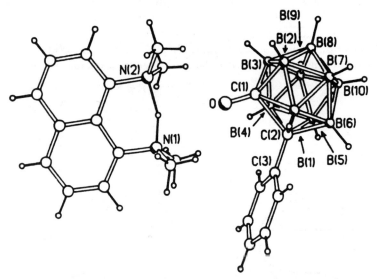

Figure 3.10. Structure of the salt $1,8\text{-}C_{10}H_6(NMe_2)_2H^+[O(Ph)C_2B_{10}H_{10}]^-$.

not correspond simply to changes in the calculated charges on the boron atoms (or the BH units) at the sites in question, there is evidence of a broad correlation between the chemical shift changes, $\Delta\delta$, and changes, Δp, in the polarity of these cages with respect to the pseudo 5-fold axis of symmetry through the boron atoms in question (Figure 3.9).

Structural evidence that electronic charge has delocalized from the *exo* oxygen atom of the anion $[1,2\text{-O(Ph)C}_2\text{B}_{10}\text{H}_{10}]^-$ into the cluster has been obtained by an X-ray crystallographic study of the proton sponge salt $\text{C}_{10}\text{H}_6(\text{NMe}_2)_2\text{H}^+ [1,2\text{-O(Ph)C}_2\text{B}_{10}\text{H}_{10}]^-$ (Figure 3.10), which revealed the carbon-oxygen bond to the *exo* oxygen atom to be only 1.25 Å long, as appropriate for what is effectively a double bond. The link between the two carbon atoms in the icosahedral cluster is long (2.00 Å). These features are readily intelligible if one treats the anion as derived from a *nido*-shaped $[\text{PhCB}_{10}\text{H}_{10}]^-$ anion interacting with a neutral carbonyl group, CO (itself a source of two electrons—the carbon "lone pair"—for skeletal bonding), near the vacant vertex. The frontier orbitals involved in their interactions are shown in Figure 3.11, with orbitals as labeled by Mingos [39]. Like a cyclopentadienide

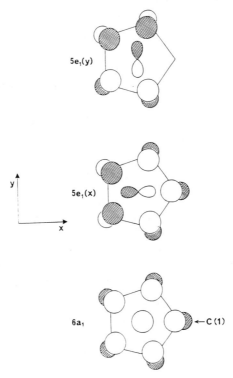

Figure 3.11. Frontier orbitals involved in bonding the CO unit in the anion $[\text{O(Ph)C}_2\text{B}_{10}\text{H}_{10}]^-$.

anion $C_5H_5^-$, the carborane residue has three orbitals through which it can interact with an atom approaching the vacant icosahedral vertex, one ($6a_1$) of suitable symmetry to accept the carbonyl group lone pair, the others of suitable symmetry for back π-bonding to the π^* orbitals of the carbonyl group, though the lower energy of the one ($5e_1x$) to which the cage carbon atom contributes makes it a far worse donor than that ($5e_1y$) to which only boron atoms contribute. It is this lower energy of the $5e_1x$ orbital that is responsible for the long C–C link. The $5e_1y$ orbital is primarily responsible for back π-bonding to the carbonyl group, which is effectively a μ_5 bridging ligand.

This particular compound shows the extent to which *exo* multiple bonding can influence the skeletal bonding of a boron cluster, causing it to start to open up in the manner familiar as the structural consequence of adding electrons, and indicating, through the antipodal effect, how the polarity along the axis bearing the substituent is affected. Further studies of this type promise to reveal more details of the distribution of the skeletal electrons in boron cluster derivatives.

REFERENCES

1. Dickerson, R. E. and Lipscomb, W. N., *J. Chem. Phys.*, **27**, 212 (1957).

2. Lipscomb, W. N., *Adv. Inorg. Chem. Radiochem.*, **1**, 117 (1959).

3. Lipscomb, W. N., *Boron Hydrides*, W. A. Benjamin, New York, 1963.

4. Williams, R. E., *Inorg. Chem.*, **10**, 210 (1971).

5. Williams, R. E., *Adv. Inorg. Chem. Radiochem.*, **18**, 66 (1976).

6. Rudolph, R. W. and Pretzer, W. R., *Inorg. Chem.*, **11**, 1974 (1972).

7. Wade, K., *Chem. Commun.*, 792 (1971).

8. Wade, K., *Electron Deficient Compounds*, Nelson, London, 1971.

9. Wade, K., *Adv. Inorg. Chem. Radiochem.*, **18**, 1 (1976).

10. Mingos, D. M. P., *Nature (London) Phys. Sci.*, **236**, 99 (1972).

11. Eberhardt, W. H., Crawford, B. L. Jr. and Lipscomb, W. N., *J. Chem. Phys.*, **22**, 989 (1954).

12. Lipscomb, W. N., *J. Chem. Phys.*, **25**, 38 (1956).

13. Lipscomb, W. N., in: Boron Hydride Chemistry (E. L. Muetterties, Ed.), Chap. 2, p. 39, Academic Press, New York, 1975.

14. Stanton, J. F., Lipscomb, W. N. and Bartlett, R. J., in: *Boron Chemistry* (IMEBORON 6) (S. Hermanek, Ed.), p. 74, World Scientific, New Jersey, 1987.

15. Epstein, I. R. and Lipscomb, W. N., *Inorg. Chem.*, **10**, 1921 (1971).

16. Lipscomb, W. N., *Inorg. Chem.*, **18**, 2328 (1979).

17. O'Neill, M. E. and Wade, K., *Polyhedron*, **3**, 199 (1984).

18. Guest, M. F. and Hillier, I. H., *Mol. Phys.*, **26**, 435 (1973).

19. Dixon, D. A., Kleier, D. A., Halgren, T. A., Hall, J. H. and Lipscomb, W. N., *J. Am. Chem. Soc.*, **99**, 6226 (1977).

20. Cavanaugh, M. A., Fehlner, T. P., Stramel, R., O'Neill, M. E. and Wade, K., *Polyhedron*, **4**, 687 (1985).

21. Gillespie, R. J., Porterfield, W. W. and Wade, K., *Polyhedron*, **6**, 2129 (1987).

22. Stone, A. J., *Inorg. Chem.*, **20**, 563 (1981).

23. Stone, A. J. and Alderton, M. J., *Inorg. Chem.*, **21**, 2297 (1982).

24. Stone, A. J., *Polyhedron*, **3**, 1299 (1984).

25. Middaugh, R. L., in: *Boron Hydride Chemistry* (E. L. Muetterties, Ed.), Chap. 8, p. 273, Academic Press, New York, 1975.

26. Longuet-Higgons, H. C., *Q. Rev. Chem. Soc. (London)*, **11**, 121 (1957).

27. Dunks, G. B. and Hawthorne, M. F., in: *Boron Hydride Chemistry* (E. L. Muetterties, Ed.), Chap. 11, p. 383, Academic Press, New York, 1975.

28. Hawthorne, M. F. and Andrews, T. D., *Chem. Commun.*, 443 (1965).

29. Grimes, R. N., *Metal Interactions with Boron Clusters*, Plenum Press, New York, 1982.

30. Grimes, R. N., in: *Comprehensive Organometallic Chemistry* (G. Wilkinson, F. G. A. Stone and E. W. Abel, Eds.) Vol. 4, p. 459, Pergamon Press, Oxford, 1982.

31. Porterfield, W. W., Stephenson, I. R. and Wade, K., in: *Boron Chemistry* (IMEBORON 6) (S. Hermanek, Ed.), p. 3, World Scientific, New Jersey, 1987.

32. Grimes, R. N., *Carboranes*, Academic Press, New York, 1970.

33. Hermanek, S., Plesek, J., Stibr, B. and Grigor, V., *J. Chem. Soc., Chem. Commun.*, 561 (1977).

34. Hermanek, S., Jelinek, T., Plesek, J., Stibr, B., Fusek, J. and Mares, F., in: *Boron Chemistry* (IMEBORON 6) (S. Hermanek, Ed.), p. 26, World Scientific, New Jersey, 1987.

35. Reed, D., *J. Chem. Res.*, 198 (1984).

36. Venable, T. L., Hutton, W. C. and Grimes, R. N., *J. Am. Chem. Soc.*, **106**, 209 (1984).

37. Nöth, H. and Wrackmeyer, B., *NMR Spectroscopy of Boron Compounds*, Springer, Munich, 1978.

38. Armstrong, D. R., Perkins, P. G. and Stewart, J. J. P., *J. Chem. Soc. A*, 3674 (1971).

39. Mingos, D. M. P., *J. Chem. Soc., Dalton Trans.*, 602 (1977).

CHAPTER 4

Ab initio Geometries and Chemical Shift Calculations for Neutral Boranes and Borane Anions

M. Buehl and P. von R. Schleyer

4.1. INTRODUCTION

After a quarter century of pioneering effort, Alfred Stock succeeded in describing a new class of unusually fascinating and important molecules, the boron hydrides [1–4]. "No guides for the studies were available from valence, mechanistic or theoretical chemistry" [2a], and Stock was not able to deduce a single structure correctly. In their 1942 review, Schlesinger and Burg sum-

Electron Deficient Boron and Carbon Clusters, Edited by George A. Olah, Kenneth Wade, and Robert E. Williams.
ISBN 0-471-52795-5 © 1991 John Wiley & Sons, Inc.

marized the situation, "numerous structures have been proposed for the higher boron hydrides, but knowledge of these chemical properties is still too meager to give support to any of these suggestions" [3].

Lipscomb's formulation of a consistent description of this class of molecules required both the development of experimental techniques (single crystal X-ray diffraction at low temperature) and the extension of valence theory to multicenter bonding. Lipscomb's combination of experiment and theory, and his insights into the nature of the bonding in these molecules was demonstrated by his impressive predictions of new structures. Lipscomb also employed ^{11}B-NMR as a major tool to investigate the nature of the boranes [4]. The symmetry and the fluxional character were revealed even though no detailed geometrical information could be deduced.

As Beaudet emphasizes in his recent review *The Molecular Structures of Boranes and Carboranes* [5], accurate structural data are essential for a refined knowledge of the bonding in such electron deficient molecules. His comprehensive critical analysis of experimental structural data, based on X-ray analysis, microwave-spectroscopy and gas phase electron diffraction was hindered by "certain incongruities", for example, omissions in the literature. Often insufficient data was presented to define a molecular structure completely. Beaudet not only summarized the available literature data, but he presented recommended sets of cartesian coordinates for 31 molecules, including the principal boranes. Particular attention was called to "unresolved accurate molecular structure problems". Beaudet felt that these could be resolved from gas phase structure determinations; the possibility that highly accurate structures could be calculated ab initio was not mentioned.

The contributions of molecular orbital theory to the understanding of the structures of the boron hydrides has been profound. Lipscomb and his associates developed extended Hückel and PRDDO methods largely for that purpose [6]. Later, his group employed ab initio theory to the calculation of the smaller boron hydrides, but these studies were limited to the use of small basis sets for geometry optimizations. The present availability of ever faster computers and more efficient quantum chemistry programs (e.g., Gaussian-88 and CADPAC) facilitate the study of even medium sized boranes at adequately high levels of theory. We demonstrate in the present paper that the results are at least as accurate as those which have been obtained experimentally.

Our work takes advantage of a new method of structural determination and assessment which employs IGLO (individual gauge for localized orbitals) chemical shift calculations [7, 8]. This combined ab initio/IGLO/NMR approach has been used with considerable success for the elucidation of the accurate structures of carbocations [9]. We now present a summary of our results on boranes and borane anions using the same approach. The IGLO chemical shifts calculated for several theoretical and experimental structure candidates for a given molecule are compared with the δ^{11}B data measured in solution. The degree of agreement between IGLO and experiment is used as a criterion to evaluate the accuracy of the competing structural models. As we show below, the results are gratifying in their excellence. The success of this approach also is a tribute to Lipscomb, who was interested very early in devising practical methods for the calculation of NMR chemical shifts [10].

4.2. METHODS, BASIS SETS, AND GEOMETRIES

The geometries were optimized ab initio at the Restricted Hartree Fock (RHF) level using the GAUSSIAN82 [11], GAUSSIAN86 [12], and CADPAC [13] programs with standard basis sets [14, 15] 3-21G, 6-31G* and in one case 6-31G**. Electron correlation was included in terms of Moller–Plesset second-order perturbation theory [16]. CADPAC was employed for most of the MP2/6-31G* optimizations using analytical gradients. In some cases single point calculations at higher levels of electron correlation (up to MP4sdtq) were performed.

The chemical shifts were calculated using the IGLO method [7, 8]. Huzinaga [17] gaussian lobe basis sets, recommended by the Bochum group [7b], were employed which were contracted as follows:

Basis DZ	B	7s3p	contracted	to	[4111,21]		
	H	3s	contracted	to	[21]		
Basis II′	B	9s5p1d	contracted to	[51111,311,1]	p-coefficient	0.5	
	H		same as DZ				

In the calculations, absolute chemical shieldings (with respect to the bare nucleus) are computed. Since a calculation for the $BF_3.OEt_2$ standard used experimentally is very cumbersome, B_2H_6 was used as the primary reference. The δ values were converted to the usual $BF_3.OEt_2$ scale using the experimental value of $\delta(B_2H_6) = 16.6$ ppm [18].

The experimental geometries which also were employed for IGLO calculations were taken mostly from Beaudet's compilation [5].

4.3. RESULTS

We considered a set of six neutral and three boron hydride anions for which experimental NMR data are available: B_2H_6 (**1**), B_4H_{10} (**2**), B_5H_9 (**3**), B_5H_{11} (**4a,b**), B_6H_{10} (**5**), B_6H_{12} (**6**), $B_2H_7^-$ (**7a–c**), $B_3H_8^-$ (**8a,b**), and $B_4H_9^-$ (**9**) (for numbering see Figure 4.1). The general performance of theoretically calculated geometries and chemical shifts is evaluated first, and then individual cases are discussed separately.

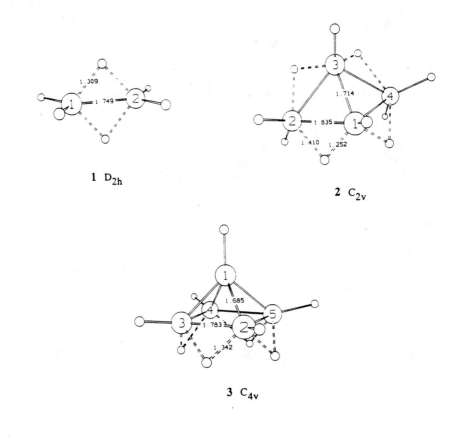

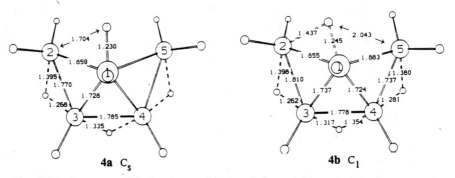

Figure 4.1. Structures of the boranes and borane anions of this study. The distances refer to the MP2/6-31G* optimized geometries.

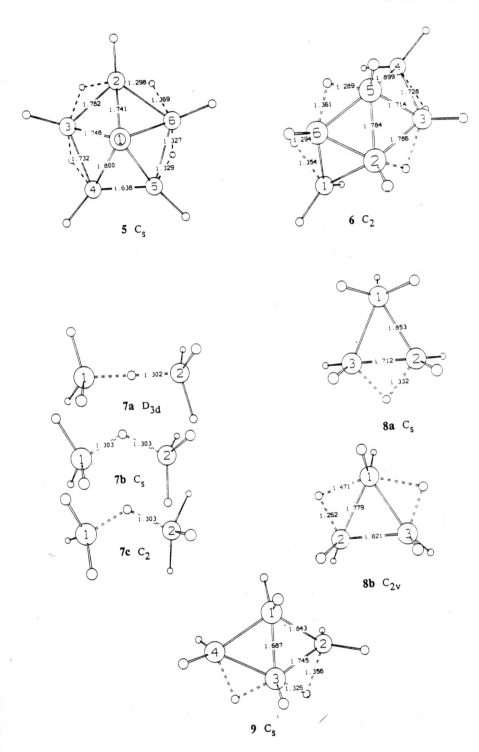

5 C_s

6 C_2

7a D_{3d}

7b C_s

7c C_2

8a C_s

8b C_{2v}

9 C_s

TABLE 4.1. Geometric parameters for 1–9 (bond distances in Å).

Compound	Level of theory	Geometry			

B_2H_6

		B1B2	B1Hb		
1	3-21G	1.785	1.315		
	6-31G*	1.779	1.316		
	MP2/6-31G*	1.749	1.309		
	Expt. (MW)[a]	*1.743*	*1.314*		
	Expt. (GED)[b]	*1.775*	*1.339*		

B_4H_{10}

		B1B2	B1B3	B1H13	B2H13
2	3-21G	1.902	1.733	1.248	1.430
	4-31G[c]	1.904	1.734	1.249	1.432
	6-31G*	1.893	1.741	1.247	1.423
	MP2/6-31G*	1.835	1.714	1.252	1.410
	MP2/6-31G**	1.838	1.715	1.245	1.405
	Expt. (MW)[d]	*1.854*	*1.718*	*1.428*	*1.425*
	Expt. (GED)[e]	*1.856*	*1.704*	*1.315*	*1.484*

B_5H_9

		B1B2	B2B3	B2H23	
3	3-21G	1.709	1.827	1.349	
	6-31G*	1.702	1.811	1.345	
	MP2/6-31G*	1.685	1.783	1.342	
	Expt. (MW)[f]	*1.690*	*1.803*	*1.352*	

B_5H_{11} [g]

		B1B2	B1B3	B1B4	B1B5	B2B3	B3B4	B4B5
4a	3-21G	1.934	1.774	1.774	1.934	1.779	1.870	1.779
	6-31G*	1.917	1.770	1.770	1.917	1.777	1.860	1.777
	MP2/6-31G*	1.859	1.728	1.728	1.859	1.770	1.785	1.770

4b	3-21G	1.893	1.795	1.772	1.952	1.835	1.852	1.762			
	6-31G*	1.896	1.761	1.758	1.939	1.856	1.830	1.750			
	MP2/6-31G*	1.855	1.737	1.724	1.883	1.810	1.778	1.737			
	Expt. (X-ray)[h]	*1.870*	*1.720*	*1.720*	*1.870*	*1.760*	*1.770*	*1.760*			
	Expt. (GED)[i]	*1.891*	*1.741*	*1.741*	*1.891*	*1.812*	*1.760*	*1.812*			
		B1Hb	**B2Hb**	**B5Hb**	**B2H23**	**B5H45**	**B3H23**	**B4H45**	**B3H34**	**B4H34**	
4a	3-21G	1.215	1.872	1.872	1.446	1.446	1.244	1.244	1.330	1.330	
	6-31G*	1.211	1.828	1.828	1.435	1.435	1.250	1.250	1.335	1.335	
	MP2/6-31G*	1.230	1.704	1.704	1.395	1.395	1.268	1.268	1.335	1.335	
4b	3-21G	1.229	1.549	1.993	1.477	1.472	1.236	1.254	1.323	1.336	
	6-31G*	1.235	1.459	2.134	1.443	1.410	1.243	1.266	1.318	1.351	
	MP2/6-31G*	1.245	1.437	2.043	1.398	1.380	1.262	1.281	1.317	1.345	
	Expt. (X-ray)[h]	*1.090*	*1.670*	*1.670*	*1.300*	*1.300*	*1.220*	*1.220*	*1.180*	*1.180*	
	Expt. (GED)[i]	*1.327*[j]	*1.594*	*1.898*	*1.394*	*1.394*	*1.274*	*1.274*	*1.335*	*1.335*	
B_6H_{10}		**B1B2**	**B1B3**	**B1B4**	**B2B3**	**B3B4**	**B4B5**	**B2H23**	**B3H23**	**B3H34**	**B4H34**
5	3-21G	1.756	1.773	1.863	1.819	1.773	1.629	1.291	1.377	1.312	1.375
	6-31G*	1.761	1.765	1.844	1.824	1.760	1.637	1.292	1.376	1.314	1.345
	MP2/6-31G*	1.741	1.748	1.800	1.782	1.732	1.638	1.298	1.369	1.327	1.329
	Expt. (X-ray)[k]	*1.736*	*1.753*	*1.795*	*1.794*	*1.737*	*1.596*	*1.32*	*1.48*	*1.31*	*1.35*
	Expt. (MW)[l]	*1.774*	*1.762*	*1.783*	*1.818*	*1.710*	*1.654*				
B_6H_{12}		**B1B2**	**B1B6**	**B2B3**	**B2B5**	**B2B6**	**B1H16**	**B6H16**	**B2H23**	**B3H12**	
6a	3-21G	1.918	1.804	1.842	1.747	1.775	1.464	1.245	1.375	1.306	
	6-31G*	1.932	1.779	1.831	1.759	1.762	1.419	1.263	1.365	1.316	
	3-21G	1.939	1.761	1.829	1.792	1.741	1.388	1.273	1.291	1.357	
	6-31G*	1.941	1.747	1.825	1.798	1.738	1.376	1.283	1.291	1.357	
	MP2/6-31G*	1.899	1.728	1.786	1.784	1.714	1.354	1.294	1.289	1.361	
	Expt. (GED)[q]	*1.778*	*1.913*	*1.699*	*1.821*	*1.777*	*1.416*	*1.200*	*1.308*	*1.308*	
	MP2/6-31G*	1.899	1.728	1.786	1.784	1.714	1.354	1.294	1.289	1.361	

TABLE 4.1. *(continued)*

B₂H₇⁻ (7)

Compound	Level of theory	7a B1Hb	7b B1Hb	7b B2Hb	7b B–H–B	7c B1Hb	7c B–H–B
$B_2H_7^-$	6-31G*[m]	1.329	1.329	1.329	141.2°	1.328	140.8°
	MP2/6-31G*[m]	1.303	1.303	1.303	127.0°	1.302	126.7°
	MP2/6-31G**		1.297	1.297	127.3°	1.298	126.4°
	Expt. (X-ray)[n]		*1.00*	*1.27*	*136.4°*		
	Expt. (Neutron-Diff.)[o]		*1.21*	*1.32*	*127.2°*		

B₃H₈⁻ (8)

Compound	Level of theory	8a B1B2	8a B2B3	8a B2H23	8b B1B2	8b B2B3	8b B1H12	8b B2H23
$B_3H_8^-$	3-21G	1.962	1.728	1.338	1.810	1.923	1.576	1.255
	6-31G*	1.940	1.718	1.337	1.798	1.907	1.543	1.254
	MP2/6-31G*	1.853	1.712	1.332	1.779	1.821	1.471	1.262
	Expt. (X-ray)[p]				*1.77*	*1.80*	*1.50*	*1.20*
	Expt. (X-ray)[r]				*1.784*	*1.832*		

B₄H₉⁻ (9)

Compound	Level of theory	B1B2	B2B3	B1B3	B2H23	B3H23
$B_4H_9^-$	3-21G	1.863	1.790	1.732	1.393	1.309
9	6-31G*	1.884	1.765	1.713	1.366	1.333
	MP2/6-31G*	1.843	1.745	1.687	1.356	1.326

a[23c]. e[29]. i[35]. m[44]. q[41].
b[62]. f[31]. j Fixed. n[45].
c[30]. g[32]. k[38]. o[46].
d[28]. h[33]. l[39]. p[48].

4.3.1. Evaluation: Ab Initio Geometries and IGLO Calculations, General Performance

The structural parameters for **1–9** are summarized in Table 4.1 and are compared with the experimental geometries, when these are available. It is apparent that geometry optimizations at the HF/3-21G level tend to overestimate most B–B bond distances considerably (up to ca. 5% too long with respect to the gas phase data). To a lesser extent, this also holds for the HF/6-31G* optimized structures. However, correlation effects are known to be essential for multicenter bonding [14], and the geometries optimized at the correlated MP2/6-31G* level show the best agreement with the experimental results [19].

However, in most cases there is some disagreement between theory and experiment with regard to the location of the bridging hydrogens. The most refined theoretical structures usually show these Hs to be bound more closely to the boron atoms than is found experimentally. However, the exact experimental location of these bridge hydrogens often is very difficult—even in the gas phase—due to large vibrational motions. Our experience with carbocations [9] showed that the position of hydrogens involved in multicenter bonding has considerable influence on the calculated chemical shifts. As is shown below, this is also true for the boron hydrides.

Although the use of p-type polarization functions on hydrogen is known to provide an improved description of bridging Hs in carbocations [9], this appears to be less important for the bridged boron hydrides [20] (also see our results on B_4H_{10}). Hence, our standard highest level optimizations have been carried out at MP2/6-31G*. Since the standard 6-31G** basis set requires three additional p-functions on all hydrogens, the use of this basis would have been very costly.

The IGLO calculated ^{11}B chemical shifts are presented in Table 4.2 together with the experimental NMR data. Figures 4.2a–f show plots of the calculated versus the experimental ^{11}B chemical shifts (see Table 4.3 for statistical analyses). In Figure 4.2a, the IGLO values are calculated for experimental geometries; the agreement with the measured chemical shifts is acceptable though not very good. In Figure 4.2b and c, the geometries employed in the IGLO calculations were optimized ab initio at Hartree–Fock levels using the 3-21G and the 6-31G* basis sets. The overall agreement of IGLO δ^{11}B with experiment is only a little better with the 6-31G* geometries, but these perform about as well as the experimental geometries (compare Figures 4.2a and c). Figure 4.2d shows the same plot for the MP2/6-31G* optimized geometries. At this level of theory, the agreement is excellent. Figure 4.2c illustrates the rather small additional improvement of employing a larger basis set (basis II') in the IGLO calculations: the slope of the least squares fit is 1.01, the standard deviation 1.4, the maximum deviation 2.2 ppm, and the correlation coefficient is 0.999.

We conclude that the performance of the 3-21G and 6-31G* geometries in the IGLO calculations is generally satisfactory (though with significant deviations

TABLE 4.2. IGLO δ^{11}B chemical shifts (in ppm rel. BF$_3$.OEt$_2$).

Level of theory	δ^{11}B (ppm)				
	B$_2$H$_6$ 1	**B$_4$H$_{10}$ 2**		**B$_5$H$_9$ 3**	
		B$_{1,3}$	B$_{2,4}$	B$_1$	B$_{2-5}$
DZ//Expt.[a]	16.6	-37.6	-3.9	-52.9	-9.0
DZ//3-21G	15.9	-41.6	-2.5	-55.0	-9.6
DZ//6-31G*	16.0	-41.7	-2.5	-55.9	-10.7
DZ//MP2/6-31G*	15.6	-42.4	-6.0	-55.2	-11.5
DZ//MP2/6-31G**		-42.7	-6.1		
II'//MP2/6-31G*	15.4	-40.0	-5.3	-55.1	-12.8
Experiment[b]	*16.6*	*-41.8*	*-6.9*	*-53.1*	*-13.4*

B$_5$H$_{11}$

	B$_1$	B$_2$	B$_5$	ϕB$_{2,5}$	B$_3$	B$_4$	ϕB$_{3,4}$
DZ//Expt. (X-ray)	-79.7			5.7			-5.0
DZ//Expt. (GED)[c]	-58.1	6.5	15.8	11.1	-7.5	1.4	-3.1
4a DZ//3-21G	-55.2			23.7			-2.1
DZ//6-31G*	-57.8			20.4			-1.7
DZ//MP2/6-31G*	-63.6			11.6			-2.0
II'//MP2/6-31G*	-61.3			15.2			0.3
4b DZ//3-21G	-51.6	14.5	23.3	18.8	-13.3	6.0	-3.7
DZ//6-31G*	-51.6	7.4	18.4	12.8	-15.5	8.8	-3.4
DZ//MP2/6-31G*	-55.8	1.4	12.8	7.1	-14.0	9.1	-2.6
II'//MP2/6-31G*	-53.9	2.2	13.7	8.0	-13.5	10.3	-1.6
Experiment[d]	*-55.3*			*7.4*			*0.5*

	B$_6$H$_{10}$ 5				**B$_6$H$_{12}$ 6b**		
	B$_1$	B$_2$	B$_{3,6}$	B$_{4,5}$	B$_{1,4}$	B$_{3,6}$	B$_{2,5}$
DZ//Expt.[k]	-45.3	-8.4	26.6	28.1	14.1	41.1	-13.3
DZ//3-21G	-52.6	-7.6	18.0	22.8	14.5	22.5	-22.1
DZ//6-31G*	-53.2	-7.4	18.5	21.5	12.9	21.6	-22.4
DZ//MP2/6-31G*	-53.4	-7.2	18.0	19.5	9.6	22.9	-23.0
II'//MP2/6-31G*	-51.5	-7.4	18.8	19.9	10.1	23.5	-22.8
Experiment[e]	*-51.8*	*-6.5*	*18.6*	*18.6*	*7.9*	*22.6*	*-22.6*

B$_2$H$_7^-$

	7a	7b	7c
DZ//Expt.(X-ray)[f]		-38.7	
DZ//Expt. (Neutron-Diff)[g]		-26.2	
DZ//6-31G*	-13.3	-17.3	-17.3
DZ//MP2/6-31G*	-14.7	-22.9	-23.1
DZ//MP2/6-31G**		-23.3	-23.2
II'//MP2/6-31G*		-24.5	-24.7
Experiment[h]		*-24.6*	

B$_3$H$_8^-$

	8a			**8b**		
	B$_1$	B$_{2,3}$	ϕB	B$_1$	B$_{2,3}$	ϕB
DZ//3-21G	-41.5	-1.1	-14.6	1.4	-37.1	-24.3
DZ//6-31G*	-42.7	-1.5	-15.9	-2.0	-38.4	-26.2
DZ//MP2/6-31G*	-48.4	-7.4	-21.1	-10.0	-42.5	-31.6
II'//MP2/6-31G*	-50.6	-7.5	-21.9	-9.3	-43.5	-32.0
Experiment[i]						*-29.8*

B$_4$H$_9^-$ 9

	B$_1$	B$_{2,4}$	B$_3$
DZ//3-21G	-54.7	-4.5	-1.8
DZ//6-31G*	-52.5	-8.4	2.9
DZ//MP2/6-31G*	-54.9	-9.9	0.9
II'//MP2/6-31G*	-55.9	-11.0	3.0
Experiment[j]	*-54.5*	*-10.2*	*0.8*

[a]Unless otherwise noted, the experimental geometries employed in the IGLO calculations were taken from [5].
[b]B$_2$H$_6$ [18]; B$_4$H$_{10}$ [34d]; B$_5$H$_9$ [63]. [c][35]. [d][33]. [e]B$_6$H$_{10}$ [37]; B$_6$H$_{12}$ [34d]. [f][45]. [g][46]. [h][47]. [i][64]. [j][54]. [k]B$_6$H$_{12}$ [41].

TABLE 4.3. Statistical analysis[a] of the performance of IGLO calculations (see Figure 4.2a–e).

IGLO basis //geometry	Slope	Correlation coefficient	Deviation[b]		Plots figure
			Largest	Standard	
DZ//Exptl.	1.08	0.984	18.5	5.3	4.2a
DZ//3-21G	1.01	0.980	13.1	5.5	4.2b
DZ//6-31G*	0.99	0.989	10.8	4.0	4.2c
DZ//MP2/6-31G*	1.01	0.999	3.1	1.2	4.2d
II'//MP2/6-31G*	1.01	0.999	2.2	1.4	4.2e

[a]Results of a least squares analysis, IGLO vs. experimental ^{11}B chemical shifts.
[b]In ppm.

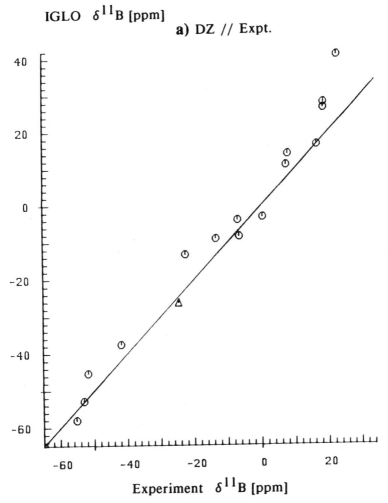

Figure 4.2. Comparison between IGLO calculated and experimental δ^{11}B chemical shifts at various levels of theory (a)–(e) (denoted as "IGLO basis // Geometry"; circles, neutral boranes **1–6**; triangles, borane anions **7–9**).

IGLO δ^{11}B [ppm]

b) DZ // 3-21G

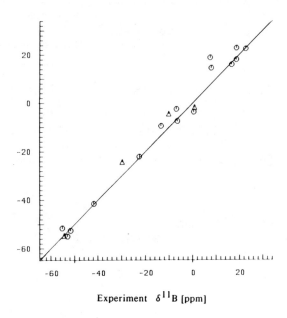

Experiment δ^{11}B [ppm]

IGLO δ^{11}B [ppm]

c) DZ // 6-31G*

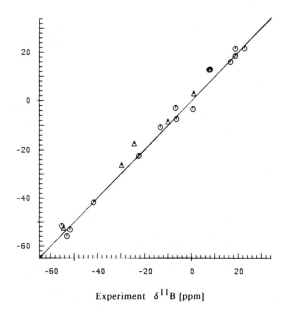

Experiment δ^{11}B [ppm]

Figure 4.2. (*Continued*)

124

IGLO δ^{11}B [ppm]

d) DZ // MP2/6-31G*

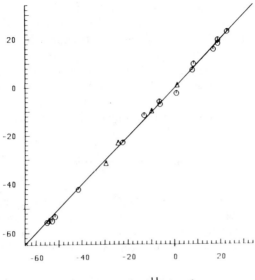

Experiment δ^{11}B [ppm]

IGLO δ^{11}B [ppm]

e) II'//MP2/6-31G*

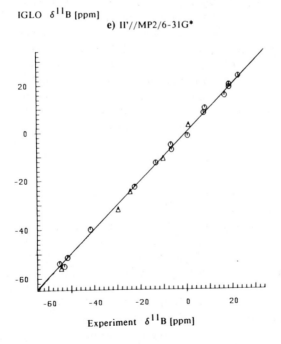

Experiment δ^{11}B [ppm]

Figure 4.2. (*Continued*)

125

in some cases) and at 6-31G* is comparable to that of the experimental geometries. The best agreement between theoretical and experimental chemical shifts is found for the MP2/6-31G* optimized structures. In view of the rather small individual deviations (the largest is only 2.2 ppm with basis II'), geometries calculated at this level should be very reliable. For the IGLO calculations, the DZ basis set performs nearly as well as the II' basis, and can be employed for most purposes.

In the following, the results for the individual compounds are discussed in more detail.

4.3.2. B_2H_6 (1)

Diborane is certainly the most important of all boron hydrides. It required many years after Stock's first preparation in 1913 [21] before a more convenient synthesis became available [3, 22] and the structure was established with certainty [23]. The discovery of the hydroboration reaction [24] has had a large impact on synthetic chemistry.

The confirmation of the bridged molecular structure of diborane required a more refined theoretical description of this electron deficient molecule. The development of the three-center bonding concept [25] and the generalization and application to the boranes [26] was a major contribution to the success of MO theory. Numerous theoretical calculations on diborane have been reported [27] recently at very high levels of sophistication [27d].

As mentioned above, B_2H_6 serves as the primary reference in the IGLO chemical shift calculations: the experimental B_2H_6 structure [23c] was taken as the standard. The IGLO values for the theoretical structures are in close agreement with experiment, without showing a pronounced dependence on the optimization level (Table 4.2). Hence, this molecule is well suited to be used as the "theoretical standard". A more detailed analysis of the IGLO results for B_2H_6 in terms of localized orbital contributions has been given [7b].

4.3.3. B_4H_{10} (2)

The "butterfly" structure of B_4H_{10} with overall C_{2v} symmetry is well established [28]. The geometrical parameters of the *boron framework* calculated at MP2/6-31G* level agree well with those obtained from microwave spectroscopy (MW) [28] and gas phase electron diffraction (GED) [29]. However, these two experimental methods do not agree in the location of the bridging hydrogens. The MW structure shows nearly identical $B1-H_b$ and $B2-H_b$ distances. The ab initio structures show rather unsymmetric B-H-B bridges with closer contacts to B1 and B3 (closer than given by the GED study). This is in agreement with calculations reported by other groups [30].

The IGLO $\delta^{11}B$ values for all structures are in reasonable agreement with experiment. The best fit, however, is found for the MP2 optimized geometries.

For this relatively small and symmetric molecule, the effect of employing a larger basis set (6-31G** which includes polarization functions on hydrogen) in the optimization could be probed. The geometrical parameters calculated at the MP2/6-31G* and MP2/6-31G** levels do not show any significant differences. The same is true for the chemical shifts calculated for these geometries. Hence, as pointed out above, polarization functions on hydrogen do not appear to be needed in the optimizations [20].

4.3.4. B_5H_9 (3)

Neither the calculated geometrical parameters nor the chemical shifts show a pronounced dependence on the optimization level. While Beaudet's experimental structure (MW) [31] gives good agreement for δ of B_1, the chemical shifts for B_{2-5} fit better for those calculated using the theoretical structures. The positions of the bridging hydrogens—which should be important in determining the chemical shifts of the basal boron atoms—could not be located experimentally with the same accuracy as the other atoms due to the large vibrational amplitude [31]. Hence, the calculated geometries probably are more reliable.

4.3.5. B_5H_{11} (4)

In a preliminary communication, we reported ab initio geometry optimizations and IGLO calculations for B_5H_{11} [32] which favored the slightly lower energy C_1 form over the C_s form. An early X-ray structure [33] and the NMR spectra [34] for B_5H_{11} indicated C_s symmetry (cf. **4a**), whereas the unique bridging hydrogen prefers an unsymmetric location (cf. **4b**) in the gas phase (GED) [35]. As reported earlier [32, 36], **4b** is the minimum structure and **4a** corresponds to a transition state for the tautomerization of H_b. The barrier for this scrambling was calculated to very low (about 1 kcal/mol with MP4sdtq/6-31G*), suggesting a rather fluxional character of the molecule. The best (theoretical) structure for **4b**—optimized at the MP2/6-31G* level—shows the critical hydrogen H_b to be involved in a perfectly normal, unsymmetrical B–H–B bond. Hence, the "styx" formulation [4] of B_5H_{11} should be 4112 rather than 3203.

The IGLO $\delta^{11}B$ values calculated for the ab initio structures are very sensitive to the "quality" of the geometry. The degree of agreement improves with increasing level of sophistication employed in the geometry optimization; the chemical shifts of **4b** steadily approach the experimental values. The MP2/6-31G* geometry gives data in excellent agreement with experiment. On the other hand, neither of the **4a** structures result in "satisfactory" IGLO values. For the MP2/6-31G* geometry, which should perform best, the deviations are up to 8 ppm (for B_1). The greatest deviation encountered for the other boranes in this study is only ca. 2 ppm at the same level. Hence, the IGLO calculations indicate that **4a** does not exist in solution.

Interestingly, the individual chemical shifts of B_3 and B_4 in unsymmetrical **4b** are calculated to be very different (-13.5 and $+10.3$ ppm), even though these atoms are not directly involved with the "critical" hydrogen bridge. This emphasizes the general importance of the exact positions of bridging hydrogens. That this pronounced chemical shift difference to a large extent is due to the unsymmetric location of H_b can be seen in the IGLO results for the experimental GED structure. In this refinement the whole molecule with the exception of H_b was constrained to have C_s symmetry (cf. Table 4.1). Even though the critical hydrogen is "displaced" to a smaller extent from the symmetric position, the calculated δ values of B_3 and B_4 also differ appreciably (-7.5 and $+1.4$ ppm).

4.3.6. B_6H_{10} (5)

The fluxionality of this molecule at room temperature is due to rapid tautomerization of the bridging hydrogens [37]. The five basal boron atoms become equivalent on the NMR time scale and only one resonance is observed for B_{2-6}. At lower temperature, this single peak is split into two resonances. These have been assigned to the static C_s structure 5 found in the solid state [38] and in the gas phase [39]. Since this symmetry requires three nonequivalent basal boron atoms, two of these ($B_{3,6}$ and $B_{4,5}$) must resonate at nearly identical frequencies. The IGLO values for the ab initio geometries are in excellent agreement with the experimental chemical shifts. Indeed very similar $\delta^{11}B$ values are calculated for $B_{3,6}$ and $B_{4,5}$. In contrast to earlier predictions [40], the significant difference in the chemical shifts between these two sets of borons and the other basal boron B_2 is reproduced well by the IGLO calculations.

4.3.7. B_6H_{12} (6)

A gas phase electron diffraction (GED) study for this molecule has been reported recently [41], which established the molecular C_2 symmetry and confirmed a 4212 type (styx notation) structure. In his very recent *ab initio* investigation [42], McKee noted considerable discrepancies between the geometrical parameters found experimentally and those calculated *ab initio* at the 3-21G and 6-31G* levels. We now use IGLO calculations in order to decide among these candidates [43].

Beginning our optimizations, we were not aware of the GED structure. We found two minima at 3-21G and 6-31G* levels that possess similar structures (6a and 6b, key geometric parameters in Table 4.1). But these differ appreciably in energy; 6a was ca. 27 kcal/mole less stable than 6b but optimized to a structure corresponding to 6b at the MP2/6-31G* level. However, this MP2/6-31G* geometry also differs appreciably from the GED structure (Table 6.1).

As noticed by McKee, the energy of the experimental structure is ca. 56 kcal/mol higher compared to the 3-21G and 6-31G* geometries of 6b (single point calculations at the MP2/6-31G* level). With respect to our MP2/6-31G* optimized geometry, this value is even higher, 59 kcal/mole! McKee showed that this enormous energy difference to some extent is due to the poor placements of the hydrogens in the GED structure [42].

The IGLO $\delta^{11}B$ values calculated using the GED geometry do not agree at all with the experimental chemical shifts (the largest deviation is 18 ppm; see Table 4.2). The chemical shifts calculated for the theoretical "false minima" **6a** also show pronounced deviations (up to ca. 11 ppm). The **6b** structures perform better, and the $\delta^{11}B$ values calculated for the MP2/6-31G* geometry are in excellent accord with experiment.

In the light of the generally very good performance of IGLO calculations for boranes, this excellent agreement suggests that the MP2/6-31G* geometry for **6b** resembles the actual structure of B_6H_{12} in solution very closely. Unfortunately, we must conclude that the experimental GED structure is not accurate.

4.3.8. $B_2H_7^-$ (7)

An earlier ab initio study including geometry optimizations at the MP2/6-31G* level [44] correctly showed the $B_2H_7^-$ anion to be strongly bent as is found in the solid state [45, 46]. The symmetrical C_2 form **7c** was predicted to be the minimum rather than the slightly unsymmetrical C_s form **7b**, which has one imaginary frequency and therefore should be a transition state (the linear D_{3d} structure has two imaginary frequencies denoting a "hilltop"). In addition, we optimized **7a** and **7b** at the MP2/6-31G** level, but as in the case of B_4H_{10} (see above), this does not have any significant influence on structure and nature of these forms.

The X-ray structure, however, reveals C_s symmetry with a strongly unsymmetrical B–H–B bridge [45]; the asymmetry of this bridge is somewhat less pronounced in the neutron diffraction structure [46]. But since the calculations showed the potential energy surface to be very flat, for example, the rotation of the BH_3 groups is practically free, the actual geometry in the solid state may well be dominated by crystal packing forces. In the gas phase or in solution more symmetric forms are to be expected.

The B–H–B angle seems to be the crucial geometric parameter that governs the chemical shifts of the various structures of $B_2H_7^-$. The value of this angle depends strongly on the level of theory employed in the optimization; the inclusion of electron correlation is especially important [44] (e.g., for **7c** the B–H–B angle is 140.8° with 6-31G* and 126.4° with MP2/6-31G* optimization). The ^{11}B chemical shift differences between the 6-31G* and the MP2/6-31G* geometries for **7b** and **7c** are significant; this shows that optimization with correlation is essential here. For these bent forms, the IGLO ^{11}B chemical shifts (−22.9 and −23.1 ppm, respectively) are in close agreement with the experimental value (−24.6 ppm) [47] whereas linear **7a** shows a clear-cut deviation (−14.7 ppm). The bent B–H–B moiety found in the solid state must be retained in solution. However, no distinction can be made between **7b** and **7c** on this basis, since the chemical shifts of these are nearly identical.

An IGLO calculation using the experimental X-ray structure shows rather poor agreement with experiment. This structure suffers from uncertainties in the hydrogen locations which are crucial for the chemical shifts. On the other hand, the neutron diffraction derived geometry, although unsymmetric, gives satisfactory IGLO results.

4.3.9. $B_3H_8^-$ (8)

Two structural alternatives were considered: **8a** with C_s and **8b** with C_{2v} symmetry, the latter has been found in crystal structures [48]. The relative energies of these have already been examined by McKee and Lipscomb [49] at various levels of theory, but the geometries only were optimized at the 3-21G level. Reoptimization at 6-31G* and MP2/6-31G* shortens nearly all bond distances. The MP2 geometry is in good agreement with the X-ray structure.

The IGLO results show a clear trend: decreasing boron-boron distances are causing an upfield shift for the $\delta^{11}B$ values. A comparison of the average chemical shifts of the MP2 geometries with the experimental value (-29.8 ppm) shows a good agreement for **8b** (-31.6 ppm) but worse correspondence for **8a** (-21.1 ppm). This indicates that the C_{2v} form of the solid state also is present in solution. The individual chemical shifts of the static form **8a** (-50.6, -7.5 ppm, basis II') lie in the same range as those found experimentally for Lewis-base adducts B_3H_7L which possess C_s symmetry (e.g., -52.6, -6.4 ppm for L=CO [50]; -51.6, -11.2 ppm for L=PH_3 [51]).

The relative energies of **8a** and **8b** were estimated at higher levels of electron correlation including correction for zero point energies (Table 4.4). At Hartree–Fock levels, **8a** is a minimum and is lower in energy than **8b** which is a transition state at that level. Electron correlation reverses the energetic sequence, and in order to confirm the nature of the stationary points frequency calculations were carried out at MP2/6-31G*. As expected, at this level **8b** is the minimum and **8a** is a transition state. A detailed mechanism for hydrogen scrambling—for which **8a** is the transition state—was proposed by Lipscomb et al. [52]. Our best estimation for this barrier, about 0.9 kcal/mol, agrees with the calculations reported by McKee and Lipscomb [49] and is consistent with the fluxional

TABLE 4.4. Absolute[a] and relative[b] energies for $B_3H_8^-$ **8a** and **8b**.

Level of theory	E_{rel}^b		
	8a	**8b**	(**8b**=0)
3-21G//3-21G	−78.20468	−78.20238	−1.4
6-31G*//6-31G*	−78.65909	−78.65803	−0.7
MP2(FU)/6-31G*//MP2/6-31G*	−78.96521	−78.96816	1.9
MP3(FC)/6-31G*//MP2/6-31G*	−78.98812	−78.99047	1.5
MP4SDTQ/6-31G*//MP2/6-31G*	−79.00549	−79.00779	1.4
ZPE(MP2/6-31G*)[c]	50.7(1)	51.3(0)	
Final			0.87

[a] In a.u.
[b] kcal/mol.
[c] Scaled by 0.95 as recommended in [65].

character of this molecule [53]. Despite this small energetic differences, however, the geometrical changes involved in the internal rearrangement are quite large. This indicates that the potential energy surface is very flat.

4.3.10. $B_4H_9^-$ (9)

No X-ray structure is known for a $B_4H_9^-$ salt. A structure (9) has been proposed on the basis of the NMR-spectrum [54]. The C_s symmetry suggested by NMR cannot be represented by a single styx structure and therefore "fractional three-center BBB-bonds" [55] have been proposed.

We carried out geometry optimizations in C_s symmetry including polarization functions and electron correlation; frequency calculations showed this structure to be a minimum. For the other boranes and borane anions in this study, the geometries optimized at MP2/6-31G* are in rather good accord with experiment. Hence, the structural parameters for $B_4H_9^-$ at this level should also be reliable. This is confirmed by the IGLO results for the MP2/6-31G* structure which are in excellent agreement with the experimental chemical shifts.

4.3.11. Hydrogen Chemical Shifts

Several problems are associated with the theoretical calculation of ^1H chemical shifts [8]; the rather small range of proton NMR shifts requires a relatively very high level of accuracy in order to produce meaningful results. Furthermore, the hydrogens are located at the "surface" of the polyhedral boranes; hence, a much greater susceptibility to solvation effects is to be expected.

The ^1H IGLO values calculated for the MP2/6-31G* geometries are presented in Table 4.5. The results obtained with IGLO basis sets DZ and II' show pronounced differences, even though the hydrogen basis for both is identical. No polarization functions for H were included, since this would increase computer time drastically without significant improving the calculated chemical shifts for the "heavy" nuclei [56]. Despite this shortcomings, the IGLO ^1H chemical shifts of the terminal hydrogens are not too far from the experimental values. The relative order of the proton resonances in most cases is reproduced correctly. The highfield shifts of the bridging hydrogens, however, consistently are overestimated in the calculations (too strongly shielded by ca. 1–3 ppm).

In summary, theoretical ^1H chemical shifts calculated at this level are not accurate enough to permit decisive assignments.

4.4. DISCUSSION

The ^{11}B chemical shifts for the molecules included in this study cover a range of about 80 ppm. The effects involved have been discussed in length in the literature [4b, 57]. The IGLO method permits an analysis in terms of contributions from individual localized orbitals. However, simple relationships are not apparent.

TABLE 4.5. IGLO ^{1}H chemical shifts^a (in ppm rel. TMS).

IGLO basis	Compounds

B$_2$H$_6$ 1, B$_4$H$_{10}$ 2, B$_5$H$_9$ 3

IGLO basis	B$_2$H$_6$ 1		B$_4$H$_{10}$ 2				B$_5$H$_9$ 3		
	H1	H12	H1	H2(e)	H2(a)	H12	H1	H2	H23
DZ	5.1	−1.2	2.3	3.1	3.1	−2.7	1.4	4.3	−5.1
II'	3.8	−2.0	1.1	2.2	2.3	−2.9	0.3	2.6	−4.8
Expt.^{b}	3.95	−0.53	1.34	2.26	2.46	−1.38	0.53	2.49	−2.28

B$_5$H$_{11}$

	Hb	H1(a)	H2(a)	H2(e)	H3	H23	H34
4a DZ	0.6	−0.1	4.1	4.6	4.5	−2.3	−4.1
II'	0.2	−1.5	3.0	3.2	3.1	−2.7	−4.6
4b^c DZ	−1.0	0.5	4.0	4.2	4.5	−2.1	−3.8
II'	−1.1	−0.7	3.0	3.0	3.1	−2.5	−3.7
Expt.^{d}	2.09	0.23	3.47	3.91	3.82	−0.64	−1.64

B$_6$H$_{10}$ 5

	H1	H2	H3	H4	H23	H34
DZ	0.0	4.1	6.3	5.6	−2.4	−4.4
II'	−1.4	2.9	4.6	4.3	−2.5	−4.6
Expt.^{e}	−1.27	3.15	4.58	4.24	0.14	−2.14

B$_6$H$_{12}$ 6

	H1(a)	H1(e)	H2	H3	H16	H23
DZ	4.2	4.5	2.4	6.3	−2.5	−3.1
II'	3.2	3.3	1.2	4.6	−2.8	−3.0
Expt.^{f}	4.2	5.1	2.1	5.1	−0.2	−0.2

B$_2$H$_7^-$ 7a, 7b, 7c

	B$_2$H$_7^-$ 7a		7b		7c	
	H1	H12	H1	H12	H1	H12
DZ	2.2	−8.3	1.6	−6.4	1.5	−6.3
II'			0.7	−7.1	0.7	−7.1
Expt.^{g}					1.2	−4.4

B$_3$H$_8^-$ 8a, 8b

	B$_3$H$_8^-$ 8a				8b			
	φH1	φH2	H23	φH	H1	H2	H12	φH
DZ	−0.5	2.6	−4.5	1.7	2.2	0.1	−2.9	0.2
II'	−1.2	1.7	−4.6	−0.2	1.5	−0.8	−3.1	−0.4
Expt.^{h}								0.3

B$_4$H$_9^-$ 9

	H1(a)	H1(e)	H2(a)	H2(e)	H3	H23
DZ	−1.3	0.0	1.9	2.3	4.1	−4.8
II'	−2.2	−0.4	0.9	1.2	2.5	−4.9
Expt.^{i}	−1.71	0.67	1.80	1.80	2.63	−2.73

^{a}MP2/6-31G* geometries.
^{b}B$_2$H$_6$ [66]; B$_4$H$_{10}$ [34a]; B$_5$H$_9$ [63].
^{c}Average values.
^{d}[34c]. ^{e}[37]. ^{f}[34d]. ^{g}[47]. ^{h}[53a]. ^{i}[54].

The atomic charge is the most obvious property to which the chemical shift might be related. Several correlations of $\delta^{11}B$ with total or π charge densities have been reported [58, 59]. But atomic charges are not observable properties and mostly have been calculated earlier at semi-empirical levels employing Mulliken population analysis (MPA). The shortcomings of MPA are widely accepted.

We employed another method, natural population analysis (NPA) [60] which generally has the advantage of showing little basis set dependence. NPA employs localized orbitals which are "improved" in order to include as much of the electron density as possible. The results of these two methods (MPA and NPA) for compounds **1–9** (6-31G* basis set for MP2/6-31G* geometries) are compared in Table 4.6 and are correlated with the IGLO chemical shifts [61] (Basis II') in Figure 4.3.

The NPA charges are considerably more negative than those obtained from the Mulliken analysis. It may seem unusual for a boron to bear nearly half a negative charge (e.g., B_1 in $B_4H_9^-$ **9**). However, the NPA data seem to be more consistent in correlating the ^{11}B chemical shifts since there is a common trend for all species. This is not the case for the Mulliken charges, where two different sets of points for the neutral and the anionic species are apparent; the boron atoms of the anions are calculated to be more positive than "expected" from the chemical shifts. The anomalously positive Mulliken charges of the apical borons

TABLE 4.6. Mulliken (MPA) and natural population analysis (NPA) charges^a.

Compound		MPA	NPA	Compound		MPA	NPA
1 B_2H_6		−0.14	−0.065	**5** B_6H_{10}	B_1	−0.013	−0.303
					B_2	−0.144	−0.187
2 B_4H_{10}	$B_{1,3}$	−0.179	−0.341		$B_{3,6}$	−0.129	−0.047
	$B_{2,4}$	−0.096	−0.103		$B_{4,5}$	−0.077	−0.151
3 B_5H_9	B_1	0.038	0.092	**6** B_6H_{12}	$B_{1,4}$	−0.78	−0.87
	B_{2-5}	−0.157	−0.111		$B_{2,5}$	−0.033	−0.092
					$B_{3,6}$	−0.146	−0.274
B_5H_{11}	B_1	−0.241	−0.487				
4a	$B_{2,5}$	−0.066	−0.040	**7c** $B_2H_7^-$		0.033	−0.199
	$B_{3,4}$	−0.089	−0.171				
				$B_3H_8^-$	B_1	−0.106	−0.489
4b	B_1	−0.193	−0.406	**8a**	$B_{2,3}$	0.053	−0.173
	B_2	−0.092	−0.094				
	B_5	−0.064	−0.053	**8b**	B_1	0.050	−0.155
	B_3	−0.141	−0.217		$B_{2,3}$	−0.061	−0.446
	B_4	−0.053	−0.135				
				9 $B_4H_9^-$	B_1	−0.080	−0.459
					$B_{2,4}$	−0.032	−0.179
					B_3	0.044	−0.192

^{a}Calculated with 6-31G* basis set for MP2/6-31G* geometries.

in B_5H_9 and B_6H_{10} appear to be artifacts of the method; the NPA charges are nearly in the expected range.

The relationship between $\delta^{11}B$ and the NPA charges, however, seems to be very rough. This is plausible since the electron distribution in the ground state is less important for the paramagnetic contributions σ_p, which dominate the chemical shielding. Unfortunately, the interpretation of this paramagnetic part (which is associated with a large chemical shift anisotropy) is not straightforward.

The least squares fit for all points gives a slope of 178 ppm per electron for the NPA charges (correlation coefficient 0.932, standard deviation 9.9 ppm). The quality of the fit is only fair and the slope is far from an estimate deduced on an empirical basis (ca. 65 ppm per electron) [57].

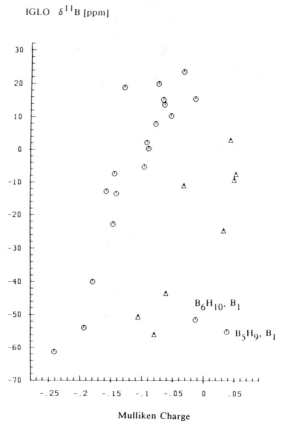

Figure 4.3. Correlation of ^{11}B chemical shifts with atomic charges obtained (a) from Mulliken and (b) from natural population analysis (NPA). Circles, neutral boranes **1–6**, triangles, borane anions **7–9**.

IGLO δ^{11}B [ppm]

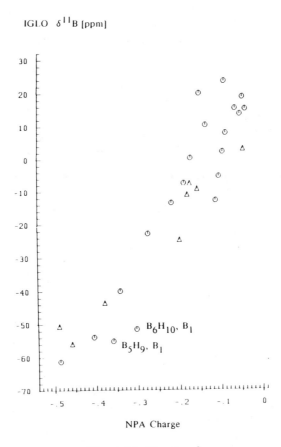

Figure 4.3. (*Continued*)

4.5. CONCLUSION

The ^{11}B chemical shifts of several neutral and anionic binary boron hydrides have been calculated ab initio using the IGLO method. The agreement between theoretical and experimental values is excellent provided accurate geometries are employed (e.g., optimized ab initio with heavy atom polarized basis sets and with inclusion of electron correlation).

For the fluxional molecules, B_5H_{11} and $B_3H_8^-$, **4b** and **8b** are strongly supported by the IGLO calculations to be the minima. For $B_4H_9^-$ (**9**), which is not yet structurally characterized, the proposed geometry is confirmed.

A pessimistic opinion stated only two years ago, "... theoretical calculations of these NMR characteristics are of little importance because they cannot be performed as accurately as necessary for the structural elucidation of boron compounds" [57], is no longer valid. A certain limitation, however, is that this accuracy—at least in the case of polyhedral boranes—can only be achieved for geometries optimized at a correlated level (these optimizations can be rather expensive, for example 50,000 CPU seconds for the MP2/6-31G* optimization of B_6H_{12} on a CRAY-YMP4/432).

ACKNOWLEDGMENTS

We thank in particular Professor W. Kutzelnigg and his associates, M. Schindler and U. Fleischer, for the development of the IGLO program, the preparation and release of the Convex version, and extensive discussions and advice. This work was supported by the Deutsche Forschungsgemeinschaft, Fonds der Chemischen Industrie, Volkswagenwerk and the Convex Computer Corporation. Computer time was provided by the KFA Jülich. A grant of the Studienstiftung des deutschen Volkes for M. B. is gratefully acknowledged.

REFERENCES

1. Stock, A., *Hydrides of Boron and Silicon*, Cornell University Press, Ithaca, NY, 1933.
2. (a) Muetterties, E. L. (Ed.), *The Chemistry of Boron and its Compounds*, Wiley, New York, London, Sydney, 1967; (b) Muetterties, E. L. and Knoth, W. H., *Polyhedral Boranes*, New York, London, 1968; (c) Muetterties, E. L. (Ed.), *Boron Hydride Chemistry*, Academic Press, New York, San Francisco, London, 1975.
3. Schlesinger, H. I. and Burg, A. B., *Chem. Rev.*, **31**, 1 (1942).
4. (a) Lipscomb, W. N., *Boron Hydrides*, Benjamin, New York, 1963; (b) Eaton, G. R. and Lipscomb, W. N., *NMR Studies of Boron Hydrides and Related Compounds*, Benjamin, New York, Amsterdam, 1969.
5. Beaudet, R. A., in: *Advances in Boron and the Boranes* (J. F. Liebmann, A. Greenberg and R. E. Williams, Eds.), Chap. 20, p. 417, Verlag Chemie, Weinheim, 1988.
6. (a) Hoffman, R. and Lipscomb, W. N., *J. Chem. Phys.*, **36**, 2179 (1962); (b) Halgren, T. A. and Lipscomb, W. N., *J. Chem. Phys.*, **58**, 1569 (1973).
7. (a) Kutzelnigg, W., *Isr. J. Chem.*, **19**, 193 (1980); (b) Schindler, M. and Kutzelnigg, W., *J. Chem. Phys.*, **76**, 1919 (1982).
8. For a review of IGLO applications see Kutzelnigg, W., Schindler, M. and Fleischer, U., *Adv. Magn. Reson. Chem.*, to be published.
9. (a) Schindler, M., *J. Am. Chem. Soc.*, **109**, 1020 (1987); (b) Bremer, M., Schleyer, P. v. R., Schötz, K., Kausch, M. and Schindler, M., *Angew Chem.*, **99**, 795 (1987); (c) Schleyer, P. v. R., Laidig, K. E., Wiberg, K. B., Saunders, M. and Schindler, M., *J. Am. Chem. Soc.*, **110**, 300 (1988); (d) Saunders, M., Laidig, K. E., Wiberg, K. B. and Schleyer, P. v. R., *J. Am. Chem. Soc.*, **110**, 7652 (1988); (e) Schleyer, P. v. R., Carneiro, J. W. de M., Koch, W. and Raghavachari, K., *J. Am. Chem. Soc.*, **111**, 5475 (1989); (f) Bremer, M., Schleyer, P. v. R. and Fleischer, U., *J. Am. Chem. Soc.*, **111**, 1147 (1989); (g) Bremer, M., Schötz, K., Schleyer, P. v. R., Fleischer, U., Schindler, M., Kutzelnigg, W. and Pulay, P., *Angew. Chem.*, **101**, 1063 (1989); (h) Schleyer, P. v. R., Koch, W., Liu, B. and Fleischer, U., *J. Chem. Soc., Chem. Commun.* 1098 (1989).
10. (a) Laus, E. A., Stevens, R. M. and Lipscomb, W. N., *J. Am. Chem. Soc.*, **94**, 8699 (1972); (b) Hall Jr., J. H., Marynick, D. S. and Lipscomb, W. N., *J. Am. Chem. Soc.*, **96**, 770 (1974).

11. Binkley, J. S., Frisch, M. J., DeFrees, D. J., Krishnan, R., Whiteside, R. A., Schlegel, H. B., Fluder, E. M. and Pople, J. A., *Gaussian82*, Carnegie-Mellon Chemistry Publishing Unit, Pittsburgh, PA 15213.

12. Frisch, M. J., Binkley, J. S., Schlegel, H. B., Raghavachari, K., Melius, C. F., Martin, R. L., Stewart, J. J. P., Bobrowicz, F. W., Rohlfing, C. M., Kahn, L. R., Defrees, D. J., Seeger, R., Whiteside, R. A., Fox, D. J., Fleuder, E. M. and Pople, J. A., *GAUSSIAN86*, Carnegie-Mellon Quantum Chemistry Publishing Unit, Pittsburgh PA, 1986.

13. Amos, R. D. and Rice, J. E., *CADPAC: The Cambridge Analytical Derivatives Package*, Issue 4.0, Cambridge, 1987.

14. Hehre, W., Radom, L., Schleyer, P. v. R. and Pople, J. A., *Ab Initio Molecular Orbital Theory*, Wiley, New York, 1986.

15. Clark, T., *A Handbook of Computational Chemistry*, Wiley, New York, 1985.

16. Binkley, J. S. and Pople, J. A., *Int. J. Quantum Chem.*, **9**, 22 (1975) and references cited therein.

17. Huzinaga, S., *Approximate Atomic Wave Functions*, University of Alberta, Edmonton, 1971.

18. Onak, T. P., Landesman, H. L., Williams, R. E. and Shapiro, I., *J. Phys. Chem.*, **63**, 1533 (1959).

19. The calculated structural parameters are equilibrium values; direct comparisons of these theoretical r_e to experimental r_o, r_s or r_z geometries should always be made with caution, cf. Harmony, M. D., Laurie, V. W., Kuczkowski, R. L., Schwendeman, R. H., Ramsay, D. A., Lovas, F. J. and McLafferty, W. I., *J. Phys. Chem. Rev. Data*, **8**, 619 (1979).

20. Not even energies are affected very much by inclusion of p-functions on hydrogen, see e.g., a study of B_3H_9: Stanton, J. F., Lipscomb, W. N., Bartlett, R. J. and McKee, M. L., *Inorg. Chem.*, **28**, 109 (1989).

21. Stock, A. and Friederici, K., *Ber. Dtsch. Chem. Ges.*, **46**, 1959 (1913).

22. Schlesinger, H. J., Brown, H. C., Hoekstra, H. R. and Rapp, L. R., *J. Am. Chem. Soc.*, **75**, 199 (1953).

23. (a) Price, W. C., *J. Chem. Phys.*, **15**, 614 (1947); **16**, 894 (1948); (b) Hedberg, K., Schomaker, V., *J. Am. Chem. Soc.*, **73**, 1482 (1951); (c) Duncan, J. L., Harper, *Mol. Phys.*, **51**, 371 (1984).

24. See e.g., Brown, H. C., *Boranes in Organic Chemistry*, Cornell University Press, Ithaca, NY, 1972.

25. Longuet-Higgins, H. C., *J. Chim. Phys.*, **46**, 275 (1949).

26. Eberhardt, W. H., Crawford Jr., B. L. and Lipscomb, W. N., *J. Chem. Phys.*, **22**, 989 (1954).

27. See e.g. (a) Switkes, E., Stevens, R. M., Newton, M. D. and Lipscomb, W. N., *J. Chem. Phys.*, **51**, 2085 (1969); (b) Laws, E. A., Stevens, R. M. and Lipscomb, W. N., *J. Am. Chem. Soc.*, **94**, 4461 (1972); (c) DeFrees, D. J., Raghavachari, K., Schlegel, H. B., Pople, J. A. and Schleyer, P. v. R., *J. Phys. Chem.*, **91**, 1857 (1987); (d) Stanton, J. F., Barlett, R. J. and Lipscomb, W. N., *Chem. Phys. Lett.*, **138**, 525 (1987).

28. Simmons, N. P. C. and Beaudet, R. A., *Inorg. Chem.*, **20**, 533 (1981).

29. Dain, C. J., Downs, A. J., Laurenson, G. S. and Rankin, D. W. H., *J. Chem. Soc., Dalton Trans.*, **4**, 472 (1981).

30. Morris-Sherwood, B. J. and Hall, M. B., *Chem. Phys. Lett.*, **84**, 194 (1981).

31. Schwoch, D., Burg, A. B. and Beaudet, R., *Inorg. Chem.*, **16**, 3219 (1977).

32. Schleyer, P. v. R., Bühl, M., Fleischer, U. and Koch, W., *Inorg. Chem.*, **29**, 153 (1990).

33. (a) Levine, L. R. and Lipscomb, W. N., *J. Chem. Phys.*, **21**, 2087 (1953); **22**, 614 (1954); (b) Moore, E. B. G., Dickerson, R. E. and Lipscomb, W. N., *J. Chem. Phys.*, **27**, 209 (1957).

34. (a) Leach, J. B., Onak, T., Spielman, Rietz, R. R., Schaeffer, R. and Sheddon, G. L., *Inorg. Chem.*, **9**, 2170 (1970); (b) Clouse, A. O., Moody, D. C., Rietz, R. R., Roseberry, T. and Schaeffer, R., *J. Am. Chem. Soc.*, **95**, 2496 (1973); (c) Onak, T. and Leach, J. B., *J. Am. Chem. Soc.*, **92**, 3513 (1970); (d) Jaworiwsky, I. S., Long, J. R., Barton, L. and Shore, S. G., *Inorg. Chem.*, **18**, 56 (1979).

35. Greatrex, R., Greenwood, N. N., Rankin, D. W. and Robertson, H. E., *Polyhedron*, **6**, 1849 (1987).

36. McKee, M. L., *J. Phys. Chem.*, **93**, 3426 (1989).

37. Brice, V. T., Johnson II, H. D. and Shore, S. G., *J. Chem. Soc., Chem. Commun.*, 1128 (1972); *J. Am. Chem. Soc.*, **95**, 6629 (1973).

38. Hirschfeld, F. L., Eriks, K., Dickerson, R. E., Lippert Jr., E. L. and Lipscomb, W. N., *Chem. Phys.*, **28**, 56 (1958).

39. Schwoch, D., Don, B. P., Burg, A. B. and Beaudet, R. A., *J. Am. Chem. Soc.*, **83**, 1465 (1979).

40. Epstein, I. R., Tossell, J. A., Switkes, E., Stevens, R. M. and Lipscomb, W. N., *Inorg. Chem.*, **10**, 171 (1971).

41. Greatrex, R., Greenwood, N. N., Millikan, M. B., Rankin, D. W. H. and Robertson, H. E., *J. Chem. Soc. Dalton Trans.*, **11**, 2335 (1988).

42. McKee, M. L., *J. Phys. Chem.*, **94**, 435 (1990).

43. Bühl, M., Schleyer, P. v. R., *Agnew. Chem.*, **102**, (1990), in press.

44. Raghavachari, K., Schleyer, P. v. R. and Spitznagel, G., *J. Am. Chem. Soc.*, **105**, 5917 (1983).

45. Shore, S. G., Lawrence, S. H., Watkins, M. I. and Bau, R., *J. Am. Chem. Soc.*, **104**, 7669 (1982).

46. Khan, S. I., Chiang, M. Y., Bau, R., Koetzle, T. F., Shore, S. G. and Lawrence, S. H., *J. Chem. Soc., Dalton Trans.*, **9**, 1753 (1986).

47. Hertz, R. K., Johnson, H. D. and Shore, S. G., *Inorg. Chem.*, **12**, 1875 (1973).

48. (a) Peters, C. R. and Nordman, C. E., *J. Am. Chem. Soc.*, **82**, 5758 (1960). (b) Deisenroth, H. J., Sommer, O., Binder, H., Wolfer, K. and Frei, B., *Z. Anorg. Allgem. Chem.*, **571**, 21 (1998).

49. McKee, M. L. and Lipscomb, W. N., *Inorg. Chem.*, **21**, 2846 (1982).

50. Glore, J. D., Rathke, J. W. and Schaeffer, R., *Inorg. Chem.*, **12**, 2175 (1973).

51. Bishop, V. L. and Kodama, G., *Inorg. Chem.*, **20**, 2724 (1981).

52. Pepperberg, M., Dixon, D. A., Lipscomb, W. N. and Halgren, T. A., *Inorg. Chem.*, **17**, 587 (1978).

53. (a) Marynick, D. and Onak, T., *J. Chem. Soc. A*, 1160 (1970); (b) Beall, H., Bushweller, C. H., Dewkett, W. J. and Grace, M., *J. Am. Chem. Soc.*, **92**, 3484 (1970).

54. Rammel, R. J., Johnson II, H. D., Jaworlwsky, I. S. and Shore, S. G., *J. Am. Chem. Soc.*, **97**, 5395 (1975).

55. Lipscomb, W. N., *Acc. Chem. Res.*, **6**, 257 (1973).

56. Fleischer, U., unpublished calculations.

57. Hermanek, S., Jelinek, T., Plesek, J., Stibr, B., Fusek, J. and Mares, F., in: *Boron Chemistry, Proceedings of the 6th IMEBORON* (S. Hermanek, Ed.), pp. 26–73, World Scientific, Singapore, 1987.

58. Kroner, J. and Wrackmeyer, B., *J. Chem. Soc., Faraday Trans. II*, **72**, 2283 (1976).

59. Kroner, J., Nölle, D. and Nöth, H., *Z. Naturforsch.*, **28b**, 416 (1973).

60. Reed, A. E., Weinstock, R. B. and Weinhold, F., *J. Chem. Phys.*, **83**, 735 (1985).

61. We did not employ the experimental $\delta^{11}B$ values, because the shifts of the individual borons for the fluxional molecules **4b** and **8b** have not been measured.

62. Kuchitsu, K., *J. Chem. Phys.*, 49, 4456 (1968).

63. Tucker, P. M., Onak, T. and Leach, J. B., *Inorg. Chem.*, **9**, 1430 (1970).

64. Paine, R. T., Fukushima, E. and Roeder, S. B. W., *Chem. Phys. Lett.*, **32**, 566 (1976).

65. DeFrees, D. J. and McLean, A. D., *J. Chem. Phys.*, **82**, 333 (1985).

66. Gaines, D. F., Schaeffer, R. and Tebbe, F., *J. Phys. Chem.* 67, 1937 (1963).

Appendix: Cartesian Coordinates of the Boranes and Borane Anions 1–9 (Optimized ab initio in the Given Symmetry at the MP2/6-31 G* Level)

$$B_2H_6 \ D_{2h} \ (1)$$

B	0.000000	0.000000	−0.874551
B	0.000000	0.000000	0.874551
H	0.000000	0.974453	0.000000
H	0.000000	−0.974453	0.000000
H	−1.038300	0.000000	1.454494
H	1.038300	0.000000	−1.454494
H	−1.038300	0.000000	−1.454494
H	1.038300	0.000000	1.454494

$$B_4H_{10} \ C_{2v} \ (2)$$

B	−0.857152	0.000000	−0.464034
B	0.000000	−1.386290	0.380067
B	0.857152	0.000000	−0.464034
B	0.000000	1.386290	0.380067
H	0.000000	−1.417325	1.576060
H	0.000000	1.417325	1.576060
H	−1.359193	0.000000	−1.537622
H	1.359193	0.000000	−1.537622
H	0.000000	−2.416574	−0.217948
H	0.000000	2.416574	−0.217948
H	−1.321460	−0.907666	0.262745
H	1.321460	−0.907666	0.262745
H	−1.321460	0.907666	0.262745
H	1.321460	0.907666	0.262745

$$B_5H_9 \ C_{4v} \ (\mathbf{3})$$

B	0.000000	0.000000	0.974594
B	0.000000	1.260839	−0.143736
B	1.260839	0.000000	−0.143736
B	0.000000	−1.260839	−0.143736
B	−1.260839	0.000000	−0.143736
H	0.000000	0.000000	2.157013
H	0.000000	2.438026	0.001463
H	2.438026	0.000000	0.001463
H	0.000000	−2.438026	0.001463
H	−2.438026	0.000000	0.001463
H	−0.948150	0.948150	−1.040277
H	−0.948150	−0.948150	−1.040277
H	0.948150	−0.948150	−1.040277
H	0.948150	0.948150	−1.040277

$$B_5H_{11} \ C_1 \ (\mathbf{4b})$$

B	0.004694	−0.257252	0.790186
B	1.507176	−0.743455	−0.182835
B	0.878403	0.951750	−0.100100
B	−0.897362	0.875028	−0.145472
B	−1.556627	−0.731510	−0.149165
H	−1.450378	1.785480	0.377167
H	1.385983	1.889029	0.417998
H	0.032661	−0.201335	1.974858
H	0.322184	−1.371674	0.333939
H	−2.595956	−0.749810	0.433738
H	−1.318011	−1.605007	−0.925944
H	2.486725	−0.873449	0.482189
H	1.403996	−1.436491	−1.156315
H	0.022518	1.348687	−1.018944
H	−1.626476	0.319984	−1.040343
H	1.655327	0.421778	−0.941414

B_6H_{10} C_s (5)

B	− 0.019415	0.851942	0.000000
B	1.468055	− 0.052984	0.000000
B	0.376343	− 0.105796	− 1.407081
B	− 1.251619	− 0.172750	− 0.819014
B	− 1.251619	− 0.172750	0.819014
B	0.376343	− 0.105796	1.407081
H	− 0.021811	2.039109	0.000000
H	2.578746	0.362822	0.000000
H	0.737524	0.178424	− 2.501679
H	0.737524	0.178424	2.501679
H	− 2.084003	0.154858	1.601660
H	− 2.084003	0.154858	− 1.601660
H	1.352315	− 0.942928	− 0.938386
H	1.352315	− 0.942928	0.938386
H	− 0.485677	− 1.113343	− 1.362148
H	− 0.485677	− 1.113343	1.362148

B_6H_{12} C_2 (6)

B	1.434763	1.102916	1.976328
B	0.891962	0.000000	0.529177
B	− 0.070815	− 1.438668	0.970624
B	− 1.434763	− 1.102916	1.976328
B	− 0.891962	0.000000	0.529177
B	0.070815	1.438668	0.970624
H	1.552806	− 0.077542	− 0.453609
H	− 1.552806	0.077542	− 0.453609
H	− 0.042119	− 2.372658	0.238557
H	0.042119	2.372658	0.238557
H	− 1.448968	− 0.408780	2.947295
H	1.448968	0.408780	2.947295
H	− 2.313246	− 1.890058	1.814164
H	2.313246	1.890058	1.814164
H	− 0.301337	− 1.812998	2.187852
H	0.301337	1.812998	2.187852
H	1.166392	− 0.986515	1.311626
H	− 1.166392	0.986515	1.311626

$B_2H_7^-$ C_2 (**7c**)

B	0.000000	1.163039	0.005545
B	0.000000	−1.163039	0.005545
H	0.000000	0.000000	0.581949
H	0.539581	1.765070	0.908135
H	−0.539581	−1.765070	0.908135
H	−1.167701	1.460615	0.144444
H	1.167701	−1.460615	0.144444
H	0.644763	1.180245	1.026940
H	−0.644763	−1.180245	1.026940

$B_3H_8^-$ C_{2v} (**8b**)

B	0.000000	0.000000	0.978050
B	0.000000	0.910573	−0.549859
B	0.000000	−0.910573	−0.549859
H	0.000000	−1.422628	0.603732
H	0.000000	1.422628	0.603732
H	1.027402	1.323453	−1.033933
H	−1.027402	1.323453	−1.033933
H	1.027402	−1.323453	−1.033933
H	−1.027402	−1.323453	−1.033933
H	1.018853	0.000000	1.631224
H	−1.018853	0.000000	1.631224

$B_4H_9^-$ C_s (**9**)

B	0.672212	−0.723403	0.000000
B	−0.256404	−0.092566	1.462085
B	0.000000	0.824146	0.000000
B	−0.256404	−0.092566	−1.462085
H	1.878132	−0.664249	0.000000
H	0.196424	−1.824506	0.000000
H	0.751086	1.758781	0.000000
H	0.300896	0.343831	2.440723
H	0.300896	0.343831	−2.440723
H	−1.230189	−0.774312	1.661969
H	−1.230189	−0.774312	−1.661969
H	−0.882038	1.006443	0.973630
H	−0.882038	1.006443	−0.973630

Skeletal Rearrangements in Clusters: Some New Theoretical Insights Involving Lipscomb's Diamond-Square-Diamond Mechanism

D. M. P. MINGOS and D. J. WALES

5.1. INTRODUCTION: THE DIAMOND-SQUARE-DIAMOND PROCESS

It is now more than twenty years since Lipscomb proposed the diamond-square-diamond (DSD) process to account for rearrangements in boranes and carboranes [1]. It is a tribute to his imagination that this process still plays a major role in modern theories of skeletal rearrangements in clusters. Such rearrangements have recently been analyzed using Stone's Tensor Surface Harmonic (TSH) theory [2–5]. The new approach enables cluster rearrangement processes

Electron Deficient Boron and Carbon Clusters, Edited by George A. Olah,
Kenneth Wade, and Robert E. Williams.
ISBN 0-471-52795-5 © 1991 John Wiley & Sons, Inc.

which have an orbital crossing, and are therefore "forbidden" in the Woodward–Hoffmann sense [6], to be identified very simply. The new theory makes extensive use of the DSD process (illustrated in Figure 5.1), in which an edge common to two triangular faces of the cluster skeleton breaks and a new edge is formed perpendicular to it.

DSD rearrangements, or combinations of several concerted DSD processes, have been proposed to rationalize fluxional processes and isomerizations of boranes, carboranes, and metalloboranes [7, 8]. King [9] has used topological considerations to distinguish between inherently rigid clusters (which contain no *degenerate* edges) and those for which one or more DSD processes are *geometrically* possible. An edge is said to be degenerate if a DSD rearrangement in which it is broken leads to a product having the same cluster skeleton as the starting material. For example; the three equatorial edges of $B_5H_5^{2-}$ are degenerate and the remainder are not. Alternative DSD processes are distinguished by the connectivities of the four vertices around the diamond in question. Hence the DSD process for the trigonal bipyramid (shown in Figure 5.2) is described as 44(33) because an edge connecting two four-coordinate equatorial vertices is broken while an edge is made between the two three-coordinate apical vertices. A general DSD process is denoted by $\alpha\beta(\gamma\delta)$, although this does not always specify the process unambiguously. For a degenerate DSD rearrangement to be geometrically possible, the connectivities of the new cluster vertices must match those of the original skeleton and we must have $\alpha + \beta = \gamma + \delta + 2$. In the *closo*-borane series only $B_5H_5^{2-}$, $B_8H_8^{2-}$, $B_9H_9^{2-}$, and $B_{11}H_{11}^{2-}$ satisfy these requirements. The maximum point group symmetry that the transition state for a DSD process in a deltahedron can have is C_{4v} if the cluster has $4p + 1$ vertices, for example, $B_5H_5^{2-}$ and $B_9H_9^{2-}$; these cases have been described as "pseudo-rotations" by Mingos and Johnston. For $B_8H_8^{2-}$ and $B_{11}H_{11}^{2-}$ the highest symmetry geometry possible for the single DSD process is C_{2v}; in this case the boron framework undergoes a pseudo-reflection.

King's approach was partially successful in that all the structures he predicted to be rigid are found experimentally to be nonfluxional. However, some of the above molecules in the $B_nH_n^{2-}$ series with topologically possible

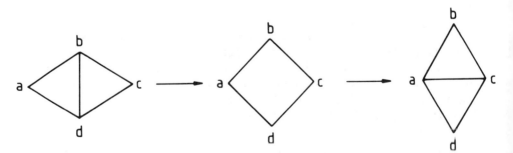

Figure 5.1. The diamond-square-diamond process.

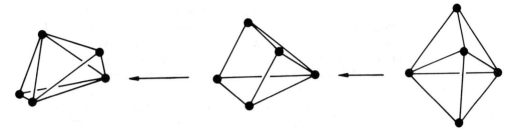

Figure 5.2. The single DSD process for $B_5H_5^{2-}$. Notice the square-based pyramidal transition state.

single DSD rearrangements are not fluxional on the NMR time scale. They are the ones which have $4p+1$ vertex atoms.

TSH theory enables us to explain why the DSD process is so favorable, and also provides a framework for the derivation of some powerful orbital symmetry selection rules. For example, we can show that transition states between *closo* boranes or carboranes with a single atom on a principal rotation axis of order three or more generally have an orbital crossing [2]. The effect of lower symmetry environments, for example, when there is a single atom on a two-fold axis, has also been considered. These results may be extended to show that an orbital crossing will generally result if a mirror plane through the critical face is retained throughout the process [3]. If a C_2 axis is retained, however, then there is an avoided crossing and the process is "allowed".

Using the above rules, and the additional criterion that multiple DSD processes are likely to be less favorable than the single DSD process, it is possible to rationalize the whole range of rearrangement rates of the *closo* boranes and carboranes. We have also discussed the application of the rules to rearrangements of transition metal clusters, and suggested some new mechanisms for these species [3]. Second-order Jahn–Teller analysis may be used to confirm the selection rules deduced from the TSH theory analysis [4]. In this review the various threads are brought together, demonstrating that these results are complementary. Lipscomb's DSD process and the related SDDS mechanism [3] feature prominently in this review.

5.2. INTRODUCING TENSOR SURFACE HARMONIC THEORY

The *Debor Principle*, which emerged in the early 1970s from the work of Williams [10], Wade [11] and Rudolph [12], first systematized the correlation between structure and electron counts for *closo*, *nido* and *arachno* boranes. The *Isolobal Principle*, which allows transition metal clusters to be understood in terms of the simpler borane structures, has also proved very useful [13]. The term *Polyhedral Skeletal Electron Pair Theory* was later introduced to cover the correlation between electron count and cluster structure [14]. Its development has often been facilitated by means of Extended Hückel calculations [15]. More recently Stone's Tensor Surface Harmonic (TSH) theory [16] has been used to provide a firmer theoretical foundation for these generalisations [17–20].

In the TSH model, linear combinations of atomic orbitals are formed using the eigenfunctions for the particle on-a-sphere problem (the spherical harmonics) and the tensor surface harmonics. By analogy with the relation between the Hückel wavefunctions for cyclic polyenes and the eigenfunctions for the particle on-a-ring problem, we expect these linear combinations to be reasonable first-order approximations to the actual cluster orbitals. Many useful results may be derived using this formalism, particularly the rationalization of cluster electron counts [16].

The theory is essentially an approximate symmetry classification. A cluster is treated as if it were spherical, and the appropriate symmetry labels for spherical symmetry are the angular momentum quantum numbers L and M, as in an atom. There is also a parity classification: functions may be changed in sign or unchanged by inversion. The prototype functions are the spherical harmonics, Y_{LM}, which have even or odd parity under inversion for L even or odd, respectively.

TSH theory classifies the basis functions of each cluster atom into σ, π, and δ orbitals which have 0, 1, and 2 nodal planes containing the radius vector from the center of the cluster to the atom, respectively. Cluster orbitals are formed from the σ basis functions σ_i (where the index i labels the cluster atoms) using the values of spherical harmonics evaluated at the cluster vertices as expansion coefficients. That is, the cluster orbital ψ_{LM}^{σ} is a linear combination of the form

$$\psi_{LM}^{\sigma} = \sum_i Y_{LM}(\theta_i, \phi_i)\sigma_i$$

where θ_i and ϕ_i are the spherical polar coordinates of the cluster atom. The sets of ψ_{LM}^{σ} with $L = 0, 1, 2, \ldots$ are denoted $S^{\sigma}, P^{\sigma}, D^{\sigma}, \ldots$, and the whole set is collectively denoted L^{σ}. For a real cluster these labels do not describe exact symmetry properties, but the approximate classification is very valuable nevertheless. If the molecular orbitals are expressed in terms of the ψ_{LM}^{σ}, rather than the original atomic σ orbitals, we find that functions of different L or M do not mix with each other very strongly. To this extent the ψ_{LM}^{σ} are good approximations to the molecular orbitals themselves.

To deal with the π functions we derive from the surface harmonics two sets of *vector surface harmonics* $\mathbf{V}_{LM}$ and $\bar{\mathbf{V}}_{LM}$, defined by

$$\mathbf{V}_{LM} = \nabla Y_{LM}, \quad \bar{\mathbf{V}}_{LM} = \mathbf{r} \times \mathbf{V}_{LM} = \mathbf{r} \times \nabla Y_{LM}$$

From each Y_{LM} we obtain in this way two vector functions, both of which are tangential to the surface of the sphere. $\mathbf{V}_{LM}$ has the same parity as the parent Y_{LM}, and is an even vector surface harmonic, while $\bar{\mathbf{V}}_{LM}$ has the opposite parity (i.e., it changes sign under inversion if L is even, and is unchanged if L is odd) and is an odd vector surface harmonic.

The direction and magnitude of the even vector surface harmonic $\mathbf{V}_{LM}$ at each vertex i are used to give the direction and magnitude of a π-orbital contribution to a π-type cluster orbital ψ_{LM}^{π} which has the same parity as the parent Y_{LM}. Similarly $\bar{\mathbf{V}}_{LM}$ yields a cluster orbital $\bar{\psi}_{LM}^{\pi}$ with the opposite parity. There is no vector surface harmonic with $L = 0$, because Y_{00} is constant and its derivative is zero, so there is no S^{π} cluster orbital, but there are $P^{\pi}, D^{\pi}, \ldots$ orbitals derived from the even vector surface harmonics, and $\bar{P}^{\pi}, \bar{D}^{\pi}, \ldots$ orbitals derived from the odd ones. These two sets are denoted generically by the symbols L^{π} and $\bar{L}^{\pi}$, respectively.

An important feature is that there is a pairing relation between the orbitals ψ_{LM}^{π} and $\bar{\psi}_{LM}^{\pi}$. Each of these may be obtained from the other by rotating the π-orbital contribution of each atom by $90°$ about the radius vector to that atom, all in the same direction. Repetition of this procedure returns the original function except for a change of sign. This pairing operation also converts bonding interactions into antibonding ones and vice versa [16, 21]. Indeed, it is usually the case that the L^{π} orbitals are bonding and the $\bar{L}^{\pi}$ ones are antibonding, especially for deltahedral clusters, in which all the faces are triangular. The usual electron count for *closo* clusters follows from the fact that the occupied orbitals are usually the n orbitals of the bonding L^{π} set (or combinations of these with the L^{σ} orbitals of the same symmetry) together with the strongly bonding S^{σ} orbital.

In certain structures, however, it happens that there is a degenerate pair of orbitals that are paired with each other, and since bonding and antibonding characteristics are exchanged by the pairing operation, both of them must be nonbonding. This occurs in tetrahedral molecules with an odd number of sets of four equivalent atoms, and in molecules in which there is an odd number of atoms on a symmetry axis of order 3 or more [18]. This results in a modification of the normal electron-counting rules: if the nonbonding pair is occupied, then there are $n + 2$ occupied cluster orbitals, and if it is vacant, there are only n. A total of $n + 1$ is unfavorable, because that would require the degenerate pair to be only half-occupied. Clusters with a unique atom on a principal rotation axis are described as *polar* [22]. Such clusters also feature in the theory of vertex rearrangements because the change in the fronter orbital pattern may correspond to an orbital crossing.

5.3. TSH THEORY ANALYSIS OF SKELETAL REARRANGEMENTS

TSH theory allows the rapid identification of orbital symmetry forbidden vertex rearrangement processes, which exhibit a crossing of occupied and virtual molecular orbitals. Since the occupied cluster orbitals of deltahedral boranes basically consist of the even L^{π} set and the S^{σ} orbital. Hence, the identification of the symmetries of the occupied molecular orbitals is easy for both the starting material and the product. Reduction in symmetry may then be performed to the point group containing the symmetry elements retained throughout the process. If the symmetries of the sets of bonding orbitals referred to this point group are not equivalent then a crossing must occur and the rearrangement will be energetically less favorable.

Of course, not all orbital symmetry allowed rearrangements will be energetically favorable; clearly, if no symmetry elements are preserved throughout a given process then there cannot be a cross-over, but this is no guarantee that the energy barrier to the process will be small.

Detailed considerations allow us to deduce two theorems for identifying degenerate rearrangements which necessarily involve orbital crossings.

THEOREM 5.1. A crossing is found to occur if the proposed transition state has a single atom lying on a principal rotation axis of order three or more.

This is the case when the degenerate edge lies opposite a single atom, as it does in $B_5H_5^{2-}$ and $B_9H_9^{2-}$. In fact we noted above that the DSD process could only pass through a C_{4v} "transition state" for *closo*-boranes, $B_nH_n^{2-}$, with $n = 4p + 1$ atoms, that is, $B_5H_5^{2-}$ and $B_9H_9^{2-}$. These orbital crossings were first noted by Gimarc and Ott [23].

The proof of the above theorem is straightforward. The p^π orbitals of the unique atom lying on the principal axis span an E representation if the molecular point group is axial and the rotation axis is of order three or more. The total number of E representations in the $L^\pi/\bar{L}^\pi$ set is then odd [24] and one of them must be nonbonding. Consider the transformation properties of the p^θ and p^ϕ orbitals of each set of n equivalent atoms lying equidistant to the n-fold principal rotation axis of a cluster with C_{nv} point group symmetry. These orbitals clearly span the regular representation of C_{nv} (where the character equals the order of the point group for the identity operation and zero otherwise), so that any E-type irreducible representation must occur an even number of times in the representation spanned by the nonpolar atoms. The TSH pairing principle then implies that one pair must be self-conjugate and nonbonding. Such structures therefore have a doubly degenerate, nonbonding pair of E orbitals at the HOMO–LUMO level which corresponds to the crossover of two molecular orbitals at this geometry, as illustrated in Figure 5.3 for the single DSD processes in $B_5H_5^{2-}$ and $B_9H_9^{2-}$.

In the intermediate geometry for the single DSD process the frontier orbitals of the open face are self-conjugate (Figure 5.4), that is, they are interchanged by the TSH parity operation. If the cluster has $4p + 1$ atoms ($p = 1, 2, \ldots$) these orbitals are degenerate, but otherwise they need not be. Theorem 5.1 enables us to identify mechanisms where there must be an orbital crossing due to a degeneracy at the HOMO–LUMO level.

A more detailed examination enables us to identify cases where there must be a crossing in lower point group symmetries. These occur when the pairing principle causes a change in the number of occupied orbitals with a given parity

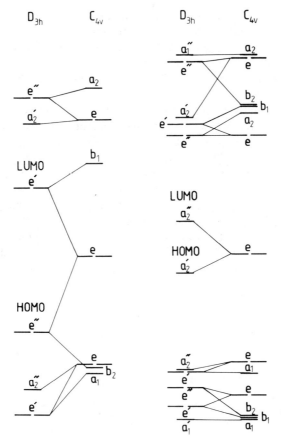

Figure 5.3. Correlation diagrams for the forbidden single DSD processes in $B_5H_5^{2-}$ (left) and $B_9H_9^{2-}$ (right). Only the L^π cluster orbitals are considered and the units of energy are parameterized so that the energy scales and origins of the two diagrams are different [19].

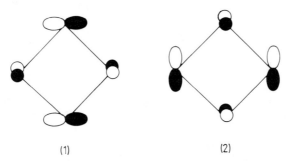

Figure 5.4. The frontier orbitals in the open face of the single DSD transition state. Both orbitals are antisymmetric with respect to a C_2 rotation about an axis through the center of the face. Under reflection in vertical and horizontal mirror planes (1) is antisymmetric and symmetric, respectively, while (2) behaves in the opposite fashion.

under reflection. The TSH theory parity operation corresponds to the pseudo-scalar irreducible representation of the point group which is symmetric to all proper rotations and antisymmetric to improper rotations and reflections. Therefore L^π and $\bar{L}^\pi$ orbitals related by the parity operation have opposite parities with respect to reflection. The high symmetry case (above) can also be discussed in these terms.

The frontier orbitals in the transition state (Figure 5.4) have complementary characteristics with respect to reflection in the mirror planes which pass through opposite vertices of the open square face. If the open face is squeezed along one of these mirror planes then the two parity related orbitals are split apart. Squeezing across the same mirror plane, however, causes splittings of the two components in the opposite sense and hence there is an orbital crossing in the correlation diagram (Figure 5.5) if a mirror plane is retained throughout a DSD process. If instead a C_2 axis is retained then there is an avoided crossing because the frontier orbitals are both antisymmetric under this operation. Note that these symmetry elements must be present for the whole cluster skeleton, not just the open face. Hence we arrive at our second theorem.

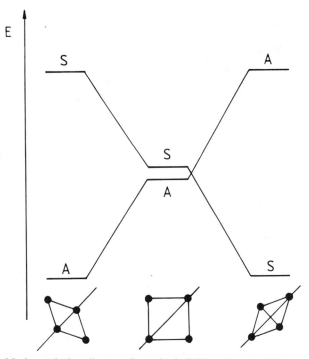

Figure 5.5. Orbital correlation diagram for a single DSD process in which a mirror plane through two of the critical atoms is retained throughout. An orbital crossing occurs due to the change in bonding character of the parity related orbitals.

THEOREM 5.2. A degenerate DSD process in which a mirror plane is retained throughout involves an orbital crossing and is therefore orbitally "forbidden".

In fact we can generalize this result further to identify the conditions under which the orbitals belonging to the L^π frontier set can cross their $\bar{L}^\pi$ partners. The point groups which may be conserved throughout fall into three distinct classes [25]:

1. L^π and $\bar{L}^\pi$ always transform according to different IRs. Groups of this type include the inversion and/or a σ_h mirror plane, for example, C_i, C_s, C_{2v}, O_h, etcetera.
2. L^π and $\bar{L}^\pi$ always transform according to the same IRs. This is the case for point groups such as C_n and D_n where the parity operator transforms as A_1.
3. Some L^π and $\bar{L}^\pi$ transform in the same way while others do not. Groups of this type include C_{nv} $(n \geqslant 3)$ and S_{4n}.

For a DSD process, mirror planes or a C_2 axis may be conserved, and in the point groups C_s and C_2 the parity operator transforms as A'' and A, respectively. Hence if a mirror plane is conserved, parity related L^π and $\bar{L}^\pi$ always span different IRs (and can cross), while if a C_2 operation only is conserved, they must span the same IR (and cannot cross). We would also expect these results to extend to cases where the symmetry elements are only approximate, but in a less rigorous manner. Only a C_2 axis is retained in the single DSD processes for $B_8H_8^{2-}$ and $B_{11}H_{11}^{2-}$ and consequently these molecules are stereochemically nonrigid.

We may also apply these orbital symmetry selection rule to nondegenerate rearrangements in which the starting and finishing clusters are different, so long as they both have $n + 1$ skeletal electron pairs. For example, in the single DSD process which interconverts 1,2-$C_2B_3H_5$ to 1,5-$C_2B_3H_5$ a mirror plane is retained throughout and there is a crossing [23]. The analysis may be extended to multiple DSD processes, whether they be concerted or stepwise. For example, consider a degenerate, concerted double DSD process in which a mirror plane is preserved throughout and passes through both open faces in the transition state. If one edge is broken across the mirror plane and the other is broken simultaneously in the plane then there is no orbital crossing. This follows because of the complementary nature of the frontier orbitals in the two open faces which preserves the total number of occupied orbitals with S and A parity under the reflection by means of two avoided crossings. In contrast, if the two edges are made and broken parallel to one another (both in or both across the mirror plane) then the number of S or A parity orbitals in the occupied set changes by plus or minus two.

5.4. APPLICATION TO BORANES

As discussed above, the single DSD process for $B_5H_5^{2-}$ is "forbidden" by orbital symmetry. Some alternative mechanisms have been considered but none are expected to be particularly favorable. For $B_6H_6^{2-}$ there is no low-order DSD process, and this molecule is not fluxional on the NMR time-scale. The two possible carborane isomers are both known [26], which is also circumstantial evidence that there is a significant barrier to rearrangement. Otherwise we might expect only the most stable carborane to be isolatable.

In contrast both $B_8H_8^{2-}$ and $B_{11}H_{11}^{2-}$ have orbital symmetry-allowed single DSD processes, and both exhibit fluxionality on the NMR time-scale [27, 28]. Only one carborane isomer has been prepared in each case [29, 30], although for $C_2B_9H_{11}$ this conclusion is based on NMR evidence alone. For $B_7H_7^{2-}$, $B_9H_9^{2-}$, and $B_{10}H_{10}^{2-}$ there is no evidence of fluxionality on the NMR time-scale [27]. However, only one of the six possible carborane isomers $C_2B_7H_9$ has been isolated [31], while three of the $C_2B_8H_{10}$ [32] and two of the $C_2B_5H_7$ [33] carborane isomers are known. For all three species there are orbital symmetry-allowed concerted double DSD processes, although this requires the two open faces to share an edge in the case of $B_{10}H_{10}^{2-}$ and $B_7H_7^{2-}$. Furthermore, in each case a mirror plane would be retained in the first step of a stepwise double DSD rearrangement. For the pentagonal bipyramid, for example, the 54(44) DSD process results in a capped octahedral C_{3v} geometry with a single atom on the principal axis as Onak and co-workers have noted [34].

For the icosahedron all three carboranes, $C_2B_{10}H_{12}$, are known; the 1,2 isomer (carbon atoms adjacent) may be converted to the 1,7 by heating to 470°C, and the 1,7 isomer may be converted to the 1,12 (carbon atoms at opposite ends of a diameter) by heating to 630°C. The high temperatures indicate high activation energy barriers for these rearrangements. A sextuple DSD mechanism has been proposed to account for the first interconversion, and a relative

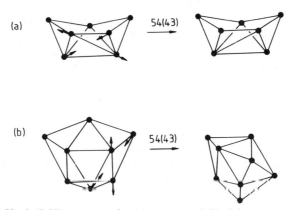

Figure 5.6. Single DSD processes for (a) seven- and (b) eight-vertex *nido*-boranes.

Figure 5.7. Concerted double DSD process for the nine-vertex *nido*-skeleton.

rotation of layers, which is actually equivalent to a pentuple DSD, for the second [7]. Both processes involve several DSD rearrangements, and are therefore expected to be less favorable than the single DSD possible for $B_8H_8^{2-}$ and $B_{11}H_{11}^{2-}$, and the double DSD possible for $B_9H_9^{2-}$. Furthermore, ab initio calculations indicate that neither geometry is a true transition state [2].

The above theories may also be applied to the *nido*-boranes and carboranes to predict the relative energy barriers to skeletal rearrangements in these species, assuming that any rearrangement must leave the open face essentially unchanged because of the bridging protons. The 7- and 8-vertex *nido* geometries have symmetry-allowed single DSD rearrangements, both described as 54(43) in King's notation [9] (Figure 5.6). The 9-vertex *nido* geometry has a symmetry-allowed double DSD rearrangement which is equivalent to concurrent 54(53) and 54(44) processes in which the two open faces share an edge (Figure 5.7). The 10- and 11-vertex *nido* geometries have no low-order topologically allowed DSD processes available, and are therefore expected to have larger barriers to rearrangement. However, the pentuple DSD which is topologically possible for the 11-vertex system does not have a TSH symmetry-forced crossing because both starting material and transition state have a single atom on the principal axis. The prediction for the relative order of the energy barriers of these *nido*-boranes and the corresponding carboranes to skeletal rearrangements is therefore:

7,8-vertex *nido* < 9-vertex *nido* ≪ 10,11-vertex *nido* geometries

5.5. THE SDDS MECHANISM

The presence of a single square face in an otherwise deltahedral cluster opens up the possibility of a new low energy rearrangement mechanism. Such a process may be written as a square→diamond, diamond→square or SDDS mechanism. This concerted process is illustrated for the square-based pyramid and the capped square antiprism in Figure 5.8. If a mirror plane is retained throughout then this mechanism involves two orbital crossings if one edge lies in the plane and the other lies across it. If the edges are parallel, as in Figure 5.8, then there are two avoided crossings and the process is orbital symmetry-allowed. There is no TSH symmetry-forced crossing, since the starting and finishing molecules

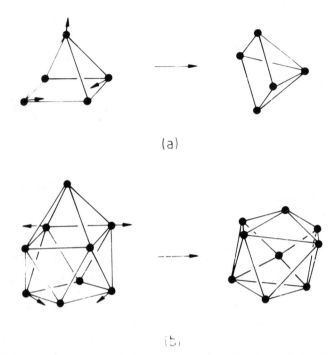

Figure 5.8. The square→diamond, diamond→square (SDDS) mechanism for a square-based pyramid (a), and a capped square-antiprism (b).

both have a single atom on the four-fold principal axis. Hence this may be a favorable process for *nido* clusters. For main group clusters, however, the presence of bridging hydrogen atoms around the open face means that this mechanism would involve a much greater perturbation to the structure of the cluster, and is therefore not likely to occur. It could, however, be significant in accounting for the behavior of $B_8H_8^{2-}$ in solution [35] where a fluxional bicapped trigonal prismatic geometry may be present. The SDDS mechanism is allowed by orbital symmetry for the latter species, and may account for its nonrigidity.

As for the DSD mechanism it is possible to understand these results from simple orbital arguments, and we will consider the case of the square-based pyramid [36]. This geometry also arises for the single DSD process in $B_5H_5^{2-}$ as discussed above, and the reaction coordinates are compared in Figure 5.9. We have already described how the two components of the nonbonding *e* pair of orbitals are split apart in opposite senses for the DSD process, leading to an orbital crossing for a cluster with six skeletal electron pairs. The structural perturbation for the SDDS process is considerably greater, and leads to mixing with the π orbitals of the vertex atom, as shown in Figure 5.10. This demonstrates two key points:

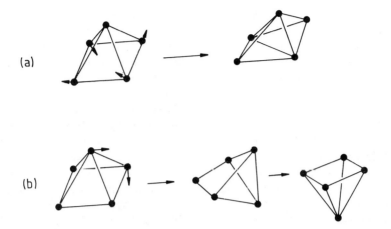

Figure 5.9. Sketches of the geometry changes for the square-based pyramid, (a) as the intermediate in the $B_5H_5^{2-}$ single DSD process and (b) as the starting point for the SDDS process in for example, $C_5H_5^+$.

1. The mixing of each e component in the open face with an e component from the apex implies that there are two avoided crossings.
2. The mixing also means that the two components are not greatly split and end up degenerate again in the product.

These conclusions clearly agree with the alternative analysis above. Also note that the square-based pyramid is probably the least favorable case for the SDDS mechanism as the structural changes involved are relatively large.

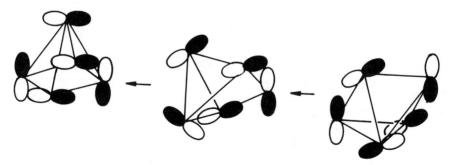

Figure 5.10. The evolution of one of the e components in the SDDS process for the square-based pyramid. The other component may be obtained simply by applying the pairing operation to each orbital.

5.6. GEOMETRICAL SELECTION RULES

The geometrical symmetry selection rules are not concerned with the conservation of *orbital* symmetry, but with the fact that any reaction mechanism must follow a vibrational normal mode of the cluster at every point [37]. In fact all of the processes considered above are geometrically allowed because normal modes of the appropriate symmetry exist. However, it is possible to eliminate some candidate structures as transition states using geometrical selection rules such as the McIvor–Stanton theorems [38] or the formulation due to Rodger and Schipper [39].

Murrell and Laidler [40] first emphasized the requirement that a true transition state must normally have precisely one negative "force constant"; that is one normal mode with an associated imaginary frequency. This follows from a straightforward consideration of the multidimensional potential energy surface; if there were two normal modes with negative force constants then the "transition state" would be a hill rather than a saddle point, and there would almost certainly be a lower energy pathway connecting reactants and products. McIvor and Stanton extended this group theoretical analysis and derived several theorems concerning the transformation of the transition vector and the effect of transition state symmetry operations on the reactants and products.

Their method may be applied as follows. We assume a particular geometry for the "transition state" and sketch the displacements of the atoms to their equilibrium positions in the products and reactants. These unlabeled displacement vectors can be regarded as the objects acted upon the symmetry operations of the transition state point group. If it happens that a symmetry operation of the putative transition state converts the product displacements into reactant displacements then the transition vector must be antisymmetric under this operation. (The transition vector lies in the direction of the critical normal mode with the negative force constant.) There are various other restrictions on the transformation properties of the transition vector which apply if a symmetry operation maps a reactant onto an equivalent structure [38].

For our purposes the most important result of Stanton and McIvor's work is the antisymmetric nature of the transition vector under operations which interconvert reactant and product displacements. This has an important corollary that a transition state cannot have an odd order symmetry element linking the product and reactant displacements [38]. (A symmetry element has odd order if it must be repeated an odd number of times to achieve the identity operation.) However, this method can only rule out a candidate structure as a transition state for a particular process. It cannot determine whether the structure may be a true transition state for a different reaction, or indeed an intermediate with no negative force constants.

In Rodger and Schipper's approach the stationary permutation-inversion operations of the product and reactant molecules are considered [39]. With the assumption that all the common stationary operations are preserved throughout the reaction path it is possible to deduce the symmetry of the transition state, if there is one, for concerted processes which involve only permutations of identical atoms (homoconversions). In fact we can summarize their results for such cases as follows: a transition state for a concerted homoconversion along a maximal symmetry path must have an even order operation which is not present in the product or reactant. This is a necessary but not a sufficient condition because the operation in question must also be a product generator operation [39]. However, it provides a simple first criterion for examining possible transition state geometries.

For heteroconversions, in which the product is not simply related to the reactant by a permutation of identical nuclei, the geometrical selection rules for concerted processes are much less restrictive. Rodger and Schipper have tabulated the vibrational modes which relate particular point groups, and the presence or absence of modes of the required type is now the determining factor. In fact their tabulation shows that there is very often a mode (usually E-type) which breaks all the symmetry, and this may help to explain why so many transition states are calculated to have C_1 symmetry [41]. Exceptions arise for point groups with no E modes such as C_{2v}.

5.7. APPLICATION TO CLUSTER REARRANGEMENTS

We consider three specific examples using the McIvor–Stanton approach. In each case two DSD processes result in a permutation of the vertices of the starting geometry. The single DSD process leading from the pentagonal bipyramid to the capped octahedron is shown in Figure 5.11. The capped octahedron has C_{3v} symmetry and the reactant displacements are clearly mapped onto the product displacements by a C_3 rotation. Since this operation has odd order we conclude that the capped octahedron cannot be a transition state for the stepwise double DSD process. It would be impossible to find a transition vector antisymmetric to this operation: such a species would also have to be antisymmetric under C_3^3 which is the identity operation. This does not preclude the possibility that this structure may be a metastable intermediate, although orbital symmetry considerations suggest otherwise [3].

The second example is the single DSD process for the bicapped square antiprism (Figure 5.12). Again this leads to a C_{3v} intermediate structure for which the reactant and product displacements are related by a C_3 rotation. Hence we can again conclude that this geometry cannot be a true transition state for the nonconcerted double DSD process. There is no such prohibition on the concerted double DSD, however. Furthermore, in both the above examples a mirror plane through the critical face is conserved during each single DSD process. However, this does not necessarily lead to an orbital crossing because the steps do not lead back immediately to the starting geometry. Instead each intermedate structure has a nonbonding e pair of orbitals at the HOMO–LUMO level, arising from the single atom on the three-fold rotation axis [2].

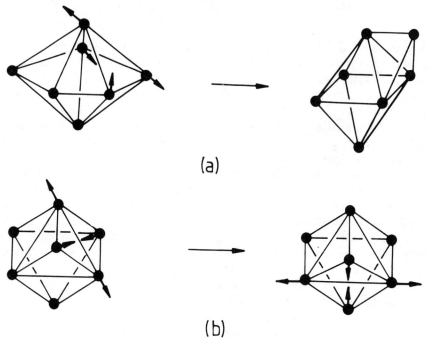

(a)

(b)

Figure 5.11. (a) The single DSD process leading from the pentagonal bipyramid to the capped octahedron for which (b) the reactant and product displacements are related by a C_3 rotation.

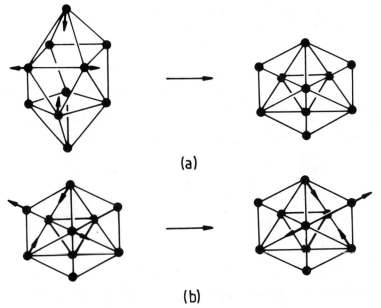

(a)

(b)

Figure 5.12. (a) A single DSD process leading from the bicapped square antiprism to a C_{3v} deltahedron for which (b) the reactant and product displacements are related by a C_3 rotation of the transition state point group.

There is bound to be a significant rise in the total orbital energy for this open shell configuration because the doubly occupied e pair is nonbonding.

In contrast the single DSD process illustrated in Figure 5.13 for the tricapped trigonal prism leads to a C_{2v} symmetry deltahedron. In this case the point group of the intermediate geometry contains no odd order operations. The reactant and product displacements are related by a C_2 operation and one of the reflections, while both sets of displacements are mapped onto themselves under reflection in the other mirror plane. Hence the transition vector must be antisymmetric under the former operations and symmetric under the latter. It must therefore transform as either B_1 or B_2, depending upon the choice of axes.

5.8. THE SECOND-ORDER JAHN–TELLER EFFECT

Here we examine the DSD mechanism using the symmetry properties of the terms arising in a perturbation theory analysis of the energy of a distorted molecule [42]. Consider distortion along a normal mode of symmetry species S_i.

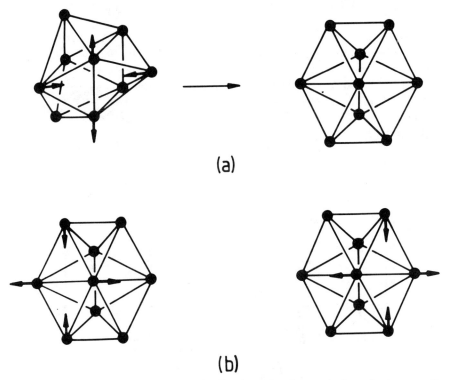

Figure 5.13. (a) A single DSD process leading from the tricapped trigonal prism to a C_{2v} symmetry deltahedron and (b) the reactant and product displacements.

Using first- and second-order perturbation theory, the energy of the distorted system may be written as [42]

$$E(S_i) = E_0 + S_i \int \psi_0 \hat{\mathscr{H}}_i \psi_0 \, d\tau$$

$$+ \frac{1}{2} S_i^2 \left[\int \psi_0 \hat{\mathscr{H}}_{ii} \psi_0 \, d\tau - 2 \sum_m' \frac{|\int \psi_0 \hat{\mathscr{H}}_i \psi_m \, d\tau|^2}{\Delta E_{m0}} \right],$$

where ψ_0 is the ground-state wavefunction of the undistorted geometry, ΔE_{m0} is the zeroth-order energy separation between the states ψ_0 and ψ_m of the undistorted geometry, $\hat{\mathscr{H}}_i = (\partial \hat{\mathscr{H}}/\partial S_i)_0$ and $\hat{\mathscr{H}}_{ii} = (\partial^2 \hat{\mathscr{H}}/\partial S_i^2)_0$. E_0 is the unperturbed energy of the system and $S_i \int \psi_0 \hat{\mathscr{H}}_i \psi_0 \, d\tau$ is the first-order Jahn–Teller term which is not important for nondegenerate electronic states because it vanishes. The term in square brackets can be identified with the "force constant" for distortion along the normal mode in question and can be divided into two parts. The first part, involving the second derivative, represents the nuclear motion in a fixed electron distribution while the second term represents the "relaxation" of the electron distribution in response to the nuclear displacement. If this "force constant" is negative then the energy of the system is lowered on distortion and we have a second-order Jahn–Teller effect.

It is often productive to proceed in a nonrigorous manner and consider only the contribution of the HOMO–LUMO mixing to the second part of this term [42]. In this case one simply seeks a normal mode whose symmetry species is contained in the direct product of the symmetry species of the HOMO and the LUMO (note that the term $(\partial \hat{\mathscr{H}}/\partial S_i)_0$ transforms like S_i). Distortion along such a mode results in the HOMO and LUMO splitting further apart through their mutual admixture.

Consider the product and reactant displacements for the single DSD process sketched in Figure 5.14. If the cluster has C_{4v} symmetry it follows from the geometrical selection rules that the transition vector must be antisymmetric

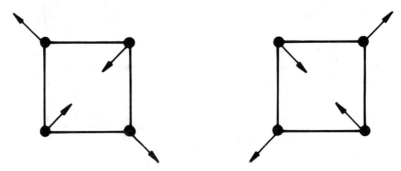

Figure 5.14. Displacements to products and reactants in the DSD process. Only the critical face is sketched for clarity.

under C_4, C_4^3, and σ_d, that is, it must transform like $x^2 - y^2$. This, in fact, simply corresponds to the displacements as sketched. Now consider descent in symmetry to the point group C_2. The HOMO and the LUMO of the hypothetical *nido* cluster both transform as B in this point group [3], and the DSD transition vector transforms as A. Since $B \otimes B = A$ we conclude that the contribution of the particular energy term considered above would be nonzero and therefore negative. In C_s, on the other hand, when a diagonal mirror plane is retained, the DSD transition vector transforms as A' while the HOMO and LUMO transform as A' and A''. Since $A' \otimes A'' = A''$ the DSD mode cannot lead to mixing of the HOMO and LUMO in this case. This fundamental difference between the C_s and C_2 point groups results from the properties of the Tensor Surface Harmonic theory parity operator, $\hat{\mathscr{P}}$, which relates L^π and $\bar{L}^\pi$ cluster orbitals [3]. $\hat{\mathscr{P}}$ is symmetric to rotations and antisymmetric to reflections, and hence in C_2 the parity related HOMO and LUMO belong to the same irreducible representation. In fact, the critical matrix element will have a nonzero component (and lead to a lowering of the energy) for vibrations which transform in the same way as $\hat{\mathscr{P}}$. The DSD vibration clearly does transform in this way in C_2, but not in C_s, and this kind of argument could also be applied to predict "mode-softening" in an equilibrium geometry.

Clearly this analysis fits in with the results of the previous sections, where we showed that a crossing occurs if a mirror plane is retained throughout, and an avoided crossing if a C_2 axis is conserved. Furthermore, if C_{2v} symmetry is conserved throughout a McIvor–Stanton analysis leads to the conclusion that the only possible symmetry species for the transition vector is A_1, which does not admit HOMO–LUMO mixing. This analysis, which pertains to the single DSD process in an icosahedron, is again in agreement with the orbital symmetry conclusion because two mirror planes are retained throughout.

5.9. SUMMARY

We have examined the interrelation of orbital and geometrical symmetry selection rules for cluster skeletal rearrangements within the framework of Stone's TSH theory. The orbital selection rules in particular have great predictive power and enable the rearrangement rates of all the *closo*-borohydrides to be rationalized. We have also shown how the selection rules for various DSD processes and the new SDDS mechanism may be understood in terms of simple orbital arguments.

The geometric selection rules are most restrictive for homoconversions along paths of maximal symmetry, in which equivalent atoms are simply permuted within the same skeleton [39]. It is interesting to note that the appearance of a new even order operation in the "transition state" will lead to an orbital crossing if the symmetry element in question is a C_4 or higher even order rotation axis [2].

We have also considered a second-order Jahn–Teller analysis of the diamond-square-diamond mechanism which results in the same conclusion as the orbital symmetry analyses: single DSD processes in which a mirror plane is conserved throughout are expected to be less favorable than those in which a C_2 rotation axis is retained.

Since the completion of this review, the DSD process has been identified as a key rearrangement mechanism in inert gas and "trapped ion" clusters. The interested reader is referred elsewhere for details [43].

ACKNOWLEDGMENT

DJW thanks Dr. A. Rodger and Dr. P. E. Schipper for some helpful discussions.

REFERENCES

1. Lipscomb, W. N., *Science (Washington DC)*, **153**, 373 (1966).

2. Wales, D. J. and Stone, A. J., *Inorg. Chem.*, **26**, 3845 (1987).

3. Wales, D. J., Mingos, D. M. P. and Zhenyang, L., *Inorg. Chem.*, **28**, 2754 (1989).

4. Wales, D. J. and Mingos, D. M. P., *Polyhedron* **8**, 1933 (1989).

5. Mingos, D. M. P. and Wales, D. J., *Introduction to Cluster Chemistry*, Prentice Hall, Englewood Cliffs, NJ, 1990.

6. Woodward, R. B. and Hoffmann, R., *Angew. Chem. Int. Ed.*, **8**, 781 (1969).

7. Wade, K., *Electron Deficient Compounds*, Nelson, London, 1971.

8. See e.g., Kennedy, J. D., in: *Progress in Inorganic Chemistry*, Vol. 34, Wiley, New York, 1986.

9. King, R. B., *Inorg. Chim. Acta*, **49**, 237 (1981).

10. Williams, R. E., *Prog. Boron Chem.*, **2**, 51 (1970); *Inorg. Chem.*, **10**, 210 (1971); *Adv. Inorg. Chem. Radiochem.*, **18**, 67 (1976).

11. Wade, K., *J. Chem. Soc., Chem. Commun.*, 792 (1971).

12. Rudolph, R. W. and Pretzer, W. R., *Inorg. Chem.*, **11**, 1974 (1972); Rudolph, R. W., *Acc. Chem. Res.*, **9**, 446 (1976).

13. Hoffmann, R., *Angew. Chem., Int. Ed. Engl.*, **21**, 711 (1982).

14. Mason, R., Thomas, K. M. and Mingos, D. M. P., *J. Am. Chem. Soc.*, **95**, 3802 (1973).

15. Hoffmann, R. and Lipscomb, W. N., *J. Chem. Phys.*, **36**, 2179 (1962); Hoffmann, R., *J. Chem. Phys.*, **39**, 1397 (1963).

16. Stone, A. J., *Molec. Phys.*, **41**, 1339 (1980); *Inorg. Chem.*, **20**, 563 (1981); Stone, A. J. and Alderton, M. J., *Inorg. Chem.*, **21**, 2297 (1982); Stone, A. J., *Polyhedron*, **3**, 1299 (1984).

17. Mingos, D. M. P. and Johnston, R. L., *Struct. Bond.*, **68**, 29 (1987).

18. Fowler, P. W. and Porterfield, W. W., *Inorg. Chem.*, **24**, 3511 (1985); Fowler, P. W., *Polyhedron*, **4**, 2051 (1985).

19. Stone, A. J. and Wales, D. J., *Molec. Phys.*, **61**, 747 (1987).

20. Wales, D. J., *Molec. Phys.*, **67**, 303 (1989).

21. Wales, D. J. (1989) submitted.

22. Johnston, R. L. and Mingos, D. M. P., *J. Chem. Soc., Dalton Trans.*, 647 (1987).

23. Gimarc, B. M. and Ott, J. J., *Inorg. Chem.*, **25**, 83, 2708 (1986); *J. Comp. Chem.*, **7**, 673 (1986).

24. Johnston, R. L. and Mingos, D. M. P., *J. Chem. Soc., Dalton Trans.*, 647 (1986).

25. Fowler, P. W., *Polyhedron*, **4**, 2051 (1985).

26. McNeill, E. A., Gallaher, K. L., Scholer, F. R. and Bauer, S. H., *Inorg. Chem.*, **12**, 2108 (1973); Shapiro, I., Keilin, B., Williams, R. E., Good, C. D., *J. Am. Chem. Soc.*, **85**, 3167 (1963); Beaudet, R. A. and Poynter, R. L., *J. Chem. Phys.*, **53**, 1899 (1970).

27. Muetterties, E. L., Hoel, E. L., Salentine, C. G. and Hawthorne, M. F., *Inorg. Chem.*, **14**, 950 (1975).

28. Muetterties, E. L., Wiersema, R. J. and Hawthorne, M. F., *J. Am. Chem. Soc.*, **95**, 7520 (1975); Tolpin, E. I. and Lipscomb, W. N., *J. Am. Chem. Soc.*, **95**, 2384 (1973).

29. Shapiro, I., Good, C. D. and Williams, R. E., *J. Am. Chem. Soc.*, **84**, 3837 (1962).

30. Rogers, H. N., Lau, K. and Beaudet, R. A., *Inorg. Chem.*, **15**, 1775 (1976); Hart, H. and Lipscomb, W. N., *Inorg. Chem.*, **7**, 1070 (1968); Berry, T. E., Tebbe, F. N. and Hawthorne, M. F., *Tetrahedron Lett.*, 715 (1965).

31. Koetzel, T. F., Scarborough, F. E. and Lipscomb, W. N., *Inorg. Chem.*, **7**, 1076 (1968).

32. Tebbe, F. N., Garrett, P. M., Youngs, D. C. and Hawthorne, M. F., *J. Am. Chem. Soc.*, **88**, 609 (1966); Rietz, R. R., Schaeffer, R. and Walter, E., *J. Organomet. Chem.*, **63**, 1 (1973); Garrett, P. M., Smart, J. C., Ditta, G. S. and Hawthorne, M. F., *Inorg. Chem.*, **8**, 1907 (1969).

33. Rietz, R. R. and Schaeffer, R., *J. Am. Chem. Soc.*, **95**, 6254 (1973); Grimes, R. N., *Carboranes*, Academic Press, New York, 1970; Onak, T. P., Gerhart, F. J. and Williams, R. E., *J. Am. Chem. Soc.*, **85**, 3378 (1963).

34. Abdou, Z. J., Abdou, G., Onak, T. and Lee, S., *Inorg. Chem.*, **25**, 2678 (1986).

35. Muetterties, E. L., Wiersema, R. J. and Hawthorne, M. F., *J. Am. Chem. Soc.*, **95**, 7520 (1973).

36. Strohrer, W.-D. and Hoffmann, R., *J. Am. Chem. Soc.*, **94**, 1661 (1972).

37. Rodger, A. and Johnson, B. F. G., *Polyhedron*, **7**, 1107 (1988).

38. McIvor, J. W. and Stanton, R. E., *J. Am. Chem. Soc.*, **94**, 8618 (1972); Stanton, R. E. and McIvor, J. W., *J. Am. Chem. Soc.*, **97**, 3632 (1975).

39. Rodger, A. and Schipper, P. E., *Chem. Phys.*, **107**, 329 (1986); *J. Phys. Chem.*, **91**, 189 (1987); *Inorg. Chem.*, **27**, 458 (1988).

40. Murrell, J. N. and Laidler, K. J., *Trans. Faraday Soc.*, **64**, 371 (1968).

41. Dewar, M. J. S., *Faraday Discuss. Chem. Soc.*, **62**, 197 (1976).

42. Bader, R. F. W., *Molec. Phys.*, **3**, 137 (1960); *Can. J. Chem.*, **40**, 1164 (1962); Pearson, R. G., *J. Am. Chem. Soc.*, **91**, 1252 (1969); *Acc. Chem. Res.*, **4**, 152 (1971).

43. Wales, D. J., *J. Chem. Phys.*, **91**, 7002 (1989).

CHAPTER 6

Some Recently Determined Borane Structures and Their Implications

N. N. GREENWOOD

This paper discusses two aspects of our current research program:

1. The detailed structure of small boranes in the gas phase, and their implications for interpreting kinetic parameters of gas-phase thermolysis reactions in the temperature range 50–150°C.

2. Some unexpected metallaborane structures and their broader implications for cluster chemistry.

6.1. STRUCTURE AND THERMOLYSIS OF SOME SMALL BORANES

A convenient starting point for the first theme is the structure of *arachno*-B_6H_{12}, the only simple borane for which dimensions are unknown from X-ray crystal structure analysis or other techniques. We have been collaborating with D. W. H. Rankin in Edinburgh, and the structure that emerged, which is one of the most complex ever done by electron diffraction, is shown in Figure 6.1 [1]. Something like this had been proposed [2] on the basis of NMR spectroscopy but, of course, there were no dimensions, and alternative structures with C_i and C_s symmetry could not be entirely eliminated. The electron diffraction results establish that the molecule has C_2 symmetry and is therefore chiral. This conformation has been described as a belt fragment of an icosahedron (which is itself generated by the fusion of two such fragments), but such a description

Electron Deficient Boron and Carbon Clusters, Edited by George A. Olah, Kenneth Wade, and Robert E. Williams.

ISBN 0-471-52795-5 © 1991 John Wiley & Sons, Inc.

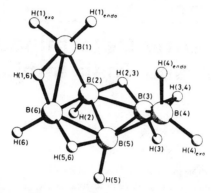

Figure 6.1. Perspective view of the structure of B_6H_{12} as obtained by gas-phase electron diffraction [1].

implies that the dihedral angles would be close to icosahedral (138.2°). In fact, the observed dihedral angles in *arachno*-B_6H_{12} are 167(2)° for the central hinge at B(2)B(5) and 128(1)° for the two outer hinges at B(2)B(6) and B(3)B(5) [1]. Alternatively, the Williams–Wade cluster geometry and electron counting rules [3] would relate the geometry of *arachno*-B_6H_{12} to that of a *closo*-dodecahedral B_8 cluster by removal of two adjacent 5-connected vertices. It is significant that the internal dihedral angles of a triangulated dodecahedron are 157° and 120°, close to those observed, suggesting that the magnitude of dihedral angles is a useful supplementary criterion for distinguishing between geometrical de-

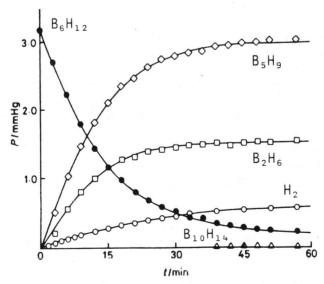

Figure 6.2. Course of thermolysis of *arachno*-B_6H_{12} at 99.4°C and an initial pressure of 3.14 mmHg [5].

scriptions of open clusters. In this connection we note that the somewhat similar structure of $hypho$-$[B_6H_{10}(PMe_3)_2]$ [4], which features a central dihedral angle of 140° and two outer dihedral angles of 154°, is better described in terms of the removal of three contiguous vertices from a $closo$-B_9^* tricapped trigonal prism (i.e., $closo$-$B_9 - 3B = hypho$-B_6) rather than as an icosahedral belt fragment. These, and other similar examples, establish the value of internal dihedral angles in helping to trace the putative parenthood of more-open cluster geometries.

The gas-phase thermolysis of $arachno$-B_6H_{12} in the temperature range 75–150°C is particularly clear (see Figure 6.2) and is very readily interpreted in terms of structure [5]. The homogeneous reaction follows first-order kinetics, the main volatile products being B_5H_9 and B_2H_6 in a 2:1 ratio. Very little H_2 or involatile solid is produced, and there is a trace of $B_{10}H_{14}$ during the later stages of the thermolysis. An Arrhenius plot over a range of temperatures leads to an activation energy, E_a, of 73.8 ± 4.5 kJ mol^{-1} and a pre-exponential factor, A, of 2.2×10^7 s^{-1}.

The main reaction can be explained by a simple two-step mechanism involving the first-order rate-determining elimination of $\{BH_3\}$ which then rapidly dimerizes to B_2H_6

$$B_6H_{12} \xrightarrow{\text{slow}} B_5H_9 + \{BH_3\}$$

$$2\{BH_3\} \xrightarrow{\text{fast}} B_2H_6$$

The deeper significance of these results can be seen by comparison with two other $arachno$-boranes that we have recently studied in detail: B_5H_{11} and B_4H_{10}. But first, a note about their structures, which turn out to be crucial to the discussion.

The structure of $arachno$-B_5H_{11} in the gas phase as determined by electron diffraction is shown in Figure 6.3. A notable feature is the absence of a vertical

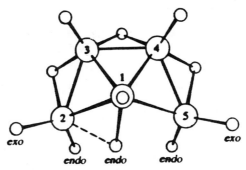

Figure 6.3. Structure of $arachno$-B_5H_{11} as determined by gas-phase electron diffraction [6]. Note the C_1 symmetry resulting from the fact that the semi-bridging H(1)$_{endo}$ atom is closer to B(2) than to B(5). This agrees with the low-temperature X-ray crystal structure [7] and the most recent theoretical study [8].

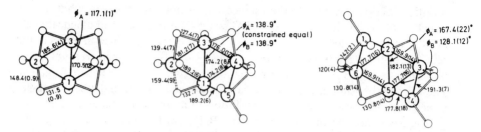

Figure 6.4. Structural relationship between the three *arachno*-boranes B_4H_{10}, B_5H_{11}, and B_6H_{12} (see text for discussion); distances in pm.

plane of symmetry through atoms $H(1)_{exo}$, $B(1)$, and $H(3,4)$ due to the semibridging nature of $H(1)_{endo}$ which is 30.5 pm closer to $B(2)$ than to $B(5)$ [159.4(9) and 189.9(9) pm, respectively] [6]. The distance $B(1)–H(1)_{endo}$ (132.7 pm) is also closer to the mean of the $B–H_\mu$ distances in the molecule than to the average $B–H_t$ distance [133.4(7) and 119.2(4) pm, respectively]. The structural parameters of *arachno*-B_6H_{12} and *arachno*-B_5H_{11} are compared with those of *arachno*-B_4H_{10} [9] in Figure 6.4. It will be noted that the structure of B_5H_{11} is related to that of B_4H_{10} by replacement of a hydrogen bridge by a BH_2 group, with simultaneous conversion of a terminal hydrogen on $B(4)$ into a bridging position between atoms $B(4)$ and $B(5)$. Hexaborane(12) is then notionally derived from B_5H_{11} by a similar modification to the opposite side of the molecule. A more detailed discussion of the observed structural trends is given in [1].

Turning next to a comparison of the kinetic parameters of these three *arachno*-boranes, it is apparent from Table 6.1 that the low activation energy and very low pre-exponential factor already noted for B_6H_{12} is precisely mirrored in the very similar data for B_5H_{11} [10]. By contrast, the activation energy for thermolysis of B_4H_{10} is some 35% higher and the pre-exponential factor some 4.5 orders of magnitude greater [11], being more typical of values

TABLE 6.1. Comparison of the kinetic parameters obtained during the thermolysis of the three *arachno*-boranes B_4H_{10}, B_5H_{11}, and B_6H_{10} in the temperature and pressure ranges indicated.

	B_4H_{10}	B_5H_{11}	B_6H_{12}
Temperature range (°C)	40–78	40–150	75–150
Pressure range (mmHg)	0.9–39	1.5–10.5	1.8–10
Order of reaction	First	First	First
E_a (kJ mol^{-1})	99.2 ± 0.8	72.6 ± 2.4	73.8 ± 4.5
A (s^{-1})	6.0×10^{11}	1.3×10^7	2.2×10^7
Reference	11	10	5

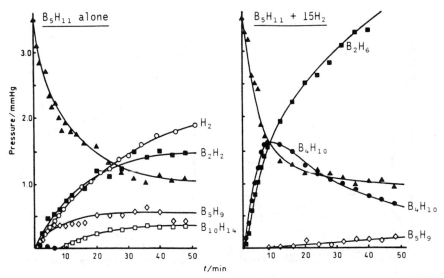

Figure 6.5. Influence of added H_2 on the thermolysis of B_5H_{11} (p_0 3.52 mmHg, T 100°C) [12]. Note the greatly increased concentration of B_2H_6 and the formation of B_4H_{10} which then itself decomposes.

normally encountered for first-order reactions. This finds a ready interpretation in terms of the structures of the three boranes and in the nature of the rate-determining step for thermolysis in each case. Thus, the structural similarity between B_5H_{11} and B_6H_{12}, which has already been mentioned, reveals the presence of an incipient $\{BH_3\}$ group, and the rate-determining step for B_5H_{11}, like that for B_6H_{12}, is believed to be the elimination of $\{BH_3\}$ from the molecule [10]

$$B_5H_{11} \longrightarrow \{B_4H_8\} + \{BH_3\}$$

The $\{BH_3\}$ then rapidly dimerizes to B_2H_6 whereas the fate of the $\{B_4H_8\}$ moiety is more complex. By contrast, the rate-determining step in the thermolysis of B_4H_{10} is the elimination of H_2 rather than $\{BH_3\}$ [11]

$$B_4H_{10} \rightleftharpoons \{B_4H_8\} + H_2$$

Consistent with these proposals, the presence of an excess of H_2 during the thermolysis of B_5H_{11} affects neither E_a or A, but *does* affect the product distribution (see Figure 6.5) [12], because H_2 reacts with $\{B_4H_8\}$ to give B_4H_{10}. The situation with B_4H_{10} itself, however, is completely different since added H_2 has a quite dramatic effect on both the rate of thermolysis and the product distribution (Figure 6.6) whilst leaving the activation energy E_a unaltered [12]. Thus a 6-fold excess of H_2 decreases the rate of thermolysis 5-fold by reducing the

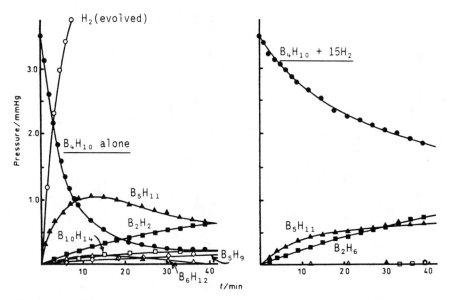

Figure 6.6. Influence of added H_2 on the thermolysis of B_4H_{10} (p_0 3.51 mmHg, T 89°C) [12]. Note the dramatic diminution in the rate of thermolysis and the complete absence of volatile higher boranes such as B_6H_{12} and $B_{10}H_{14}$; the formation of solid "polymer" is also almost completely suppressed in the presence of added H_2.

concentration of $\{B_4H_8\}$; the rate-determining step still remains the formation of $\{B_4H_8\}$ from B_4H_{10} so the activation energy is unaltered. Details of the modified product distribution can also be interpreted [12, 13].

Before leaving the structural/kinetic work on the boranes themselves and moving to some novel metallaborane structures, let me interpolate a very recent electron-diffraction tour de force by David Rankin. This concerns the structure of B_4H_8CO [14], a species long known and much studied. Detailed NMR studies by several groups (see [14] for detailed literature citations) had led to the conclusion that there were two isomers, *exo* and *endo*, but the structure of neither had been determined. We confirmed the presence of two isomers in the ratio *endo/exo* = 62/38 by NMR spectroscopy, and electron diffraction of this mixture led to the two structures shown in Figure 6.7. Interestingly, the dihedral angle at the "hinge" B(1)B(3) for the *exo*-isomer is 144(2)°, which is larger than that observed for the *endo*-isomer, namely 135(4)°. The angle B(1)–C–O is essentially 180° in both isomers but the angle B(3)–B(1)–CO is 109(2)° for the *exo*-isomer and 125(2)° for the *endo*. The only other structural data on an adduct B_4H_8L derive from an X-ray diffraction analysis of *endo*-$[B_4H_8(PF_2NMe_2)]$ which led to a dihedral angle at the B(1)B(3) "hinge" of 137° (close to that obtained by electron diffraction for *endo*-B_4H_8CO) and an angle B(3)–B(1)–P of 135° [15]. The significance of these angles and of the curious ratios of *endo/exo* concentrations observed for several phosphine ligands PF_2X (where $X = Bu^t$, OMe, SMe, Me, or CF_3) [16], still awaits detailed interpretation.

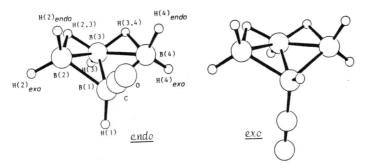

Figure 6.7. The structures of *exo-* and *endo-*B$_4$H$_8$CO as determined by electron diffraction of their gaseous mixture [14].

6.2. SOME RECENTLY DETERMINED METALLABORANE STRUCTURES

Turning now to the synthesis and structure of some novel metallaboranes, we might subtitle this section "Tales of the Unexpected".

As is clear from Figure 6.8, *nido-*B$_{10}$H$_{14}$ has four geometrically distinct types of boron atom: B(1,3), B(2,4), B(5,7,8,10), and B(6,9). Subrogation of a BH$_t$ vertex by the isolobal Ru(η^6-C$_6$Me$_6$) group, for example, might therefore lead one to predict the possibility of four isomers for *nido-*[(η^6-C$_6$Me$_6$)RuB$_9$H$_{13}$], provided that viable synthetic routes could be devised. The easiest isomer to make is the 6-Ru derivative which can be obtained in 80% yield by reaction of *arachno-*B$_9$H$_{14}^-$ with [(η^6-C$_6$Me$_6$)RuCl$_2$]$_2$ at room temperature [17]. The 5-isomer was obtained in somewhat lower yield by deprotonating *nido-*B$_{10}$H$_{14}$ to give B$_{10}$H$_{13}^-$ and then reacting this with [(η^6-C$_6$Me$_6$)RuCl$_2$]$_2$ at room temperature to give the yellow, air-stable subrogation product [5-(η^6-C$_6$Me$_6$)-*nido-*5-RuB$_9$H$_{13}$] [17].

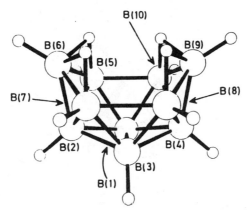

Figure 6.8. Numbering system for the boron cluster atoms in *nido-*B$_{10}$H$_{14}$.

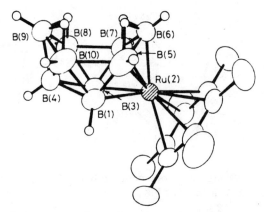

Figure 6.9. Molecular structure of $[2\text{-}(\eta^6\text{-}C_6Me_6)\text{-}nido\text{-}2\text{-}RuB_9H_{13}]$ as determined by single-crystal X-ray diffraction [19].

Examples of 2-metalladecaborane isomers are rare and, indeed, before our own work, the sole example was $[2\text{-}(\eta^5\text{-}C_5Me_5)\text{-}nido\text{-}2\text{-}CoB_9H_{13}]$ [18]. We have found that reaction of the versatile reagent $[(\eta^6\text{-}C_6Me_6)RuCl_2]_2$ with KB_6H_{11} in THF/CH_2Cl_2 at $-25°C$ gives a modest yield of the required $[2\text{-}(\eta^6\text{-}C_6Me_6)\text{-}nido\text{-}2\text{-}RuB_9H_{13}]$ as an air-stable yellow solid [19]. The mechanism is obscure but the structure is not (see Figure 6.9). The final isomer, $[1\text{-}(\eta^6\text{-}C_6Me_6)RuB_9H_{13}]$, was isolated in 9% yield as a coproduct from the same reaction mixture. However, instead of the expected *nido* structure, the compound turned out to have the more open structure shown in Figure 6.10; this features an unprecedented 7-membered open face Ru(1)B(5,6,7,8,9,10) [19]. Note that the B(5)...B(10) distance is now nonbonding (249.7 pm) compared

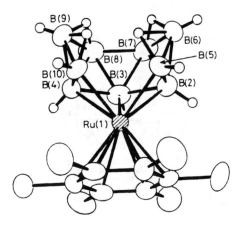

Figure 6.10. Molecular structure of $[1\text{-}(\eta^6\text{-}C_6Me_6)\text{-}1\text{-}RuB_9H_{13}]$ as determined by single-crystal X-ray diffraction [19].

with 183.5(15) pm for the corresponding distance B(7)–B(8). The structure is therefore an interesting (and we believe significant) departure from the Williams–Wade systematics [3] and it is tempting to speculate that the rupture of the B(5)...B(10) link reflects an increasing electron contribution from Ru to the cluster count. However, I hasten to add that we have, as yet, no direct evidence as to whether the oxidation state is Ru^{II} or Ru^{IV}. The structure could, in fact, be considered as a novel "*isoarachno*" 10-vertex cluster obtained by removing two adjacent vertices from a 12-membered nonicosahedral "*isocloso*" cluster of the type recently observed [20] for the $\{WC_2B_9\}$ cluster in $[Pt(PEt_3)_2(CO)_2WC_2B_9H_8Me_2(CH_2C_6H_4Me)]$.

Use of the same versatile metallaborane synthon $[(\eta^6\text{-}C_6Me_6)RuCl_2]_2$ with *arachno*-$B_{10}H_{14}^{2-}$ in MeCN (instead of with solutions of *nido*-$B_{10}H_{13}^-$ or *arachno*-$B_6H_{11}^-$ in THF/CH$_2$Cl$_2$) gives an almost quantitative yield of a unique 11-vertex *nido*-metallaborane with three bridging H atoms (Figures 6.11 and 6.12a) [21]

$$3[(\eta^6\text{-}C_6Me_6)RuCl_2]_2 + 4[NEt_4]_2[B_{10}H_{14}] + 12MeCN \longrightarrow$$

$$2[Ru(NCMe)_6][(\eta^6\text{-}C_6Me_6)RuB_{10}H_{13}]_2 + 8[NEt_4]Cl + 4HCl + 6C_6Me_6$$

It is noteworthy that, although over 100 *nido*-7-metallaundecaborane species incorporating some 20 different metals are known [22], virtually all have only two bridging H atoms in the open face (Figure 6.12b). Furthermore, since the $\{Ru(\eta^6\text{-}C_6Me_6)\}$ group is isolobal with a $\{BH_t\}$ group, the anion $[7\text{-}(\eta^6\text{-}C_6Me_6)\text{-}$

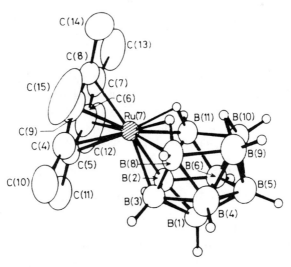

Figure 6.11. Molecular structure of the anion $[7\text{-}(\eta^6\text{-}C_6Me_6)\text{-}nido\text{-}7\text{-}RuB_{10}H_{13}]^-$ as determined by single crystal X-ray diffraction [21]. A significant feature of the structure is the presence of three bridging H atoms (two Ru–H$_\mu$–B and one B H$_\mu$ B).

 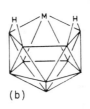 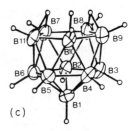

(a)   (b)   (c)

Figure 6.12. (a) Line diagram of the cluster structure of the ruthenaundecaborane anion $[7\text{-}(\eta^6\text{-}C_6Me_6)\text{-}nido\text{-}7\text{-}RuB_{10}H_{13}]^-$ emphasizing the presence of three bridging H atoms in the open face [21]; (b) the cluster structure adopted by almost all known *nido*-7-metallaundecaboranes, showing the presence of just two bridging H-atoms in the open face [22]; (c) the recently determined structure of *nido*-$B_{11}H_{14}^-$ featuring two B–H–B bridges and one BH_2 group in the open face [23].

nido-7-$RuB_{10}H_{13}]^-$ is equivalent to *nido*-$B_{11}H_{14}^-$. Interestingly, a very recent X-ray crystal structure of this latter ion shows the presence of two rather than three B–H–B bridges, the third *endo*-H atom being incorporated as part of a BH_2 group (Figure 6.12c) [23]; the ion thus retains effective C_s symmetry as does the ruthenaborane analogue (Figures 6.11 and 6.12a). The anion $[7\text{-}(\eta^6\text{-}C_6Me_6)\text{-}$ *nido*-7-$RuB_{10}H_{13}]^-$ shows bridge-proton fluxionality even at $-90°C$, with an upper limit to the free energy of activation, $\Delta G^{\ddagger}_{183}$, for the fluxional process of about $34\,kJ\,mol^{-1}$; low-temperature proton fluxionality is also a well known feature of the parent anion, $B_{11}H_{14}^-$.

In the synthesis of all the metallaboranes mentioned so far we have started with a binary borane precursor and reacted this with an organometallic species. Suppose, however, we start not with a B_{10} cluster but with a preformed MB_9 complex: for example, reaction of $[6\text{-}(\eta^5\text{-}C_5Me_5)\text{-}nido\text{-}6\text{-}IrB_9H_{13}]$ with $[(\eta^5\text{-}C_5Me_5)RhCl_2]_2$ in the presence of "proton sponge" for 2 days in dichloromethane solution at room temperature affords the novel *isonido* bimetalla 11-vertex cluster $[(\eta^5\text{-}C_5H_5)_2RhHIrB_9H_{10}]$ as shown in Figure 6.13 [24]. The compound can be derived by the simple stoichiometry of the following equation, though the observed structure implies a concurrent isomerization of the 6-irida starting material to the 5-irida analog during the course of the reaction

$$[6\text{-}(\eta^5\text{-}C_5Me_5)IrB_9H_{13}] + \tfrac{1}{2}[(\eta^5\text{-}C_5Me_5)RhCl_2]_2 + 2\ \text{Base} \longrightarrow$$

$$(\eta^5\text{-}C_5Me_5)Rh(\eta^5\text{-}C_5Me_5)HIrB_9H_{10}] + 2[\text{Base H}]^+Cl^-$$

The Rh–H–Ir and B–H–B bridges in the quadrilateral open face were detected by NMR spectroscopy. It will be noted that, since both $\{Rh(\eta^5\text{-}C_5Me_5)\}$ and $\{Ir(\eta^5\text{-}C_5Me_5)\}$ are isolobal with $\{BH\}$ and all are expected to provide two electrons towards the cluster bonding, the dimetalla product is a 24 electron $(2n + 2)$ 11-vertex species. One might therefore have expected a *closo* configuration; however, the product has neither a *closo* nor a regular *nido* structure

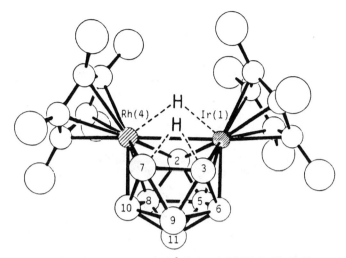

Figure 6.13. The molecular structure of $[(\eta^5\text{-}C_5Me_5)_2RhHIrB_9H_{10}]$. Due to crystallographic disorder there is a 50:50 occupancy factor between B(2) (as shown) and its equivalent position [designated B(2')] over the Rh(1)Ir(4)B(3)B(7) face. Likewise the two metal positions are not assignable between Rh and Ir and there is a similar interchange between C_5Me_5 rotamers [24].

but again adopts an *isonido* configuration formed by notional removal of a 4-connected vertex from the previously mentioned Stone *isocloso* structure [20], as illustrated in Figure 6.14. The new *isonido* geometry can be generated by distorting two overlapping "diamond" lozenges (in the upper front portion of the *closo* cluster) into a "square" face. It is as if bis-protonation of the parent *closo*-$B_{11}H_{11}^{2-}$ dianion analog has opened up the quadrilateral face, an occurrence for which there is some precedence among various other borane and heteroborane clusters.

The unexpected *isonido* structure of $[(\eta^5\text{-}C_5Me_5)_2RhHIrB_9H_{10}]$ discussed in the preceding paragraph raises the question as to whether other 11-vertex $(2n + 2)$ heteroborane clusters, which in the past have reasonably been assumed to have a *closo* octadecahedral structure, do instead have a more open structure.

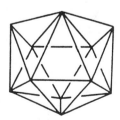  

Figure 6.14. Some *closo*- and *nido*-geometries for 11-vertex clusters.

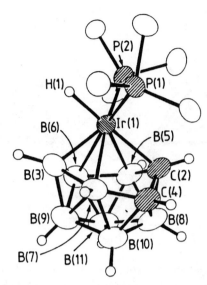

Figure 6.15. The structure of $[1,1,1\text{-}H(PPh_3)_2\text{-}isonido\text{-}1,2,4\text{-}IrC_2B_8H_{10}]$ as determined by X-ray crystallography [27]. A notable feature is the quadrilateral open face Ir(1)C(2)C(4)B(7) with the Ir(1)–C(4) distance of 277.9(8) pm being clearly nonbonding.

A good example is the iridadicarbaborane $[(PPh_3)_2HIrC_2B_8H_{10}]$ which was first reported in 1976 [25] and which has generally been assumed to be a 24-electron *closo* cluster [26]. In fact, we have recently found [27] that the compound has the *isonido* structure illustrated in Figure 6.15. The "diamond→square" relationship between the two structures is emphasized in Figure 6.16. Several other examples of 24-electron $(2n + 2)$ 11-vertex *isonido*-clusters are now emerging such as the $\{(arene)RuC_2B_8\}$ clusters exemplified by $[1\text{-}(\eta^6\text{-}MeC_6H_4\text{-}p\text{-}Pr^i)\text{-}2,4\text{-}Me_2\text{-}1,2,4\text{-}RuC_2B_8H_8]$ [28] and $[1\text{-}(\eta^6\text{-}C_6Me_6)\text{-}1,2,5\text{-}RuC_2B_8H_9\text{-}4\text{-}Br]$ [29].

Figure 6.16. The relationship between the "expected" *closo* structure and the "unexpected" *isonido* structure found for $[1,1,1\text{-}H(PPh_3)_2\text{-}1,2,4\text{-}IrC_2B_8H_8]$ [27].

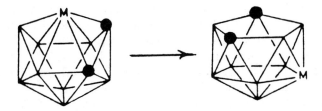

A more drastic structural modification of a 24-electron 11-vertex structure can be effected by thermolytic rearrangement of an $\{OsC_2B_8\}$ cluster at 400°C for 14 min. Under these conditions $[1-(\eta^6-C_6Me_6)-2,4-Me_2-1,2,4-OsC_2B_8H_8]$ quantitatively isomerizes to the orange air-stable *nido*-cluster $[2-(\eta^6-C_6Me_6)-8,10-Me_2-2,8,10-OsC_2B_8H_8]$ as indicated in the subjoined scheme [30]. The crystal structure of this astonishing product (Figure 6.17) indicates that it has adopted the *nido*-configuration of $B_{11}H_{14}^-$ with a five-membered $\overline{BCBCB}$ open face in which the two carbon atoms are no longer contiguous; more surprisingly, the Os atom has moved to a non-open-face five-connected cluster position well away from the two carbon atoms. This is clearly fundamentally different from the *closo* structure expected from the formal $(2n + 2)$ cluster-electron count and cannot be rationalized in terms of "slipped" *closo* or other distortions. The behavior is entirely novel and was certainly not anticipated on the basis of currently accepted theories.

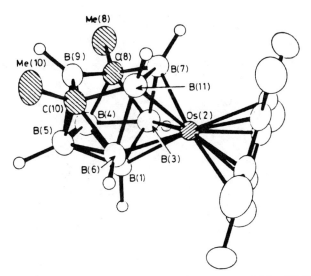

Figure 6.17. The molecular structure of *nido*-$[2-(\eta^6-C_6Me_6)-8,10-Me_2-2,8,10-OsC_2B_8H_8]$ [30]. Note the 5-connected osmium atom away from the open face, the noncontiguous carbon atoms, and the absence of bridging hydrogen atoms around the five-membered $\overline{BCBCB}$ open face.

Finally, I should like to describe our recent study of an unprecedented fluxional isomerism that is both specifically defined and extremely facile. This remarkable behavior was discovered in two new *closo* 10-vertex metalladicarba-boranes of ruthenium and rhodium, namely, $[2-(\eta^6-C_6Me_6)-closo-2,1,6-RuC_2B_7H_9]$ and $[2-(\eta^5-C_5Me_5)-closo-2,1,6-RhC_2B_7H_9]$ [31]. The compounds were obtained as air-stable, yellow crystalline solids in more than 60% yield by the reaction of *arachno*-$[4,6-C_2B_7H_{12}]^-$ with either $[(\eta^6-C_6Me_6)RuCl_2]_2$ or $[(\eta^5-C_5Me_5)RhCl_2]_2$ in $CHCl_3$ solution. X-ray diffraction analysis of the 3,9-dideuteriated isotopomer of the ruthenium compound gave the structure shown in Figure 6.18. The cluster is a straightforward bicapped square antiprism, as expected for a 10-vertex cluster based on *closo*-$1,6-C_2B_8H_{10}$, in which the $\{BH(2)\}$ vertex has been subrogated by the isolobal $\{Ru(\eta^6-C_6Me_6)\}$ unit. Proton and ^{11}B-NMR spectroscopy on solutions in CD_2Cl_2 at room temperature indicated fluxionality, but with the scrambling occurring only within each of selected site pairs. Low-temperature NMR ($< -60°C$) revealed seven separate ^{11}B and nine separate ^{1}H resonances for the $\{C_2B_7H_9\}$ unit consistent with the solid-state structure. At intermediate temperatures, interchange occurred within each of the pairs CH(1)/CH(6), BH(5)/BH(10), and BH(4)/BH(7) with $\Delta G^{\ddagger}_{272} \cong 45\,\text{kJ mol}^{-1}$, whereas each of BH(3), BH(8), and BH(10) remained in (or was converted to) a chemically identical position throughout the fluxionality.

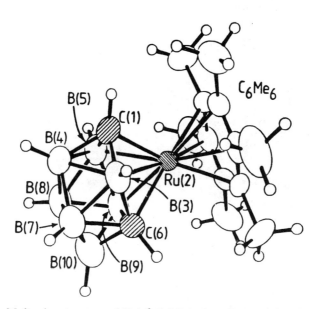

Figure 6.18. Molecular structure of $[2-(\eta^6-C_6Me_6)-closo-2,1,6-RuC_2B_7H_9-3,9-d_2]$ [31]. The compound crystallizes in space group $P\bar{1}$ with two molecules per unit cell, the crystallographic inversion center being achieved by the specific ordering of the two enantiomers (formally the 2,1,6- and 2,1,9-isomers) in the crystal.

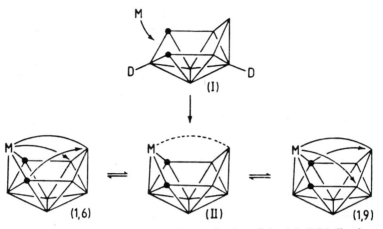

Figure 6.19. Scheme for the formation and isomerization of the 1,6- (1,9-) dicarbaruthenaborane cluster shown in Figure 6.18 (see text).

Variable temperature NMR on specifically deuteriated $[2\text{-}(\eta^6\text{-}C_6Me_6)\text{-}closo\text{-}2,1,6\text{-}RuC_2B_7H_9\text{-}3,9\text{-}d_2]$ confirmed this and showed that other more extensive site-scrambling did not occur on a longer time-scale. The rhodium analog was static at 20°C but showed equivalent fluxionality at higher temperatures, for example 70°C, with $\Delta G^{\ddagger}_{345} \cong 58\,\text{kJ}\,\text{mol}^{-1}$ [31].

These results specifically eliminate the possibility of isomerization by triangular face rotation and define the fluxional isomerization as occurring via two concerted or sequential diamond-square-diamond processes involving the Ru(2)C(1)B(5)B(9) and Ru(2)C(6)B(9)B(10) vertices. These four-atom units both involve the ruthenium atom which is known [19, 28] to induce open character among adjacent faces as discussed above. The process may therefore proceed via quadrilateral-faced *isonido* intermediates possibly linked by a transition state that approximates geometrically to a 6,9-bridged normal *nido* structure as illustrated in Figure 6.19 (structure II). In general, the activation energies for known *closo*-heteroborane cluster isomerizations are much higher than those observed here, so that temperatures above about 330°C or ultraviolet irradiation are required [32]. The specifically defined and extremely facile fluxional isomerism of these *closo*-ruthena- and -rhoda-dicarbadecaboranes is therefore without precedent.

ACKNOWLEDGMENTS

The work described in this paper has been done in collaboration with my colleagues, Dr Robert Greatrex and Dr John Kennedy. We have been greatly assisted by the mass-spectrometric abilities of Mr Darshan Singh, the NMR insights of Dr Xavier Fontaine, and the X-ray crystallographic expertise of Dr

Mark Thornton-Pett. Equally important has been the enthusiastic commitment of the graduate students and post-doctoral assistants mentioned in the detailed literature citations. Part of the work was supported by the European Office of the U.S. Army, and support has also been received throughout from the Science and Engineering Research Council, U.K. The work cited in [27, 29, 31] was done in collaboration with colleagues at the Institute of Inorganic Chemistry, Czechoslovak Academy of Sciences, 250 68 Řež near Prague, Czechoslovakia.

REFERENCES

1. Greatrex, R., Greenwood, N. N., Millikan, M. B., Rankin, D. W. H. and Robertson, H. E., *J. Chem. Soc., Dalton Trans.*, 2335 (1988).
2. Gaines, D. F. and Schaeffer, R., *Proc. Chem. Soc.*, 267 (1963); *Inorg. Chem.*, **3**, 438 (1964); Lutz, C. A., Philips, D. A. and Ritter, D. M., *Inorg. Chem.*, **3**, 1191 (1964); Collins, A. L. and Schaeffer, R., *Inorg. Chem.*, **9**, 2153 (1970); Leach, J. B., Onak, T., Spielman, J., Rietz, R. R., Schaeffer, R. and Sneddon, L. G., *Inorg. Chem.*, **9**, 2170 (1970); Clouse, A. O., Moody, D. C., Rietz, R. R., Rosebury, T. and Schaeffer, R., *J. Am. Chem. Soc.*, **95**, 2496 (1973).
3. Williams, R. E., *Adv. Inorg. Chem. Radiochem.*, **18**, 67 (1976); Wade, K., *Adv. Inorg. Chem. Radiochem.*, **18**, 1 (1976).
4. Mangion, M., Long, J. R., Clayton, W. R. and Shore, S. G., *Cryst. Struct. Commun.*, **4**, 501 (1975); Mangion, M., Hertz, R. H., Denniston, M. L., Long, J. R., Clayton, W. R. and Shore, S. G., *J. Am. Chem. Soc.*, **98**, 449 (1976).
5. Greatrex, R., Greenwood, N. N. and Waterworth, S., *J. Chem. Soc., Chem. Commun.*, 925 (1988).
6. Greatrex, R., Greenwood, N. N., Rankin, D. W. H. and Robertson, H. E., *Polyhedron*, **6**, 1849 (1987).
7. Huffman, J. C., Ph.D. Thesis, Indiana University, 1974.
8. McKee, M. L. and Lipscomb, W. N., *Inorg. Chem.*, **20**, 4442 (1981).
9. Dain, C. J., Downs, A. J., Laurenson, G. S. and Rankin, D. W. H., *J. Chem. Soc., Dalton Trans.*, 472 (1981).
10. Attwood, M. D., Greatrex, R. and Greenwood, N. N., *J. Chem. Soc., Dalton Trans.*, 385 (1989).
11. Greatrex, R., Greenwood, N. N. and Potter, C. D., *J. Chem. Soc., Dalton Trans.*, 2435 (1984); 81 (1986).
12. Attwood, M. D., Greatrex, R. and Greenwood, N. N., *J. Chem. Soc., Dalton Trans.*, 391 (1989).
13. Greenwood, N. N. and Greatrex, R., *Pure Appl. Chem.*, **59**, 857 (1987).
14. Cranson, S. J., Davies, P. M., Greatrex, R., Rankin, D. W. H. and Robertson, H. E., *J. Chem. Soc., Dalton Trans.*, 101 (1990).
15. La Prade, M. D. and Nordman, C. E., *Inorg. Chem.*, **8**, 1669 (1969).
16. Odom, J. D. and Zozulin, A. J., *Inorg. Chem.*, **20**, 3740 (1981).
17. Bown, M., Greenwood, N. N. and Kennedy, J. D., *J. Organomet. Chem.*, **309**, C67 (1986); Bown, M., Fowkes, H., Fontaine, X. L. R., Greenwood, N. N., Kennedy, J. D., MacKinnon, P. and Nestor, K., *J. Chem. Soc., Dalton Trans.*, 2597 (1988).

18. Wilczynski, E. and Sneddon, L. G., *Inorg. Chem.*, **18**, 864 (1979); Zimmerman, G. J., Hall, L. W. and Sneddon, L. G., *Inorg. Chem.*, **19**, 3642 (1980); Gromek, J. M. and Donahue, J., *Cryst. Struct. Commun.*, **10**, 871 (1981).

19. Bown, M., Fontaine, X. L. R., Greenwood, N. N., Kennedy, J. D. and MacKinnon, P., *J. Chem. Soc., Chem. Commun.*, 817 (1987).

20. Atfield, M. J., Howard, J. A. K., Jelfs, A. N. de M., Nunn, C. M. and Stone, F. G. A., *J. Chem. Soc., Chem. Commun.*, 918 (1986).

21. Bown, M., Fontaine, X. L. R., Greenwood, N. N., Kennedy, J. D. and Thornton-Pett, M., *J. Chem. Soc., Dalton Trans.*, 1169 (1987).

22. Kennedy, J. D., *Prog. Inorg. Chem.*, **32**, 519 (1984); **34**, 211 (1986) and references therein.

23. Getman, T. D., Krause, J. A. and Shore, S. G., *Inorg. Chem.*, **27**, 2398 (1988).

24. Nestor, K., Fontaine, X. L. R., Greenwood, N. N., Kennedy, J. D. and Thornton-Pett, M., *J. Chem. Soc., Chem. Commun.*, 455 (1989).

25. Jung, C. W. and Hawthorne, M. F., *J. Chem. Soc., Chem. Commun.*, 1499 (1976); *J. Am. Chem. Soc.*, **102**, 3024 (1980).

26. Baker, R. T., *Inorg. Chem.*, **25**, 109 (1986).

27. Nestor, K., Fontaine, X. L. R., Greenwood, N. N., Kennedy, J. D., Plešek, J., Štíbr, B. and Thornton-Pett, M., *Inorg. Chem.*, **28**, 2219 (1989).

28. Bown, M., Fontaine, X. L. R., Greenwood, N. N., Kennedy, J. D. and Thornton-Pett, M., *Organometallics*, **6**, 2254 (1987).

29. Unpublished collaborative work from the Leeds and Řež groups, 1987–1988.

30. Bown, M., Fontaine, X. L. R., Greenwood, N. N., Kennedy, J. D. and Thornton-Pett, M., *J. Chem. Soc., Chem. Commun.*, 1650 (1987).

31. Bown, M., Jelínek, T., Štíbr, B., Heřmánek, S., Fontaine, X. L. R., Greenwood, N. N., Kennedy, J. D. and Thornton-Pett, M., *J. Chem. Soc., Chem. Commun.*, 974 (1988).

32. See e.g., Grimes, R. N., in: *Comprehensive Organometallic Chemistry*, (Wilkinson, G., Stone, F. G. A. and Abel, E., Eds.) Part 1, Chap. 5, pp. 480–485, Pergamon, Oxford, 1982 and references cited therein; for more recent work on carbaboranes without metal vertices, see e.g., Abou, Z. J., Soltis, M., Oh, B., Siwap, G., Banuelos, T., Nam, W. and Onak, T., *Inorg. Chem.*, **24**, 2363 (1985), and related work referred to therein.

CHAPTER 7

Electron Deficiency Aspects of the Small Carboranes

T. ONAK and K. FULLER

7.1. INTRODUCTION

The label "electron deficient" has been applied to group III element compounds probably more often than to other compounds that are electronically unsaturated. A definitive opus on this subject appeared nearly two decades ago [1], and the greatest portion of this work was devoted to compounds of boron. The material presented below concentrates primarily on one aspect of polyborane chemistry, namely the electron deficiency nature and consequent chemistry of the common small $closo$ carboranes, 1,5-$C_2B_3H_5$, 1,6-$C_2B_4H_6$, and 2,4-$C_2B_5H_7$ (Figure 7.1) [2] and related species.

Many electron deficient (electron unsaturated) compounds have in common that they contain too few valence electrons to make all their two-center bonds electron-pair bonds; that is, they involve three-center, or other multicenter, bonding [1]. The label "electron deficient" is also used to refer to atoms at the positive ends of polar bonds, atoms which act as sites for nucleophilic attack by electron rich species; and that a substance truly deficient in electrons might be

Electron Deficient Boron and Carbon Clusters, Edited by George A. Olah, Kenneth Wade, and Robert E. Williams.
ISBN 0-471-52795-5 © 1991 John Wiley & Sons, Inc.

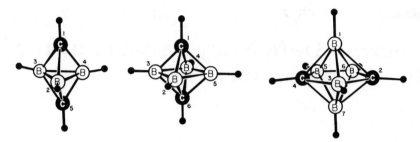

Figure 7.1. Structures of the *closo* compounds $1,5\text{-}C_2B_3H_5$, $1,6\text{-}C_2B_4H_6$, and $2,4\text{-}C_2B_5H_7$.

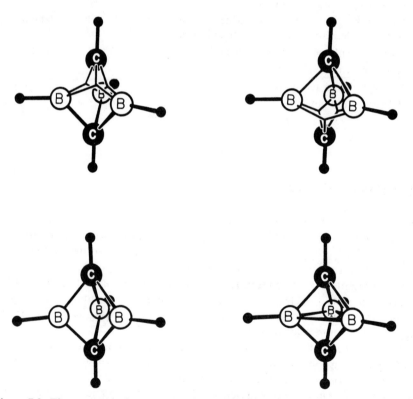

Figure 7.2. Three canonical structures of *closo*-$1,5\text{-}C_2B_3H_5$; the first two depict the six pairs of cage electrons by a combination of three- and two-center bonds; the third depicts the cage electrons as six "classical" two-center bonds only; the fourth is a ball and stick structure correctly depicting the symmetry of the molecule in which the sticks attaching the cage atoms do not formally represent the six pairs of electrons holding together the five cage atoms.

expected to function as an electron (or electron-pair) acceptor, that is, as a Lewis acid. Consider, for example, one of the simplest known small carboranes, *closo*-1,5-$C_2B_3H_5$. One can specify its bonding structure in terms of a number of combined canonical forms contributing to the actual D_{3h} structure, for example, (1) with three 3-center and three 2-center bonds for the framework bonds (Figure 7.2) (where $IED_{total} = IED_{tc} = 3$) [3], or (2) with classical 2-center bonds only, which then leaves three vacant boron orbitals (Figure 7.2) (where $IED_{total} = IED_{vo} = 3$) [3], or (3) as some (partial) combination of these (not shown). But in both (1) and (2), and with any structure that could be imagined to fall in category (3), the IED_{total} is the same, and in this case, three. With the next larger carborane, *closo*-1,6-$C_2B_4H_6$, it is not possible to draw a single reasonable canonical form which represents the known D_{4h} symmetry of this compound (Figure 7.3). As one proceeds to still "higher" carboranes this type of situation is very often encountered, and it may be advisable to draw a number of "resonance" forms to illustrate a given structural point about the compound. But no matter which canonical form is drawn, the IED stays the same for the same carborane, and in the case of 1,6-$C_2B_4H_6$ this is a value of 4 (i.e., $IED = IED_{tc} + IED_{vo} = 4$). The virtual vacant boron orbitals implied by the degree of electron unsaturation in the carboranes (e.g., the present IED concept), and construed from the various bonding schemes, play an important part in conceptualizing the acceptor chemistry of the polyhedral systems mentioned in the remainder of this chapter.

7.2. REACTIONS WITH AMINES AND PHOSPHINES

Both of the closo carboranes, 1,5-$C_2B_3H_5$ and 1,6-$C_2B_4H_6$, react rapidly with trimethylamine and trimethylphosphine to produce 1:1 adducts [5,6]. In the case of $(CH_3)_3N:C_2B_3H_5$ the structure of the adduct has not been fully established and, as suggested by the nature of the trimethylphosphine adduct

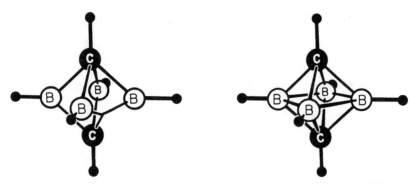

Figure 7.3. One (left) of many canonical forms of the D_{4h} closo-1,6-$C_2B_4H_6$ (right); seven pairs of the 26 bonding electrons are involved in bonding together the framework atoms in a tetragonal bipyramidal (nearly octahedral) geometry.

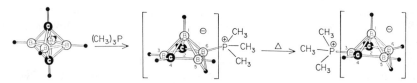

Figure 7.4. Reaction of *closo*-1,6-$C_2B_4H_6$ with trimethylphosphine and subsequent rearrangement of the initial product, 5-$(CH_3)_3$P-*nido*-2,4-$C_2B_4H_6$ to the 3-$(CH_3)_3$P-isomer.

properties, may well be polymeric in nature. Not unrelated to this chemistry is the surprising result of an ab initio calculational study [7] on the ammonia/1,5-$C_2B_3H_5$ system in which it is concluded that a 1:1 bonded adduct of these two molecules is not energetically favored over the dissociated molecules.

The structure of the $(CH_3)_3L:C_2B_4H_6$ (L = N,P) adduct shows that the *closo* cage of the startiang material, 1,6-$C_2B_4H_6$, has opened to a *nido* cluster (Figure 7.4) [6]. The $(CH_3)_3L$- group attaches, initially, to the 5-position of the pentagonal pyramid, but this adduct rearranges to an isomer in which $(CH_3)_3L-$ becomes attached to the basal boron position with more symmetry elements, the 3-position.

Trimethylamine does not react at all with the parent 2,4-$C_2B_5H_7$, but does form adducts with halo derivatives of this system (vide infra).

All of the small carboranes are found to react rapidly with primary and secondary amines to give cage cleavage products, presumably via initial Lewis base attack at a boron site of the carborane, for example [6c]

$$2,5\text{-}C_2B_5H_7 \xrightarrow{R_2NH} (Me_2NBHMe)_n \ (n=1,2) + (Me_2N)_2BH + Me_2NH:BH_3$$

$$+ (Me_2NBH_2)_2 + Me_2NBMe_2 + (Me_2N)_2BMe$$

7.3. REACTIONS WITH HYDROXYLIC COMPOUNDS

Similar to reactions of the small *closo*-cage carboranes with primary or secondary amines, these same carboranes react with alcohols, or water, to give cage degradation products. With both 1,6-$C_2B_4H_6$ and 2,4-$C_2B_5H_7$, the reaction is not nearly as fast as the reactions with primary and secondary amines, and they yield products of the type

$$1,6\text{-}C_2B_4H_6 \xrightarrow{CH_3OH} [(CH_3O)_2B]_2CH_2 + (CH_3O)_2BCH_3 + (CH_3O)_3B + H_2$$

The chief reason for the slower reaction of the hydroxylic compounds, as compared to the reactions with primary or secondary amines, is most probably that the nucleophilicity of the amines toward the cage compounds, under the conditions of the experiments, is greater than that of the hydroxyl compounds. In the case of 1,5-C$_2$B$_3$H$_5$ the alcoholysis and hydrolysis is much more rapid than with C$_2$B$_4$H$_6$ or C$_2$B$_5$H$_7$, and careful control of the reaction leads to the nearly quantitative formation of an open-chain B–C–B–C–B unit with alternating boron and carbon atoms [8]

$$C_2B_3H_5 + 5\,ROH \rightarrow (RO)_2BCH_2B(OR)CH_2B(OR)_2 + 3H_2$$

7.4. THE ACTION OF (CH$_3$)$_3$L (L = N,P) ON *B*-Cl-2,4-C$_2$B$_5$H$_6$, AND SUBSEQUENT REACTION OF THE ADDUCT(S) WITH BCl$_3$ OR CH$_2$Cl$_2$

As mentioned above, a marked enhancement of *closo*-2,4-C$_2$B$_5$H$_7$ reactivity with Me$_3$L: is observed upon placement of halogen atom on a boron atom [9]. The three isomers of *B*-Cl-2,4-C$_2$B$_5$H$_6$ (*B* = 1, 3, 5) all form 1:1 adducts with Me$_3$N and with Me$_3$P. Trimethylamine can be slowly removed from the (CH$_3$)$_3$N:ClC$_2$B$_5$H$_6$ adduct by using CH$_2$Cl$_2$

$$(CH_3)_3N:ClC_2B_5H_6 + CH_2Cl_2 \rightarrow ClC_2B_5H_6 + [(CH_3)_3NCH_2Cl]Cl$$

Attempts to remove the Me$_3$L: (L = N,P) from the adducts with the use of BCl$_3$ were unsuccessful; instead, Cl$^-$ is quantitatively removed to give *B*-Me$_3$L$^+$-*closo*-2,4-C$_2$B$_5$H$_6$ (Figure 7.5). With both the 5- and 3-Cl-carborane adducts the respective 5- and 3-Me$_3$L$^+$-*closo*-2,4-C$_2$B$_5$H$_6$ products are formed. However, when starting with the 1-Cl- isomer, the reaction sequence (i.e., adding trimethylamine to the chlorocarborane followed by the addition of BCl$_3$ to the resulting adduct) gives rise to the rearranged 3-Me$_3$L$^+$-*closo*-2,4-C$_2$B$_5$H$_6$. The nature of the *B*-Me$_3$L-*B*-Cl-closo-2,4-C$_2$B$_5$H$_6$ adducts have been of considerable interest and it may well be that they are the cage counterparts of Meisenheimer complexes found in aromatic chemistry. In the case of benzene, the introduction of electron withdrawing groups onto the ring makes it easier to carry out nucleophilic displacement reactions; and in some cases the intermediate in which both the entering group and incipient-leaving group are attached to the same carbon can be isolated (Figure 7.6). In an analogous fashion the "electron deficient" cage of C$_2$B$_5$H$_7$ may make it easier to form a *closo*-cage adduct in which both entering and eventual leaving groups are attached to the same boron (Figure 7.7). In this respect Gaussian-86 STO-3G calculations on the three *B*-Me$_3$L-*B*-Cl-*closo*-2,4-C$_2$B$_5$H$_6$ adducts are illuminating. The search for an energy minimum for the 1-Me$_3$L-1-Cl-*closo*-2,4-C$_2$B$_5$H$_6$ isomer has resulted instead in the cage opening to a *nido* system. On the other hand energy minima for the *B*-Me$_3$L-*B*-Cl-*closo*-2,4-C$_2$B$_5$H$_6$ (*B* = 3,5) isomers have been found.

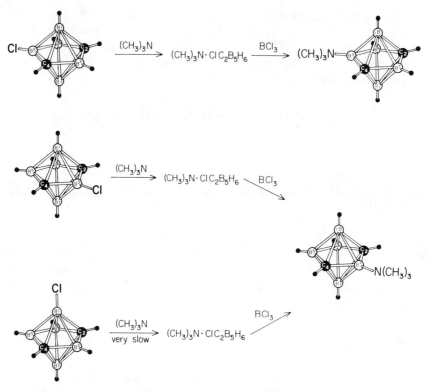

Figure 7.5. Reaction of *B*-chloro derivatives of *closo*-C$_2$B$_5$H$_7$ with trimethylamine and subsequently with boron trichloride. The two *closo* structures on the right are cations; BCl$_4^-$ is the anion.

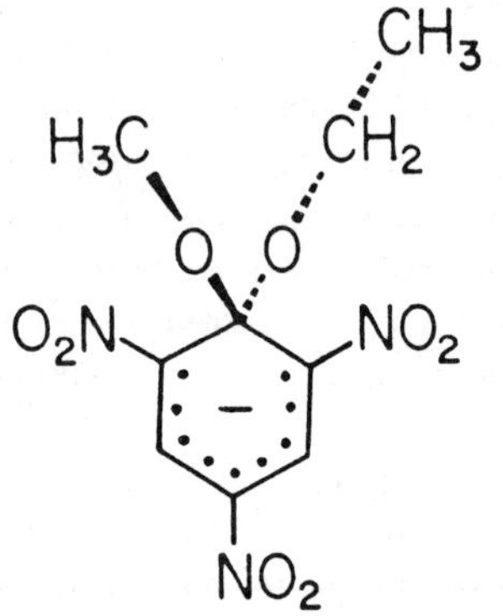

Figure 7.6. Meisenheimer complex isolated from the reaction of a substituted aromatic system with a nucleophile.

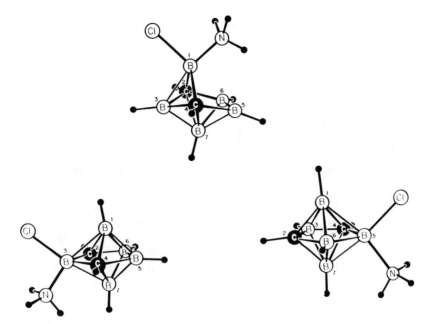

Figure 7.7. Meisenheimer analogs proposed for the trimethylamine adducts of B-Cl-$C_2B_5H_6$. Gaussian-86 STO-3G optimization of the adduct formed from 1-Cl-$C_2B_5H_6$ is not stable in the *closo* cage configuration.

7.5. SUBSTITUENT EFFECTS ON ELECTRON DEFICIENT *closo*-2,4-C₂B₅H₇ RELATIVE STABILITIES OF CARBORANE DERIVATIVES, EXPERIMENTAL, AND CALCULATIONAL

The effect of substituents on the relative stability of various isomeric mono B-halo derivatives (with their potential pi-donating abilities) on the electron deficient *closo*-2,4-$C_2B_5H_7$ has been assessed by both experimental and calculational approaches. The experimental data has been gathered from thermal equilibration results on various isomer sets [10]. Application of the MNDO semi-empirical MO method to these same isomer sets has been used to derive relative energies [11]. Both experimental results and MNDO relative energy assessments give the same order of stabilities for each of the following isomer sets:

$$3\text{-Cl-} > 5\text{-Cl-} > 1\text{-Cl-} \quad \text{(exp. and MNDO)}$$

$$3\text{-Br-} > 5\text{-Br-} > 1\text{-Br-} \quad \text{(exp. and MNDO)}$$

$$5\text{-I-} > 3\text{-I-} > 1\text{-I-} \quad \text{(exp. and MNDO)}$$

Similar correspondence is found between experimental results and MNDO calculations for the B,B'-dihalo (halogen = Cl, I) systems

$3,5$-Cl_2- > $1,3$-Cl_2- > $5,6$-Cl_2- > $1,5$-Cl_2- > $1,7$-Cl_2- (exp. and MNDO)

$5,6$-I_2- > $3,5$-I_2- > $1,5$-I_2- > $1,3$-I_2- > $1,7$-I_2- (exp. and MNDO)

Only for the dibromo system is there a minor discrepency

$3,5$-Br_2- > $5,6$-Br_2- > $1,3$-Br_2- > $1,5$-Br_2- > $1,7$-Br_2- (exp.)

$3,5$-Br_2- > $1,3$-Br_2- > $5,6$-Br_2- > $1,5$-Br_2- > $1,7$-Br_2- (MNDO)

Ab initio calculations on the three B-Cl-*closo*-$2,4$-$C_2B_5H_6$ isomers at the Gaussian-86 STO-3G level confirm that the wrong order of stability is predicted, using this basis set, for the 3- and the 5- isomers [12]. We find that this incorrect order of stabilities continues at the 3-21G level but begins to turn around at the 6-31G level (Figure 7.8). However, as the quality of the basis set is increased to this point, the 1-isomer, although continuing to be the least stable during this series of calculations, shifts to relative energy values that are increasingly disparate. Adding configuration interaction with double substitutions shifts the relative energy values closer to those experimentally observed, but a better correspondence to the experimental [10a, c] relative ΔH values is experienced upon incorporating d-orbitals for the nonhydrogen atoms (i.e., 6-31G* level). At the CID/6-31G*//6-31G and MP3/6-31G*//6-31G levels the calculated relative enthalpies are almost those, within experimental error, obtained from experimental equilibration results [12b].

These results could well suggest that all future ab initio calculations on

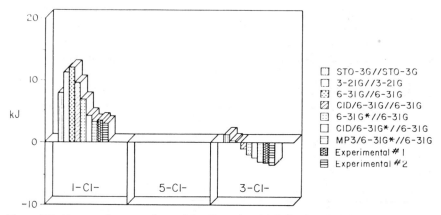

Figure 7.8. Bar-graph comparison of the Gaussian-86 derived enthalpies of the three B-Cl-*closo*-$2,4$-$C_2B_5H_6$ isomers while varying the orbital basis set. The energy value for the 5-Cl- isomer is set to zero in each case.

substituted carboranes be conducted at least to the CID/6-31G*//6-31G or MP3/6-31G*//6-31G level for useful information about relative stabilities. Unfortunately, this is not practical for $C_2B_5H_7$ derivatives with bromine or iodine because of computer time constraints.

7.6. REACTIONS WITH METAL ALKYLS AND HYDRIDES

A strong base such as butyl-lithium will abstract the carbon-attached cage hydrogens from the small *closo*-carboranes $1,6-C_2B_4H_6$ and $2,4-C_2B_5H_7$ [13]. The yield of the conjugate base anion from this reaction is not always quantitative, leading to the speculation that some butyl anion attack at a boron site may be occurring. However, to date, no products of this sort have been reported. On the other hand, action of hydride, and Me_3L: $(L = N,P,$ vide supra$)$ on some of the small *closo*-carboranes lead to products which highly suggest Lewis base attack at a boron site. The action of NaH, or LiH, on $1,6-C_2B_4H_6$ or the action of NAH on either 3- or $5-Me_3N^+-nido-2,4-C_2B_4H_6$ gives the parent $nido-2,4-C_2B_4H_7^-$ ion (Figure 7.9) [6b]. The *closo*-$2,4-C_2B_5H_7$ does not react

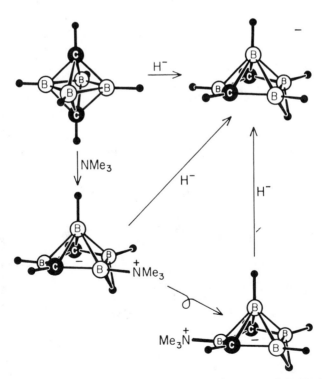

Figure 7.9. Reaction of *closo*-$1,6-C_2B_4H_6$ and $B-Me_3N^+-nido-2,4-C_2B_4H_6^-$ isomers with hydride.

with hydride but the 2,3-$C_2B_5H_7$ isomer (as a diethyl derivative) reacts with the hydride portion of $Li(BEt_3)H$ to form a *nido*-2,4-$C_2B_5H_8^-$ ion derivative [14].

7.7. REACTIONS WITH AMIDE IONS

1,6-$C_2B_4H_6$ appears to be rather impervious to attack by amides whereas 2,4-$C_2B_5H_7$ reacts with $Li[NR_2]$ in acetonitrile to eventually yield, in a quantitative fashion, *nido*-2,4-$C_2B_4H_7^-$. Along the way, two intermediates are formed and the proposed structure of one of these [15] is a five-boron cluster (Figure 7.10) which may represent the product of the initial step of the overall conversion.

7.8. HALIDE EXCHANGE REACTIONS

Net front-side displacement reactions involving *B*-halogenated derivatives 2,4-$C_2B_5H_7$ result in the replacement of larger halogens with smaller ones. For example, the reaction of 5-Br-6-Cl-*closo*-2,4-$C_2B_5H_5$ with benzyltriethylammonium chloride in dichloromethane yields 5,6-Cl_2-*closo*-2,4-$C_2B_5H_5$; and reaction of 3,5-I_2-*closo*-2,4-$C_2B_5H_5$ with tetrabutylammonium fluoride in tetrahydrofuran gives 3,5-F_2-*closo*-$C_2B_5H_5$ [16]. The rate of substitution, as the halide ion is varied, corresponds with the expected nucleophilicity trend $F^- > Cl^- > Br^-$ in nonhydroxylic solvents.

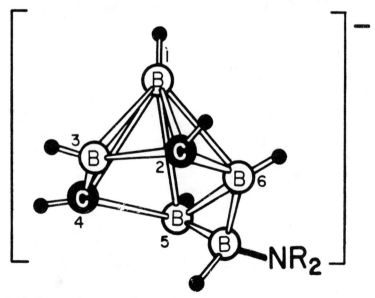

Figure 7.10. Proposed structure for an observed intermediate in the reaction of $LiNR_2$ with *closo*-2,4-$C_2B_5H_7$.

ACKNOWLEDGMENT

We thank the National Science Foundation and the MBRS-NIH (K.F.) for partial support of this study.

REFERENCES AND FOOTNOTES

1. Wade, K. *Electron Deficient Compounds*, Thomas Nelson, Camden, NJ, 1971.

2. Williams, R. E., *Adv. Inorg. Chem. Radiochem.*, **18**, 67 (1976).

3. A concept of IED (index of electron deficiency) can be defined as: $IED_{tc} = s + t$ (where tc = three-center bonds and where s and t are taken from Lipscombs styx topological treatment [4], s = three-center BHB bonds and t = three-center BBB bonds, either open or closed). For the observed stable structures of diborane(6) and pentaborane(9), $IED_{tc} = 2$ for B_2H_6 and $IED_{tc} = 5$ for B_5H_9. An additional way of defining IED is $IED_{vo} = vo$ (where vo = number of vacant 2nd shell orbitals on boron); $IED_{vo} = 1$ for BH_3 and $IED_{vo} = 2$ for $(CH_3)_2B-B(CH_3)_2$. Or, more generally, combining both, $IED_{total} = IED_{tc} + IED_{vo} = (s + t) + vo$. So two BH_3 units taken together, or B_2H_6, would have an IED_{total} of 2 whether it is considered as diborane (with two bridging BHB bonds) or as a collection of two independent BH_3 units with one vo apiece (for a total of two vo). Obviously, reacting B_2H_6, or $2 BH_3$, with H^- (or other electon donors such as other anions or amines, etc.) gives a product with a lower IED.

$$B_2H_6 \xrightarrow{\quad H^- \quad} \qquad B_2H_7^- \xrightarrow{\quad H^- \quad} 2 BH_4^-$$

IED   2 (bridging BHB $s = 2$)   1 (bridging BHB $s = 1$)   0 ($s = 0$)

In the case of carboranes, reacting a *closo* cage compound with H^- (at a boron site) normally produces a *nido* cage structure with a lowered IED, for example

$$closo\text{-}1,6\text{-}C_2B_4H_6 + H^- \quad \rightarrow \quad nido\text{-}2,4\text{-}C_2B_4H_7^-$$

IED 4   (combination of bridging   3 (combination of bridging
bonds and vacant orbitals)   bonds and vacant orbitals)

However, formal addition of two Hs to proceed from a *closo* to a *nido* configuration, or to proceed from a *nido* to an *arachno* configuration does not lower the IED.

$$closo\text{-}C_2B_4H_6 \xrightarrow{\quad 2\,(H) \quad} nido\text{-}C_2B_4H_8$$

IED 4   4

4. Lipscomb, W. N., *Boron Hydrides*, Benjamin, New York, 1963; Lipscomb, W. N., *Boron Hydride Chemistry* (E. L. Muetterties, Ed.), Academic Press, New York, 1975, pp. 39–78.

5. Burg, A. B. and Reilly, T. J., *Inorg. Chem.* **11**, 1962 (1972).

6. (a) Lockman, B. and Onak, T., *J. Am. Chem. Soc.*, **94**, 7923 (1972). (b) Onak, T., Lockman, B. and Haran, G., *J. Chem. Soc., Dalton Trans.*, 2115 (1973); (c) Lew, L., Haran, G., Dobbie, R., Black, M. and Onak, T., *J. Organomet. Chem.*, **111**, 123 (1976).

7. McKee, M. L., *Inorg. Chem.*, **27**, 4241 (1988).

8. Dobbie, R. C., Wan, E. and Onak, T., *J. Chem. Soc., Dalton Trans.*, 2603 (1975).

9. (a) Siwapinyoyos, G. and Onak, T., *Inorg. Chem.*, **21**, 156 (1982); (b) Siwapinyoyos, G. and Onak, T., *J. Am. Chem. Soc.*, **102**, 420 (1980).

10. (a) Abdou, Z. A., Abdou, G., Onak, T. and Lee, S., *Inorg. Chem.*, **25**, 2678 (1986); (b) Takimoto, C., Siwapinyoyos, G., Fuller, K., Fung, A. P., Liauw, L., Jarvis, W., Millhauser, G. and Onak, T., *Inorg. Chem.*, **19**, 107 (1980); (c) Abdou, Z. J., Soltis, M., Oh, B., Siwap, G., Banuelos, T., Nam, W. and Onak, T., *Inorg. Chem.*, **19**, 2363 (1985). (d) Ng, B., Onak, T., Banuelos, T., Gomez, F. and DiStefano, E. W., *Inorg. Chem.*, **24**, 4091 (1985).

11. (a) O'Gorman, E., Banuelos, T. and Onak, T., *Inorg. Chem.*, **27**, 912 (1988); (b) Onak, T., O'Gorman, E., Banuelos, T., Alfonso, C. and Yu, M., unpublished results.

12. (a) Beltram, G. A., Jasperse, C., Cavanaugh, M. A. and Fehlner, T. P., *Inorg. Chem.*, **29**, 329 (1990). (b) Onak, T., O'Gorman, E., Banuelos, T., Alfonso, C. and Yu, M., *Inorg. Chem.*, **29**, 335 (1990).

13. (a) Kesting, R. E., Jackson, K. F., Klusmann, E. B. and Gerhart, F. J., *J. Appl. Polym. Sci.*, **14**, 2525 (1970); (b) Williams, R. E., private communication; (c) Olsen, R. R. and Grimes, R. N., *J. Am. Chem. Soc.*, **92**, 5072 (1970); (d) Olsen, R. R. and Grimes, R. N., *Inorg. Chem.*, **10**, 1103 (1971).

14. Beck, J. S., Quintana, W. and Sneddon, L. G., *Organometallics*, **7**, 1015 (1988).

15. Abdou, Z. J., Gomez, F. and Abdou, G., *Inorg. Chem.*, **27**, 3679 (1988).

16. (a) Ng, B., Onak, T. and Fuller, K., *Inorg. Chem.*, **24**, 4371 (1985); (b) Ng, B. and Onak, T., *J. Fluorine Chem.*, **27**, 119 (1985).

CHAPTER 8

Hypho-Dithiaborane Clusters: Links Between Electron Deficient and Electron Precise Cage Systems

S. O. KANG and L. G. SNEDDON

8.1. INTRODUCTION

Most polyhedral boranes have been found to be members of the *closo-*, *nido-* or *arachno-* electronic classes, containing $n + 1$, $n + 2$ or $n + 3$ skeletal electron pairs, respectively (where n = the number of cage atoms). Although many exceptions are known, compounds which fall into these classes generally have predictable structures based on closed deltahedra or deltahedra missing one or two vertices, respectively [1]. There are many fewer members of the *hypho* electronic class, that is, clusters containing $n + 4$ skeletal electrons. Examples include, $B_6H_{10}(PMe_3)_2$ [2], $B_5H_{12}^-$ [3], $B_5H_9L_2$ (L = NR_3 or PR_3) [4,5], B_5H_9L (L = dppm, dppe, tmeda, etc.) [6,7], $B_3H_5 \cdot 3PMe_3$ [5b,8], Me_3N-CB_5H_{11} [9], $B_4H_8(PR_3)_2$ [10], $B_4H_8 \cdot$ tmeda [11], B_6H_{14} [12], $B_7H_{12}Fe(CO)_4^-$ [13], and $C_2B_6H_{13}^{1-}$ [14]. Again, simple electronic counting rules would predict that these compounds should have structures based on deltahedra missing three vertices, but actual structural and bonding patterns in the class have still not been established.

In 1977, Plešek, Heřmánek and Janoušek reported [15] the high yield synthesis of the first dithiaborane, *arachno-*6,8-$S_2B_7H_9$. This cluster, the structure of which is shown in Figure 8.1, contains both two sulfur atoms and two boron-boron bridging hydrogens on the open face and would appear to be a versatile starting material for the generation of a range of new types of hybrid

Electron Deficient Boron and Carbon Clusters, Edited by George A. Olah,
Kenneth Wade, and Robert E. Williams.
ISBN 0-471-52795-5 © 1991 John Wiley & Sons, Inc.

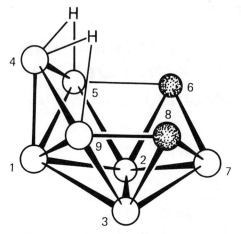

Figure 8.1. Molecular structure of *arachno*-$S_2B_7H_9$.

clusters. Since a cage-sulfur atom is a four skeletal-electron donor (isoelectronic with BH^{2-}) the incorporation of two sulfur atoms into a boron hydride framework necessitates the formation of open cage geometries suggesting that the *arachno*-$S_2B_7H_9$ cage system might also serve as a primary starting material for the syntheses of a wide range of new *hypho*-clusters. Furthermore, because of their increased number of skeletal bonding electrons and, as a result, their greater ability to form two-center two-electron, rather than multicenter, interactions, these *hypho*-clusters might well exhibit structural and bonding features which bridge those observed in electron precise and electron deficient cage systems. We have now investigated these possibilities [16–18] and describe in this chapter the syntheses and structures of a unique series of dithiaborane clusters having formal *hypho* skeletal electron counts. The relationships between these polyhedral boranes and both transition metal clusters and organic cage compounds are also discussed.

8.2. AN 11-VERTEX *hypho*-CLUSTER: *hypho*-5-CH_3-5,11,7,14-$CNS_2B_7H_9$

Our original interest in the chemistry of *arachno*-$S_2B_7H_9$ grew out of our studies of the reactions of nitriles with polyhedral borane anions. In particular, we were interested in the development of new methods by which nitrile insertion reactions could be accomplished. The reactions of neutral boron hydrides with nitriles have usually involved initial electrophilic addition at nitrogen and have resulted in the production of simple base adducts [19, 20]

$$2\,RC\equiv N + B_{10}H_{14} \rightarrow 6{,}9\text{-}(RCN)_2B_{10}H_{12} + H_2 \qquad (8.1)$$

$$RC\equiv N + 6\text{-}SB_9H_{11} \rightarrow 9\text{-}(RCN)\text{-}6\text{-}SB_9H_{11} \qquad (8.2)$$

However, the C≡N bond in a nitrile is strongly polarized and nucleophilic addition at the electropositive carbon can also occur. For example, acyl guanidines react with phenylacetonitrile to produce a 1,3,5-triazine derivative [21], presumably by a reaction sequence involving initial attack at the nitrile carbon.

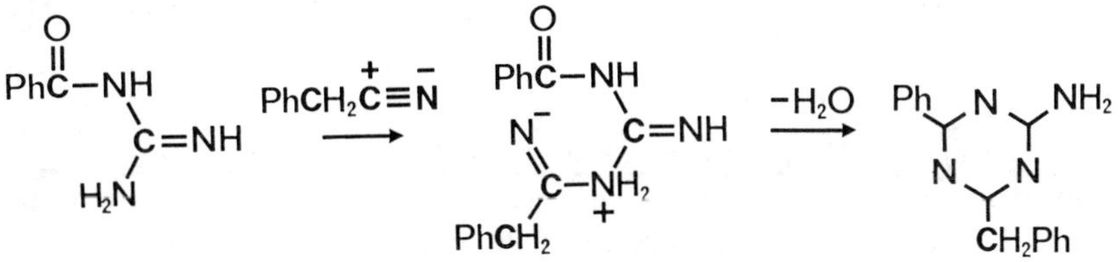

(8.3)

Likewise, it has been shown that mercaptoamines will react at the nitrile carbon to yield, upon elimination of ammonia, cyclic thiazolines [22].

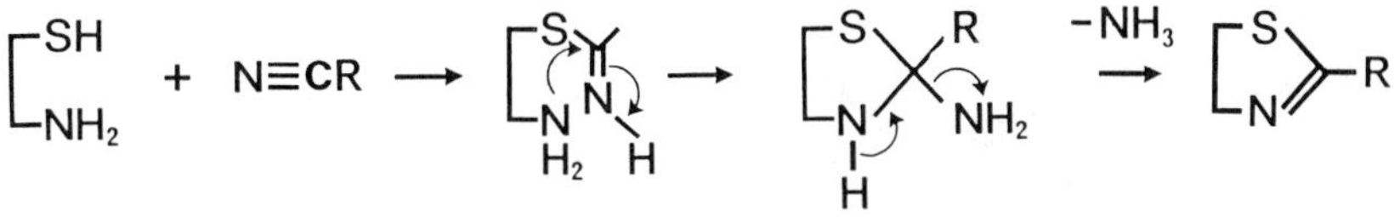

(8.4)

We felt that certain polyhedral borane anions might also be capable of undergoing similar types of cyclizations leading to the formation of new types of clusters derived from nitriles. For example, as outlined below, if a reactant borane anion is strongly nucleophilic then initial addition at a nitrile carbon could form an intermediate which might then undergo intramolecular condensation to yield either CN incorporation or further elimination of ammonia and one-carbon insertion.

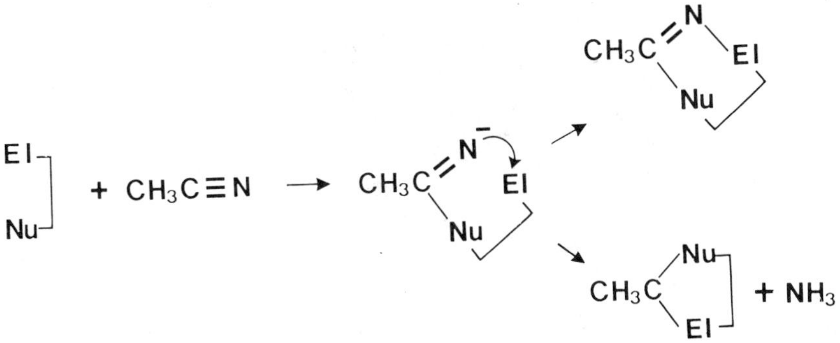

(8.5)

Indeed, we found [16] that when the *arachno*-$S_2B_7H_8^-$ anion is reacted with acetonitrile, reduction of the nitrile and insertion of the resulting imine unit into the cage occurs to yield the 11-vertex *hypho*-cluster *hypho*-5-CH$_3$-5,11,7,14-CNS$_2$B$_7$H$_8$

$$arachno\text{-}S_2B_7H_8^- + MeC{\equiv}N \xrightarrow[\text{2. H}^+]{\text{1. reflux}} hypho\text{-}5\text{-}CH_3\text{-}5,11,7,14\text{-}CNS_2B_7H_9$$

(I)   (8.6)

A single crystal X-ray structural study of **I** revealed that the compound adopts the unique *hypho*-structure shown in Figure 8.2 that contains four different (boron, sulfur, carbon, and nitrogen) main group cage substituents. As can be seen in the figure, hydroboration of the nitrile has occurred to produce an imine which bridges two borons (B4 and B12). The structure thus formed contains three open faces. There are two puckered hexagonal faces on either side of the CN group, (C5, B4, B9, S14, B12, N11) and C5, B4, B1, S7, B12, N11), and

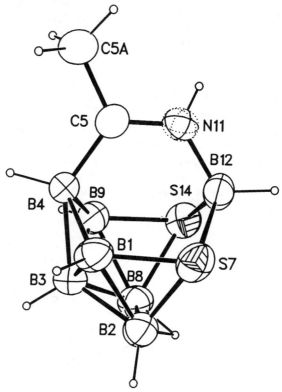

Figure 8.2. ORTEP drawing of the molecular structure of *hypho*-5-CH$_3$-5,11,7,14-CNS$_2$B$_7$H$_9$ **I**.

one planar pentagonal face containing the two sulfur atoms and three borons (B12, S14, B8, B2, S7). A single bridge hydrogen is found between B2 and B8 on the five-membered open face. The isolated boron B12 is four-coordinate sitting between the nitrogen and two sulfur atoms and has approximate tetrahedral geometry.

There are at least two different ways in which this structure may be viewed. First, if the CN group is considered to be part of the cluster, then the cluster would have 30 skeletal electrons and be an 11-vertex $n + 4$ *hypho*-cage system. Such a cluster should then have a structure based on a *closo* 14-vertex polyhedron [23] missing three vertices. Indeed, a structure consistent with that observed for **I** can be derived from a bicapped hexagonal antiprism structure, as shown in Figure 8.3. Thus, removal of three five-coordinate vertices, accompanied by additional bond breaking between positions 1 and 5, and 11 and 14 to allow the CN group to adopt a symmetrical position, generates the observed structure.

Alternatively, **I** can be viewed as hybrid molecule which contains a "nonclassical" thiaborane cluster framework substituted by a "classical" bridging exopolyhedral imine group. Supporting this interpretation are the facts that the C5–N11 bond distance is within the normal CN double bond range and the bond angles about C5 indicate formal sp^2 hybridization. Thus, the imine group would be bound to the thiaborane cage via one boron-carbon sigma bond between B4 and C5 and a nitrogen-boron dative bond between N11 and B12 and as a result would be a three electron donor to the $S_2B_7H_8$ cage framework.

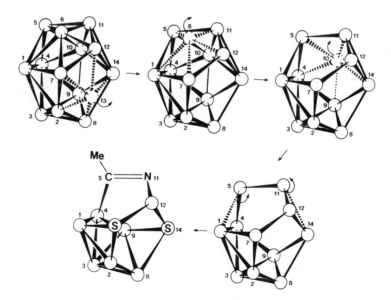

Figure 8.3. Derivation of an 11-vertex *hypho*-structure from a bicapped hexagonal antiprism by removal of three vertices.

The thiaborane cage would contain 26 skeletal electrons and therefore be a 9-vertex *hypho*-cluster. Such a cluster should have a structure based on an icosahedron missing three vertices, which is consistent with the geometry observed for the S_2B_7 framework.

A reasonable reaction sequence leading to the formation of **I** (Figure 8.4) could involve an initial nucleophilic attack of the anion at the acetonitrile carbon, followed by hydroboration of the nitrile bond. This would then generate a strongly nucleophilic imine intermediate. The most positively charged boron in the original S_2B_7 framework would be expected to be the 7-boron atom, because of its location between the two electronegative sulfur atoms. Therefore, cluster condensation driven by the formation of a nitrogen-boron dative bond between N11 and B7 could result in the rearrangement of the cage in the manner shown in the figure to yield the observed final structure containing the unique tetrahedral boron B12.

This type of reaction between a polyhedral borane anion and acetonitrile has not been observed previously; however, similar reactions have been reported in metal cluster chemistry. For example, Kaesz [24] has studied the stepwise reduction of nitriles at a metal cluster and observed that $HFe_3(CO)_{11}^-$, as well as certain other iron carbonyl anions, reacts with acetonitrile to give the

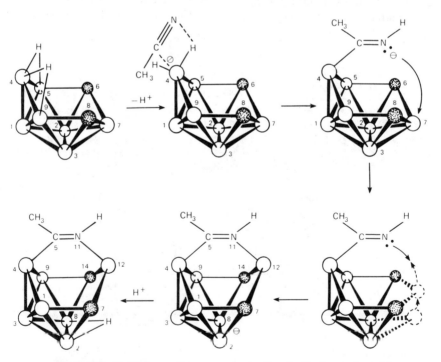

Figure 8.4. Possible reaction sequence leading to the formation of **I**.

$Fe_3(CO)_9(\mu\text{-}\eta^3\text{-MeCNH})^-$ and $Fe_3(CO)_9(\mu\text{-}\eta^3\text{-MeHCN})^-$ anions which can then be protonated to yield neutral compounds:

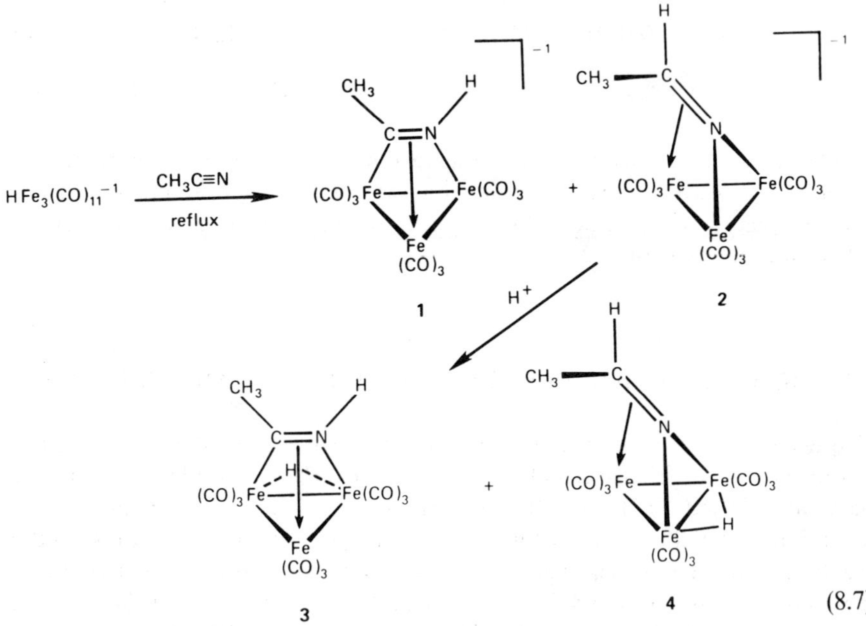

$$
HFe_3(CO)_{11}^{-1} \xrightarrow[\text{reflux}]{CH_3C\equiv N}
$$

(8.7)

It is also of interest to note that **3** can be thermally converted to **4** and that when **4** is treated with H_2 under pressure further reduction of the C–N unit occurs to give the nitrene cluster $(\mu_3\text{-RN})Fe_3(CO)_9$. **I** and its corresponding anion are the polyhedral borane analogues of **1** and **3** above, but no rearrangement of the nitrile such as observed in the formation of **2**, **4** or $(\mu_3\text{-RN})Fe_3(CO)_9$ was found. Further reduction of the C–N unit leading to carbon, rather than nitrogen cage insertion, was observed in the reaction of acetonitrile with the *arachno*-$C_2B_7H_{12}^-$ anion (isoelectronic with $S_2B_7H_8^-$) which gave the three carbon carborane *nido*-6-CH_3-5,6,9-$C_3B_7H_{10}$ **II** (Figure 8.5) as the final product.

$$
arachno\text{-}6,8\text{-}C_2B_7H_{12}^- + MeC\equiv N \xrightarrow{\text{reflux}} nido\text{-}CH_3C_3B_7H_9^- + NH_3 \tag{8.8}
$$

$$
nido\text{-}CH_3C_3B_7H_9^- + H^+ \rightarrow nido\text{-}6\text{-}CH_3\text{-}5,6,9\text{-}C_3B_7H_{10} \tag{8.9}
$$

$$
\textbf{II}
$$

Thus, both the thiaborane and carborane anions are activated for nucleophilic attack at the acetonitrile carbon with the differences in observed reactivity being related to the number of hydrogens available for reduction. In the $S_2B_7H_8^-$ anion there is only one extra hydrogen and the formation of an imine group occurs (equation 8.10), while in the $C_2B_7H_{12}^-$ anion three hydrogens are available for reaction allowing complete reduction resulting in deamination and one-carbon insertion (equation 8.11)

$$arachno\text{-}S_2B_7H_8^- + MeC\equiv N \rightarrow hypho\text{-}MeCNHS_2B_7H_7^- \qquad (8.10)$$

$$arachno\text{-}6,8\text{-}C_2B_7H_{12}^- + MeC\equiv N \rightarrow nido\text{-}MeC_3B_7H_9^- + NH_3 \qquad (8.11)$$

The above results suggest that the reactions of other polyhedral borane anions with nitriles may, depending on the number of bridging hydrogens present, yield a range of new clusters resulting from either carbon or carbon-nitrogen cage-insertions.

8.3. EIGHT AND NINE VERTEX *hypho*-DITHIABORANE CLUSTERS

The results of the nitrile reactions discussed above suggested that polyhedral borane anions might also readily attack other polarized multiple bonds and we have also found [17] that the *arachno*-$S_2B_7H_8^-$ anion will readily react (equation 8.12) with excess acetone at room temperature but, in contrast to the reaction with acetonitrile, cage-degradation (B7) results in the production of a new six-boron anion in excellent yields. Addition of PPN^+Cl^- to the reaction solution led to the isolation of the $PPN^+[S_2B_6H_9]$ salt **III**. However, it was

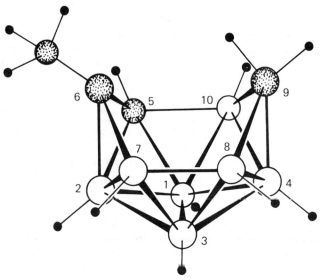

Figure 8.5. Proposed structure for *nido*-6-CH_3-5,6,9-$C_3B_7H_{10}$ **II**.

also found that in situ reaction (8.13) of the anion with excess methyl iodide results in the good yield formation of the dimethylated derivative, **IV**.

$$arachno\text{-}6,8\text{-}S_2B_7H_8^- \xrightarrow[\text{2. PPN}^+\text{Cl}^-]{\text{1. (CH}_3)_2\text{CO}} hypho\text{-}S_2B_6H_9^-\,\text{PPN}^+ \quad (8.12)$$

III

$$hypho\text{-}S_2B_6H_9^- + \text{excess CH}_3\text{I} \rightarrow hypho\text{-}2,3\text{-}Me_2S_2B_6H_8 + \text{NaI} + \text{HI} \quad (8.13)$$

IV

Thiaboranes of the formulas $S_2B_6H_9^-$ or $(CH_3)_2S_2B_6H_8$ would be $n+4$ *hypho* skeletal-electron systems (8 cage atoms and 12 skeletal electron pairs) and would be expected to adopt open cage geometries based on an octadecahedron missing three vertices. Possible *hypho*-structures for both **III** and **IV** that are derived by the removal of one six-coordinate and two five-coordinate vertices are illustrated in Figure 8.6.

The ^{11}B-NMR spectra of **III** and **IV** (Figure 8.7) have several similar features and support the structures proposed in Figure 8.6. Both spectra show four doublets of relative intensities 2:2:1:1 with the resonances at -25.2 ppm in **III**

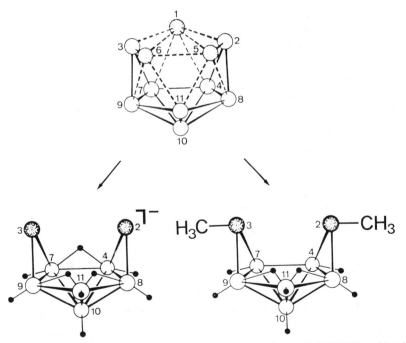

Figure 8.6. Derivation of eight-vertex *hypho*-structures for *hypho*-$S_2B_6H_9^-$ **III** and *hypho*-2,3-$Me_2S_2B_6H_8$ **IV** from an octadecahedron by removal of three vertices.

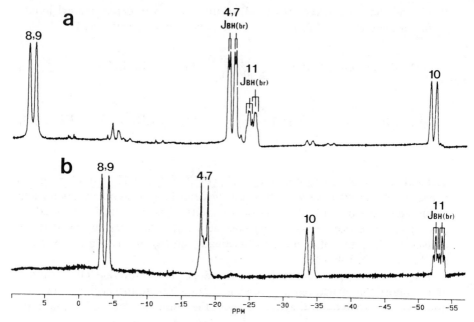

Figure 8.7. 160.5 MHz ^{11}B-NMR spectra of **III** (a) and **IV** (b).

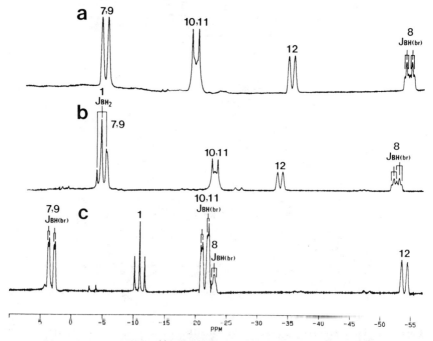

Figure 8.8. 160.5 MHz ^{11}B-NMR spectra of **V** (a), **VI** (b), and **VII** (c).

and -54.2 ppm in **IV**, further split into triplets ($J = \sim 55$ Hz and $J = \sim 65$ Hz, respectively) consistent with their assignment to the boron (B11) in each cage that is bonded to two bridging hydrogens. The presence of an additional bridge hydrogen in **III**, located between borons B4 and B7, is indicated by the doublet fine coupling ($J = \sim 50$ Hz) on the intensity two doublet at -22.2 ppm. The relative chemical shifts of the boron resonances in the spectra of these compounds, and of the analogous resonances arising from the borons in the pyramidal boron fragments of **V**, **VI**, and **VII** discussed below (Figure 8.8), are of special interest since they are highly diagnostic of their structures. Thus, **III** and **VII**, as well as the *hypho*-carborane anion $C_2B_6H_{13}^-$, [14] exhibit similar spectra and are proposed to have structures in which all of the bridging positions on the pentagonal boron face are occupied. Compounds **IV**, **V**, and **VI** also exhibit similar spectral features and are each proposed to have structures in which one boron-boron edge is unsubstituted.

As shown in the ORTEP drawing given in Figure 8.9, a single-crystal X-ray structural determination of **IV** confirms that the compound adopts a structure similar to the predicted *hypho*-geometry in Figure 8.6. The only previously structurally characterized [13] example of an eight-vertex *hypho* polyhedral borane cage is the ferraborane anion $[Fe(CO)_4B_7H_{12}]^-$. The complex was shown to have a structure similar to that observed for **IV**, consisting of a B_6H_9 pentagonal pyramid framework having three bridging hydrogens, as well as BH_3 and $Fe(CO)_4$ groups bridging nonadjacent boron-boron sites on the face. Thus, both **III** and **IV** may also be considered as dibridged derivatives of hexaborane(10).

A reaction analogous to that shown in (8.13) between *hypho*-$S_2B_6H_9$ and

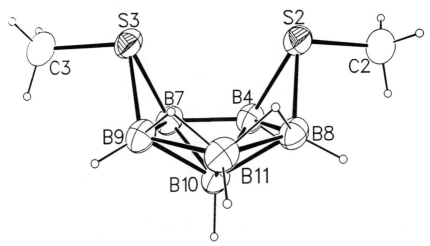

Figure 8.9. ORTEP drawing of the molecular structure of *hypho*-2,3-Me$_2$S$_2$B$_6$H$_8$ **IV**.

diiodomethane resulted in the formation of a single product which was found to be a carbadithiaborane containing a sulfur-bridging methylene group

$$hypho\text{-}S_2B_6H_9^- + CH_2I_2 \rightarrow hypho\text{-}1\text{-}CH_2\text{-}2,5\text{-}S_2B_6H_8 + HI \qquad (8.14)$$

$$\mathbf{V}$$

Based on skeletal electron counting procedures, **V** would be classified as a 9-vertex *hypho*-cage system (26 skeletal electrons). As can be seen in the ORTEP drawing in Figure 8.10, a single-crystal X-ray determination confirmed that the compound has a geometry that is consistent with this prediction. The structure can be derived, as shown in Figure 8.11, from an icosahedron by the removal of three vertices.

Sulfur-sulfur bridged methylenes have not previously been observed in dithiaboranes; however, they have been found in metal-sulfur clusters, including $CH_2S_2Fe_2(CO)_6$ [25] and $CH_2S_2(CpMoS)_2$ [26]. The methylene carbon in **V** is in a distorted tetrahedral environment, with the larger S2–C1–S5 angle (113.9(3)°) being necessitated by the presence of these three atoms in the planar pentagonal B10–B11–S5–C1–S2 ring. The C1–S2 and C1–S5 distances are comparable to those observed for the sulfur-methyl distances observed in **IV**,

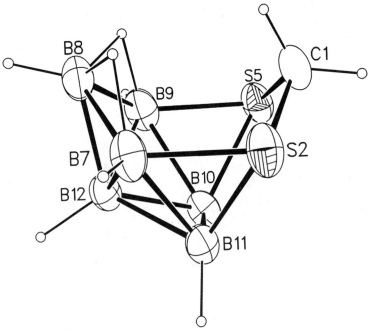

Figure 8.10. ORTEP drawing of the molecular structure of *hypho*-1-CH₂ 2,5 Me₂S₂B₆H₈ **V**.

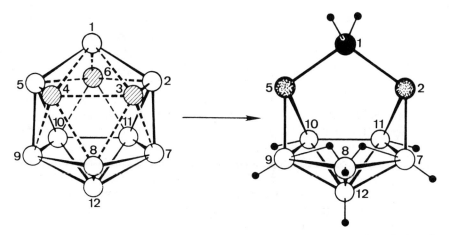

Figure 8.11. Derivation of a nine-vertex *hypho* structure for **V** from an icosahedron by removal of three vertices.

suggesting localized two-center, two-electron bonding between the methylene unit and sulfur atoms.

A wide range of 9-vertex *hypho*-clusters in which the sulfur atoms are bridged by different species isoelectronic with a CH_2 unit, such as BH_2^- or BH_3, should be possible. Shore [27] has previously demonstrated that many polyhedral borane anions will readily add BH_3 to give expanded cage products (with or without loss of H_2) when reacted with diborane or $BH_3 \cdot$ ether adducts and, indeed, it was found that reaction of the *hypho*-$S_2B_6H_9^-$ anion with $BH_3 \cdot THF$ generated *hypho*-$S_2B_7H_{10}^-$ **VI**, which is the boron analogue of **V**. Protonation of **VI** with gaseous HCl results in the formation of **VII**.

$$hypho\text{-}S_2B_6H_9^- + BH_3 \cdot THF \rightarrow hypho\text{-}S_2B_7H_{10}^- + H_2 \qquad (8.15)$$

VI

$$hypho\text{-}S_2B_7H_{10}^- + H^+ \rightarrow hypho\text{-}7,8\text{-}S_2B_7H_{11} \qquad (8.16)$$

VII

It was subsequently found that **VI** or **VII** may also be obtained by the direct reaction of *arachno*-$S_2B_7H_8^-$ or *arachno*-6,8-$S_2B_7H_9$ with $BH_3 \cdot THF$ or $NaBH_4$, respectively.

$$arachno\text{-}S_2B_7H_8^- \xrightarrow[\text{2. }H^+]{\text{1. }BH_3 \cdot THF} hypho\text{-}7,8\text{-}S_2B_7H_{11} \qquad (8.17)$$

$$arachno\text{-}S_2B_7H_9 \xrightarrow[\text{2. }H^+]{\text{1. }NaBH_4} hypho\text{-}7,8\text{-}S_2B_7H_{11} \qquad (8.18)$$

Since **VI** and **VII** are isoelectronic with **V**, similar structures are proposed as shown in Figure 8.12. These structures are also strongly supported by the spectral data, such as the ^{11}B-NMR spectra shown in Figure 8.8. The formation of *hypho*-7,8-$S_2B_7H_{11}$, rather than either $S_2B_8H_{11}$ or $S_2B_8H_9$ in (8.17) and (8.18) may again arise because of the unique reactivity of the B7 boron in *arachno*-6,8-$S_2B_7H_9$. Thus, a BH_3 unit would be expected to initially add to the $S_2B_7H_8^-$ anion at the B4–B5–S6 edge to produce $S_2B_8H_{12}^-$. In view of the cage rearrangement and degradation reactions involving B7 that were discussed above, a weakening of the cage bonding interaction with B7 might also be expected, resulting in decomposition of the framework by cleavage of this boron. This proposed reaction sequence is, of course, speculative and additional detailed studies will be required before an exact reaction mechanism can be confirmed; however, it is clear that the B7 boron appears to play a unique role in each of the above reactions.

8.4. COMPARISONS WITH CARBON AND TRANSITION METAL CLUSTERS

The series of new *hypho*-dithiaborane clusters discussed above are unprecedented in polyhedral borane chemistry; however, related cage compounds are well known in both organic and transition metal cluster chemistry. For example, **III** and **IV** are isoelectronic (in a cluster sense) with C_8H_8, and **V**, **VI**, and **VII** are isoelectronic analogs of the $C_9H_9^+$ carbocation. Several structures which have been either confirmed or proposed for organic cage fragments with the compositions of C_8H_8 or $C_9H_9^+$ are shown in Figure 8.13 [28]. Comparison of these structures with those determined for **IV** (Figure 8.9) and **V** (Figure 8.10) reveal significant differences. Most importantly, the boron atoms in both **III** and **IV** occupy either four or five coordinate positions in the cage, while the carbon

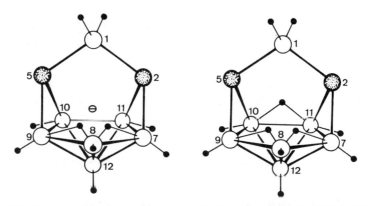

Figure 8.12. Proposed Structures for *hypho*-$S_2B_7H_{10}^-$ **VI** and *hypho*-7,8-$S_2B_7H_{11}^-$ **VII**.

C$_8$H$_8$ Cage Structures

C$_9$H$_9^+$ Cage Structures

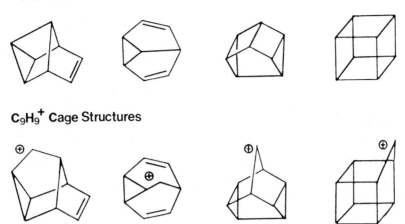

Figure 8.13. Cage structures confirmed or proposed for C$_8$H$_8$ or C$_9$H$_9^+$.

atoms in either C$_8$H$_8$ and C$_9$H$_9^+$ are in only two or three coordinate cage positions. These differences most likely arise because of the hybrid nature of the dithiaborane clusters resulting in the localization of skeletal electrons between the more electronegative elements (carbon or sulfur). For example, in **V** the electron rich bridging S–CH$_2$–S unit appears to be connected by conventional two-center, two-electron bonds, whereas the electron deficient six-boron unit adopts a configuration which favors multicenter interactions. Thus, these molecules can be described as having both classical and nonclassical portions.

Such a trend is also evident when the structures (Figure 8.14a, b) of *arachno*-6,8-S$_2$B$_7$H$_9$ and *hypho*-7,8-S$_2$B$_7$H$_{11}$ are compared. In *arachno*-6,8-S$_2$B$_7$H$_{11}$, boron B7 is bound to both sulfurs and borons B2 and B3, and is clearly part of the cage framework. In *hypho*-7,8-S$_2$B$_7$H$_{11}$, the tetrahedral boron B1 is attached to only the two sulfur atoms by what could be considered two-center, two-electron interactions. It should also be noted that the cage framework in **VII** is remarkably similar to that determined for the thiaborane fragment in *hypho*-5-CH$_3$-5,11,7,14-CNS$_2$B$_7$H$_8$ **I** (Figure 8.14c) which, as discussed above, may be described either as an 11-vertex *hypho*-cage system or a 9-vertex *hypho*-cage with a bridging imine substituent.

The above observations suggest that in these systems the transition from an electron deficient to electron precise cluster occurs by the progressive formation of two-center bonds to produce classical components, and furthermore, bring into question at what point should such a classical fragment be considered as an exopolyhedral substituent rather than a component of a cluster framework. For example, even though an eight vertex *hypho*-formulation seems reasonable for **III**, the isoelectronic compound **IV** might alternately be described as a six-vertex *nido*-system substituted by "classical" methylthiol groups, that is, *nido*-(μ-

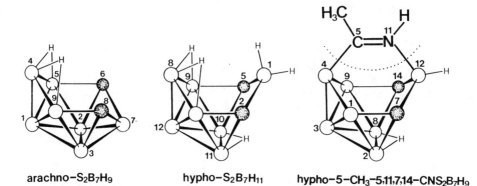

arachno–S₂B₇H₉   hypho–S₂B₇H₁₁   hypho-5–CH₃–5,11,7,14–CNS₂B₇H₉

Figure 8.14. Comparison of the structures of *arachno*-6,8-$S_2B_7H_9$, *hypho*-7,8-$S_2B_7H_{11}$, and *hypho*-5-CH$_3$-5,11,7,14-CNS$_2$B$_7$H$_8$.

MeS)$_2$B$_6$H$_8$. Clearly, more structurally characterized examples of larger cage *hypho*-clusters will be needed before the nature of the bonding interactions within these clusters can be fully understood and one focus of our ongoing research involves investigations of the synthesis and structural characterizations of new types of hybrid cluster systems.

Owing to their open structures and "classical"/"nonclassical" hybrid natures, the reactivities of the *hypho*-dithiaboranes might also be expected to be intermediate between those observed for the polyhedral boranes and other types of clusters. We have now observed similarities between their reactions and those of certain transition metal sulfur complexes containing bridging sulfur or methylthiol groups. For example, it has been shown that the anion Fe$_2$(CO)$_6$S$_2^{2-}$ can be readily generated from the complex Fe$_2$(CO)$_6$S$_2$ and that it reacts with methyl iodide as follows [29]

$$(8.19)$$

As in S$_2$B$_6$H$_9^-$, methylation occurs exclusively at the sulfur atoms consistent with a localization of negative charge at these sites in the Fe$_2$(CO)$_6$S$_2^{2-}$ anion. However, protonation of Fe$_2$(CO)$_6$S$_2^{2-}$ yields the corresponding thiol compounds Fe$_2$(CO)$_6$(SH)$_2$ [29]. In *hypho*-S$_2$B$_6$H$_9^-$ (which would be the protonated analog of the S$_2$B$_6$H$_8^{2-}$ anion) the additional proton is present not as a thiol, as found in Fe$_2$(CO)$_6$(SH)$_2$, but rather in a bridge site on the borane fragment and the anion will not react with additional acid to give the neutral compound S$_2$B$_6$H$_{10}$, but instead decomposes.

The reaction of $hypho\text{-}S_2B_6H_2^-$ with diiodomethane or $BH_3\cdot THF$ to yield the nine-vertex bridged complexes $hypho\text{-}1\text{-}CH_2\text{-}2,5\text{-}S_2B_6H_8$ **V** and $hypho\text{-}7,8\text{-}S_2B_7H_{11}$ **VI** was also suggested by similar reactions observed in metal-sulfur cluster chemistry. For example, Seyferth [30] has observed that $S_2Fe_2(CO)_6^{2-}$ will react with diiodomethane to generate the methylene bridged compound (a 5-vertex $hypho$-cluster)

$$(8.20)$$

Similarily, the reaction of $S_2Fe_2(CO)_6^{2-}$ [31] or $S_2B_6H_9^-$ [18] with $(\eta\text{-}C_5H_5)Co(Co)I_2$ results in the insertion of a $(\eta\text{-}C_5H_5)Co$ unit (a 2-electron donor) to form the corresponding five and nine vertex $arachno$-cage systems.

$$S_2Fe_2(CO)_6^{2-} + (\eta\text{-}C_5H_5)Co(Co)I_2 \rightarrow arachno\text{-}(\eta\text{-}C_5H_5)CoS_2Fe_2(CO)_6$$
$$+ 2\,LiI + CO \qquad (8.21)$$

$$hypho\text{-}S_2B_6H_9^- + (\eta\text{-}C_5H_5)Co(Co)I_2 \rightarrow arachno\text{-}(\eta\text{-}C_5H_5)CoS_2B_6H_8 + NaI$$
$$+ HI + CO \qquad (8.22)$$

The structure [32] of $arachno\text{-}(\eta\text{-}C_5H_5)CoS_2B_6H_8$ is depicted in Figure 8.15 and is seen to have a cage geometry identical to that observed for $arachno\text{-}S_2B_7H_9$ (Figure 8.1), again illustrating the close structural relationships between polyhedral boranes, metallaboranes, and traditional metal cluster systems.

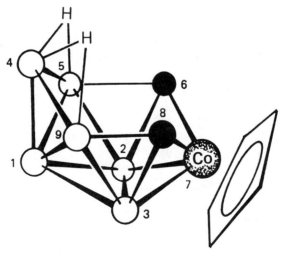

Figure 8.15. The molecular structure of $arachno\text{-}(\eta\text{-}C_5H_5)CoS_2B_6H_8$.

Numerous other bridged derivatives of $S_2Fe_2(CO)_6^{2-}$ have also been generated [29–31] by reaction of the anion with various dihalides and comparable reactions with *hypho*-$S_2B_6H_9^-$, as well as the *hypho*-$S_2B_7H_{11}^-$, *hypho*-$CH_2S_2B_7H_8^-$, and *hypho*-$CH_3CNHS_2B_7H_7^-$ anions, should now result in the production of a wide variety of new cluster systems having a continuum of bonding, structural, and reactivity properties intermediate between the electron deficient and electron precise cluster classes.

ACKNOWLEDGMENT

We thank the National Science Foundation for the support of this work.

REFERENCES

1. Wade, K. *Adv. Inorg. Chem. Radiochem.*, **18**, 1 (1976).

2. Mangion, M., Hertz, R. K., Denniston, M. L., Long, J. R., Clayton, W. R. and Shore, S. G., *J. Am. Chem. Soc.*, **98**, 449 (1976).

3. Remmel, R. J., Johnson, II, H. D., Jaworiwsky, I. S. and Shore, S. G., *J. Am. Chem. Soc.*, **97**, 5395 (1975).

4. (a) Burg, A. B., *J. Am. Chem. Soc.*, **79**, 2129 (1957); (b) Savory, C. G. and Wallbridge, M. G. H., *J. Chem. Soc., Dalton Trans.*, 179 (1973).

5. (a) Fratini, A. V., Sullivan, G. W., Denniston, M. L., Hertz, R. K. and Shore, S. G., *J. Am. Chem. Soc.*, **96**, 3013 (1974); (b) Kameda, M. and Kodama, G., *Inorg. Chem.*, **19**, 2288 (1980).

6. Alcock, N. W., Colquhoun, H. M., Haran, G., Sawyer, J. F. and Wallbridge, M. G. H., *J. Chem. Soc., Chem. Commun.*, 368 (1977).

7. Miller, N. E., Miller, H. C. and Muetterties, E. L., *Inorg. Chem.*, **3**, 866 (1964).

8. Hertz, R. K., Denniston, M. L. and Shore, S. G., *Inorg. Chem.*, **17**, 2673 (1978).

9. Duben, J., Heřmánek, S. and Štíbr. B., *J. Chem. Soc., Chem. Commun.*, 287 (1978).

10. Kodama, G. and Kameda, M., *Inorg. Chem.*, **18**, 3302 (1979).

11. (a) Colquhoun, H. M., *J. Chem. Res.*, 451 (1978); (b) Alcock, N. W., Colquhoun, H. M., Haran, G., Sawyer, J. F. and Wallbridge, M. G. H., *J. Chem. Soc., Dalton Trans.*, 2243 (1982).

12. Brellochs, B. and Binder, H., *Agnew. Chem., Int. Ed. Engl.*, **27**, 262 (1988).

13. (a) Hollander, O., Clayton, W. R. and Shore, S. G., *J. Chem. Soc., Chem. Commun.*, 604 (1974); (b) Mangion, M., Clayton, W. R., Hollander, O. and Shore, S. G., *Inorg. Chem.*, **16**, 2110 (1977).

14. Jelinek, T., Plešek, J., Heřmánek, S. and Štíbr, B., *Main Group Metal Chem.*, **10**, 397 (1987).

15. Plešek, J., Heřmánek, S. and Janoušek, Z., *Coll. Czech. Chem. Commun.*, **42**, 785 (1977).

16. Kang, S. O., Furst, G. T. and Sneddon, L. G., *Inorg. Chem.*, **28**, 2339 (1989).

17. Kang, S. O. and Sneddon, L. G., *J. Am. Chem. Soc.*, **111**, 3281 (1989).

18. Mazighi, K., Kang, S. O. and Sneddon, L. G., to be submitted.

19. Hyatt, D. E., Owen, D. A. and Todd, L. J., *Inorg. Chem.*, **5**, 1749 (1966).

20. Arafat, A., Friesen, G. D. and Todd, L. J., *Inorg. Chem.*, **22**, 3721 (1983).

21. Russell, P. B., Hitchings, G. H., Chase, B. H. and Walker, J., *J. Am. Chem. Soc.*, **74**, 5403 (1952).

22. Kuhn, V. R. and Drawert, F., *Justus Liebigs Ann. Chem.*, **590**, 55 (1954).

23. (a) Evans, W. J. and Hawthorne, M. F., *J. Chem. Soc., Chem. Commun.*, 38 (1974); (b) Brown, L. D. and Lipscomb, W. N., *Inorg. Chem.*, **16**, 2989 (1977); (c) Maxwell, W. M., Bryan, R. F. and Grimes, R. N., *J. Am. Chem. Soc.*, **99**, 4008 (1977); (d) Maxwell, W. N., Weiss, R., Sinn, E. and Grimes, R. N., *J. Am. Chem. Soc.*, **99**, 4016 (1977); (e) Pipal, J. R. and Grimes, R. N., *Inorg. Chem.*, **17**, 6 (1978); (f) Bicerano, J., Marynick, D. S. and Lipscomb, W. N., *Inorg. Chem.*, **17**, 2041 (1978); (g) Bicerano, J., Marynick, D. S. and Lipscomb, W. N., *Inorg. Chem.*, **17**, 3443 (1978).

24. (a) Andrews, M. A. and Kaesz, H. D., *J. Am. Chem. Soc.*, **99**, 6763 (1977); (b) Andrews, M. A. and Kaesz, H. D., *J. Am. Chem. Soc.*, **101**, 7238 (1979); (c) Andrews, M. A., Van Buskirk, G., Knobler, C. B. and Kaesz, H. D., *J. Am. Chem. Soc.*, **101**, 7245 (1979); (d) Andrews, M. A. and Kaesz, H. D., *J. Am. Chem. Soc.*, **101**, 7255 (1979); (e) Andrews, M. A., Knobler, C. B. and Kaesz, H. D., *J. Am. Chem. Soc.*, **101**, 7260 (1979).

25. Shaver, A., Fitzpatrick, P. J., Steliou, K. and Butler, I. S., *J. Am. Chem. Soc.*, **101**, 1313 (1979).

26. McKenna, M., Wright, L. L., Miller, D. T., Tanner, L., Haltiwanger, R. C. and DuBois, M. R., *J. Am. Chem. Soc.*, **105**, 5329 (1983).

27. (a) Johnson, II, H. D., Bruce, V. T. and Shore, S. G., *Inorg. Chem.*, **12**, 689 (1973); (b) Johnson, II, H. D. and Shore, S. G., *J. Am. Chem. Soc.*, **93**, 3798 (1971); (c) Geanangel, R. A., Johnson, II, H. D. and Shore, S. G., *Inorg. Chem.*, **10**, 2363 (1971).

28. Leone, R. E., Barborak, J. C. and Schleyer, P. v. R., in: *Carbonium Ions* (G. A. Olah and P. v. R. Schleyer, Eds.), pp. 1837–1939, Wiley, New York, 1973, and references therein.

29. Seyferth, D., Henderson, R. S. and Song, L.-C., *J. Organomet. Chem.*, **192**, Cl (1985).

30. Seyferth, D., Henderson, R. S. and Song, L.-C., *Organometallics*, **1**, 125 (1982).

31. Cowie, M., DeKock, R. L., Wagenmaker, T. R., Seyferth, D., Henderson, R. S. and Gallagher, M. K., *Organometallics*, **8**, 119 (1989).

32. Zimmerman, G. J. and Sneddon, L. G., *J. Am. Chem. Soc.*, **103**, 1102 (1981).

CHAPTER 9

C-Trimethylsilyl-Substituted Carboranes and Their Derivatives

N. S. HOSMANE and J. A. MAGUIRE

9.1. INTRODUCTION

There has been considerable interest in the synthesis and reaction chemistry of *nido*-2,3-$C_2B_4H_8$ [dicarbahexaborane(8)] and its C,C'-substituted derivatives [1]. Not only are these carborane derivatives important precursors to *closo*-carboranes, and tetracarbon carboranes, they also act as η^1, η^2, and η^5 coordinate ligands to transition metals and main group metals [1]. The $[R_2C_2B_4H_4]^{2-}$ dianion is isoelectronic with cyclopentadienide anion, $(C_5R_5)^-$, by contributing 16 valence electrons (3 and 2 electrons from each of the CR and BH groups, respectively) to the cage or ring. Consequently, the η^5-carborane complexes can be regarded as analogs of η^5-cyclopentadienyl complexes. However, there are a number of important differences. First, the greater polarizability of boron versus carbon would encourage stronger metal-carborane bonding [2,3]. Second, the C_2B_3 face of the carborane is not necessarily flat but can be slightly folded away from the capping metal group. This folding, coupled with the heteronuclear nature of the carborane bonding face, encourages a slippage of the capping metal atom away from the centroidal position above the C_2B_3 face. Slip distortion seems to be a common feature of the main group metallacarboranes and is invariably toward the boron side of the pentagonal face [4]. The last difference to be noted is that of charge; the carborane ligands bear a 2$^-$ charge while the cyclopentadienide ligands are

Electron Deficient Boron and Carbon Clusters, Edited by George A. Olah,
Kenneth Wade, and Robert E. Williams.
ISBN 0-471-52795-5 © 1991 John Wiley & Sons, Inc.

mono-anions. This increased charge should help stabilize bonding with positively charged metal or metal groups and can effect the metal oxidation state most stabilized by bonding to the carborane.

Until recently, the usual method of preparation of $C_2B_4H_8$ derivatives was that described by Hosmane and Grimes [5, 6]. This method is complicated by the use of Lewis bases such as $(C_2H_5)_3N$ and is limited to a maximum of 1–2 g of carboranes per synthesis. Despite these limitations, in terms of yield, safety, and practicality, the method is preferable to the one reported by Onak et al. [7] especially in the production of small carboranes in laboratory scale. More recently, a clean, high yield, laboratory scale preparative route to *nido*-$(Me_3Si)(R)C_2B_4H_6$ ($R = SiMe_3$, Me or H) derivatives has been reported (see Scheme I) [8]. Several aspects of these trimethylsilyl analogues of *nido*-2,3-$C_2B_4H_8$ make them attractive ligands for synthetic organometallic chemistry.

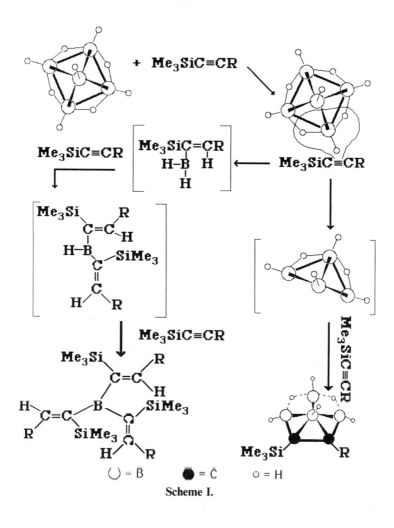

Scheme I.

The trimethylsilyl-substituted carboranes offer the following advantages: (1) they can be routinely prepared in multigram quantities as single pure isomers; (2) all are air-stable liquids and hence do not require special handling procedures; (3) they are fairly reactive liquids due to two B–H–B bridge hydrogens even in the absence of solvents; (4) their reactions generally proceed essentially quantitatively to yield carborane products; (5) they are converted to fairly air-stable products; (6) the C-SiMe$_3$ moiety is conducive to single-crystal growth; (7) the products tend to be more soluble in organic solvents than their C-H, C-alkyl, or C-aryl-substituted analogs; (8) the C-SiMe$_3$ bond is cleaved selectively; and finally (9) their chemistry seems to be more systematic than that of the alkyl or aryl analogs. Recent work, described in the next section, indicates that the C-SiMe$_3$ substituted carboranes are versatile reagents and that their reactivity toward main group metals, metalloids, and transition metals should lead to a new family of cluster compounds with interesting chemical, structural, and bonding features.

9.2. CHEMISTRY OF C$_{(cage)}$-SiMe$_3$-SUBSTITUTED DERIVATIVES OF *nido*-DICARBAHEXABORANE(8)

Several reactions that lead to the formation of tris[[(trimethylsilyl)-1-alkenyl]borane and also to several air-stable derivatives of C-trimethylsilyl-substituted C$_2$B$_4$H$_8$ have been reported (see Scheme I) [8]. Although the trivinylborane derivatives could be produced in almost quantitative yields, with the exception of *nido*-(Me$_3$Si)$_2$C$_2$B$_4$H$_6$, the *nido*-carboranes were obtained in much lower yields. Since *nido*-(Me$_3$Si)$_2$C$_2$B$_4$H$_6$ can be produced in over 73% yield (12–13 g per batch), it has been the major subject for most of the recent investigations.

The controlled reaction of liquid *nido*-(Me$_3$Si)$_2$C$_2$B$_4$H$_6$ with solid NaHF$_2$ or gaseous HCl in a pyrex-glass reactor or stainless-steel reactor at 140°C, produces *nido*-(Me$_3$Si)(H)C$_2$B$_4$H$_6$ by elimination of one of the cage-SiMe$_3$ groups while not affecting the terminal B–H and B–H–B bridge bonds [9]. The 2-(trimethylsilyl)-2,3-dicarba-*nido*-hexaborane(8) was produced in very high purity by this method so that its molecular structure could be determined by gas-phase electron diffraction [9]. The structure is shown in Figure 9.1. It should be noted that this synthesis avoids the use of expensive starting materials, such as Me$_3$SiC≡CH, that are needed in preparations that have been previously reported [8, 10]. It also should be noted that *nido*-2,3-(SiMe$_3$)$_2$-2,3-C$_2$B$_4$H$_6$, when reacted with different molar ratios of anhydrous HCl gas at elevated temperatures, can give rise to a number of carborane products (see Scheme II) [11]. For example, when the carborane and HCl were taken in a molar ratio of 1:3.3 and heated at 160–170°C for 4 days in a high-vacuum stainless-steel cylinder, the carborane products *nido*-2-(SiMe$_3$)-2,3-C$_2$B$_4$H$_7$, *nido*-2,3-C$_2$B$_4$H$_8$ (parent carborane), and *closo*-C$_2$B$_4$H$_2$Cl$_4$ were obtained in 31, 27, and 8.9% yields, respectively. When the ratio of bis(trimethylsilyl)carborane to HCl was

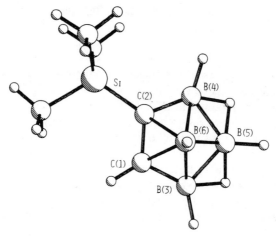

Figure 9.1. Structure of 2-(trimethylsilyl)-2,3-dicarba-*nido*-hexaborane(8) as determined by gas-phase electron diffraction.

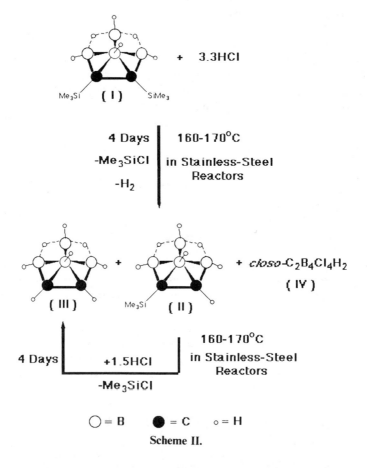

O = B   ● = C   o = H

Scheme II.

1:1.5, the parent carborane was produced in 30% yield, while with a large excess of HCl under similar conditions the reaction shown in Scheme II produced $closo$-$C_2B_4H_2Cl_4$ in 91% yield as the sole carborane product. Thus, the ratio of reactants in Scheme II determines the formation of either one particular carborane or a mixture of several carborane products and affects their respective yields. Evidently, a large excess of HCl produces B-chlorinated product, whereas an insufficient quantity of HCl yields mainly mono(trimethylsilyl)-substituted carborane. However, for the high-yield preparation of the parent C_2B_4-carborane the ideal molar ratio between the bis(C-SiMe$_3$)-substituted carborane and HCl was found to be 1:3.3 [11]. This synthetic route is of extreme interest for the following reasons: (1) this is the laboratory method of choice over the existing method of preparation of 2,3-dicarba-$nido$-hexaborane(8), which involves the gas-phase pyrolysis of a mixture of HC≡CH and B_5H_9; (2) the reaction avoids the utilization of large volume gas bulbs and Lewis bases, and the necessity of handling the volatile flammable and explosive pentaborane-acetylene mixture; (3) it does not result in the formation of a complex mixture of carboranes under specific conditions and hence eliminates the necessity of separating products by gas chromatography; and finally (4) it could produce the parent carborane in gram quantities since its precursor is routinely prepared in multigram quantities.

The $nido$-carborane 2,3-(SiMe$_3$)$_2$-2,3-$C_2B_4H_6$ undergoes direct fusion at 210°C to give $nido$-(SiMe$_3$)$_2$C$_4$B$_8$H$_{10}$ without going through a C_4B_8 metal complex intermediate or without the need of a metal catalyst [12]. The exact mechanism of the fusion process is not known, but a plausible scheme is outlined in Scheme III. This involves the high temperature formation of trimethylsilyl radical, which could then extract one of the carborane bridge-hydrogen atoms forming Me$_3$SiH and a reactive carborane fragment that could condense with another such fragment [12]. Irrespective of the mechanism, the significance of

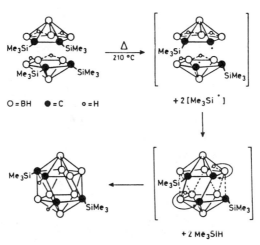

$O = BH \qquad \bullet = C \qquad o = H$

Scheme III.

the trimethylsilyl moiety in synthetic carborane transformations has been once again demonstrated.

Quite recently, the instantaneous oxidative cage closure of the small carborane has been observed in the reaction of $closo$-1-Sn-2,3-$(SiMe_3)_2$-2,3-$C_2B_4H_4$ with anhydrous $PtCl_4$ or $PtCl_2$ in $CHCl_3$ at room temperature [13]. The $closo$-carborane 1,2-$(SiMe_3)_2$-1,2-$C_2B_4H_4$ was the sole carborane product of this reaction. The tin and platinum were converted to $SnCl_2$ and elemental platinum (Pt°), respectively. No platinacarborane was identified among the products. This platinum halide induced reaction is shown in equations (1a) and (1b) [13]

$$closo\text{-1-Sn-2,3-}(SiMe_3)_2C_2B_4H_4 + PtCl_2 \xrightarrow[CHCl_3, fast]{25°C}$$

$$closo\text{-1,2-}(SiMe_3)_2\text{-1,2-}C_2B_4H_4 + SnCl_2 + Pt° \qquad (9.1a)$$

$$2closo\text{-1-Sn-2,3-}(SiMe_3)_2C_2B_4H_4 + PtCl_4 \xrightarrow[CHCl_3, fast]{25°C}$$

$$2closo\text{-1,2-}(SiMe_3)_2\text{-1,2-}C_2B_4H_4 + 2SnCl_2 + Pt° \qquad (9.1b)$$

9.3. CHEMISTRY OF MONO- AND DIANIONS OF THE $C_{(cage)}$-SiMe₃-SUBSTITUTED C_2B_4-CARBORANES

Nearly 25 years ago Onak and Dunks reported the heterogeneous reaction between the 2,3-dicarba-$nido$-hexaborane(8) and excess NaH in tetrahydrofuran (THF) to produce the sodium salt of the carborane mono-anion exclusively; attempts to produce the disodium salts of the C_2B_4-carborane dianions resulted only in failure [14]. Recent work has shown that the $nido$-2,3-$(SiMe_3)_2$-2,3-$C_2B_4H_6$ reacts with a large excess of pure NaH in THF in a method identical to that employed by Onak and Dunks [14] to produce exclusively the novel THF-solvated ion cluster, $(C_4H_8O.Na^+)_2$ [2,3-$(SiMe_3)_2$-2,3-$C_2B_4H_5^-]_2$, as shown in Scheme IV. The molecular geometry of this ion cluster was confirmed by single-crystal X-ray analysis (see Figure 9.2) [15]. The crystal packing diagram, shown in Figure 9.3, is of an extended network of dimeric $\{Na^+[(SiMe_3)_2C_2B_4H_5]^-\}_2$ ion clusters that are layered almost symmetrically on top of each other. Each cluster consists of two C_2B_4-carborane cages and two THF-solvated sodium ions with a crystallographic center of symmetry half way between the sodium ions. Each Na^+ is primarily involved in equal exo-polyhedral η^1-type bonding interactions with one of the basal borons in one cage [Na–B(5b) = 2.800(11) Å] and the C(1) atom of the other cage [Na–C(1) = 2.832(10) Å] (see Figure 9.2) [15]. The interactions with the other cage atoms are weaker. The distance of 2.57(8) Å between Na and the bridged hydrogen [H(34)] is quite short. There is also a strong interaction between the sodiums in one dimeric ion cluster $[Na^+(R_2C_2B_4H_5)^-]_2$ and the apical and unique borons in a neighboring cluster. However, all sodium-nearest neighbor

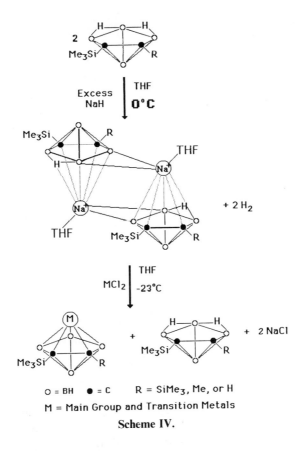

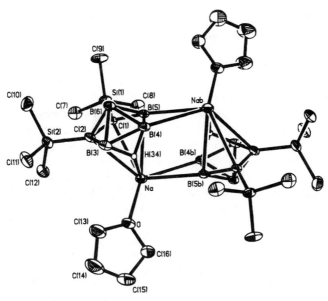

Figure 9.2. Crystal structure of the ion cluster $(C_4H_8O \cdot Na^+)_2[2,3-(SiMe_3)_2-2,3-C_2B_4H_5^-]_2$.

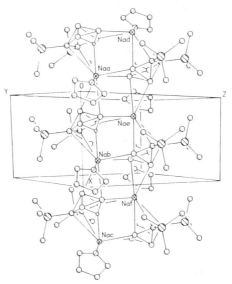

Figure 9.3. The crystal packing diagram showing the extended network of dimeric ion cluster, $(C_4H_8O \cdot Na^+)_2[2,3-(SiMe_3)_2-2,3-C_2B_4H_5^-]_2$.

distances are greater than that expected for covalent bonding and indicate that the interactions are all essentially ionic. It is clear from Figure 9.2, and even more apparent when space-filling model is used (see Figure 9.4), that the second bridge hydrogens are very well protected within the cluster [15]. The protection of this hydrogen within the cluster along with the heterogeneous nature of the reaction between NaH and *nido*-carborane could be partly responsible for the inability to produce the disodium salts of the *nido*-C_2B_4 carborane dianions. However, the ion cluster, in turn, undergoes reaction with group 14 metal halides to produce a variety of 1-(metalla)-*closo*-2,3-$C_2B_4H_4$ carborane derivatives as shown in Scheme IV.

Although the preparation of disodium salt of the *nido*-C_2B_4-carborane is not yet known, the synthesis of its first stable dianion was reported by Hosmane and coworkers in 1986 [16,17]. According to this report, *nido*-2-(SiMe$_3$)-3-(R)-2,3-$C_2B_4H_6$ (R = SiMe$_3$, Me, H) carboranes can be doubly deprotonated via reaction with NaH followed by *n*- or *t*-C_4H_9Li to give stable Na$^+$Li$^+$[2-(SiMe$_3$)-3-(R)-2,3-$C_2B_4H_4$]$^{2-}$ double salts [16,17]. The reaction is shown in Scheme V. Recently, Grimes and co-workers found that this reaction sequence is generally applicable to C-alkyl or aryl-substituted *nido*-C_2B_4-carboranes as well [18]. However, the solid-state molecular geometry of these dianions are still unknown. The formation of the sodium/lithium double salts of the *nido*-C_2B_4-carboranes could be primarily due to homogeneous nature of the reaction between the ion cluster, $(C_4H_8O \cdot Na^+)_2[2,3-(SiMe_3)_2-2,3-C_2B_4H_5^-]_2$, and BuLi in either THF or hexane as this would allow the base to reach the second bridge

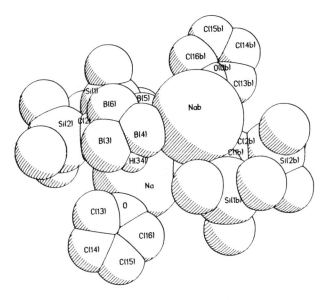

Figure 9.4. Space filling model of $(C_4H_8O \cdot Na^+)_2[2,3\text{-}(SiMe_3)_2\text{-}2,3\text{-}C_2B_4H_5^-]_2$.

hydrogens through relatively small gaps of the dimeric ion cluster for abstraction since the steric crowding around the small Li^+ cation is considerably minimum. It is important to point out that the dilithium salts of the C_2B_4-carboranes could be accomplished by the treatment of the *nido*-carborane precursor with 2 equivalents of *n*- or *t*-BuLi in hexane at low temperatures. However, the dilithium salts are highly reactive and their reactions with metal halides gave irreproducible results [17]. Thus, the discovery of the Na^+/Li^+ double salts of the C_2B_4-carborane system opened up a new era in the chemistry of metallacarboranes especially in the production of several *closo*- and *commo*-sila- and germacarboranes (see Scheme V).

9.4. CHEMISTRY OF C(cage)-SiMe3-SUBSTITUTED MAIN GROUP HETEROCARBORANES

There are several monographs [1, 19] and a number of review articles [1, 2, 20, 21] that adequately cover the earlier literature. Recent advances in the research of main group metallacarboranes containing groups 13 and 14 metals and/or metalloids covering the year 1988 have also been reviewed [4, 22]. Therefore, this section describes mainly the research reported since 1988. Earlier work will be briefly mentioned only as background to current results or for purposes of comparison.

The reaction of $Na[(SiMe_3)(R)C_2B_4H_5]$ with $SnCl_2$ in THF produces an air-sensitive THF stannacarborane intermediate, $(C_4H_8O)_2Sn(SiMe_3)(R)C_2B_4H_4$

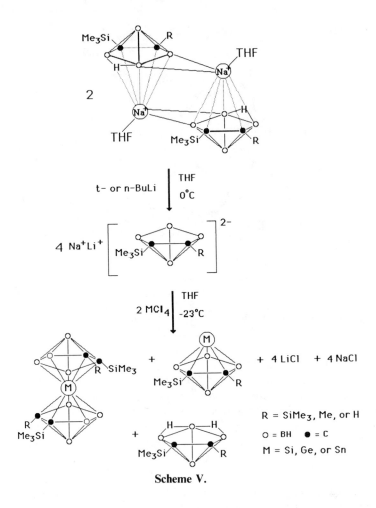

Scheme V.

(R = SiMe$_3$, Me, H), which decomposes to give THF and the corresponding stannacarborane as a sublimable off-white solid, in gram quantities (see Scheme VI) [23, 24]. The structures of these compounds show a pentagonal-bipyramidal geometry with the tin occupying an apical position [24–26]. These crystal structures provide the first unambiguous proof of how Sn(II) is incorporated into small carborane clusters. Wong and Grimes could not obtain single crystals suitable for X-ray analysis of the similar compound, Sn(R)$_2$C$_2$B$_4$H$_4$ (R = Me, H) [27]. It seems therefore that the presence of the C-SiMe$_3$ moiety is conducive to single-crystal growth. Despite the presence of a lone pair of electrons on the "bare" tin atom, the stannacarboranes do not react with BH$_3$.THF or BF$_3$ but react almost quantitatively with 2,2′-bipyridine, 2,2′-bipyrimidine, 2,2′ : 6′,2″-terpyridine, and (ferrocenylmethyl)-N,N-dimethylamine in benzene to form the electron donor-acceptor complexes as shown in Scheme VII [24, 28, 29]. The structures of these charge-transfer complexes were determined by single-crystal

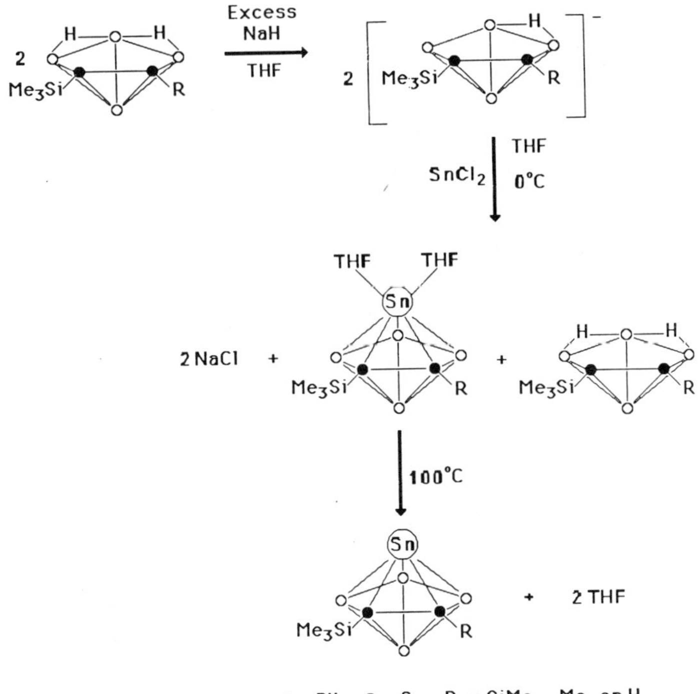

O = BH   ● = C   R = SiMe₃, Me, or H

Scheme VI.

X-ray diffraction [24, 28–32]. Complexation is accompanied by a slippage of tin from η^5 to η^3-bonding. The ^{119m}Sn Mössbauer effect spectra of these complexes exhibit quadrupole split doublets and clearly indicate that tin is formally in the +2 oxidation state [24]. The formation of these complexes shows that the apical tin atom of the stannacarboranes is behaving as a Lewis acid as in the case of the pyridine or bipyridine complexes of [(Me₅C₅)Sn⁺] [33]. Fenske-Hall and MNDO-SCF calculations were carried out on the bipyridine complexes [34, 35]. Both calculations show that complexation with bipyridine gives rise to orbitals with antibonding tin-carbon(cage) interactions, which would encourage slippage. MNDO calculations indicate that the major bonding interactions between the bipyridine molecule and the tin are through tin orbitals oriented parallel to the C_2B_3 face of the carborane. Hence, maximum tin-bipyridine bonding would be expected when the rings of the bipyridine molecule and the C_2B_3 face are essentially parallel. Repulsion between the two coordinating groups would prevent such an ideal alignment. An increased slip distortion of tin would tend to decrease ligand-ligand repulsion and yield a more favorable bipyridine orientation and stronger tin-bipyridine bonding. In general, one would expect a decrease in base-carborane dihedral angle and an increase in slip distortion on forming stronger tin-base adducts [35]. This has been generally

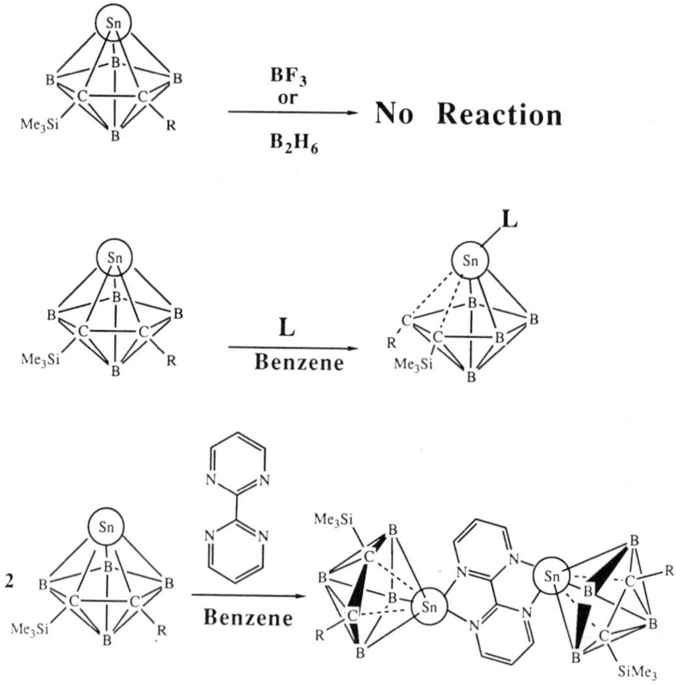

R = Me₃Si , Me , H
B = BH
L = Mono-, Bi-, or Tri-dentate Lewis Base

Scheme VII.

borne out by experiment. Adducts with monodentate bases show less slip distortion than found in the bipyridine complexes as does the complex with the weaker bipyrimidine base [24, 28–32].

The synthetic potential of the stannacarboranes has been demonstrated in the preparation of *closo*-1,2-$(SiMe_3)_2$-1,2-$C_2B_4H_4$ [13], and also in the production of *closo*-osmacarboranes [36, 37], *commo*-bis(germacarborane) [38] and B-Ge^{IV}Cl$_3$-substituted *closo*-germacarboranes [39, 40]. These reactions are given in equations (9.1)–(9.4).

$$3\,[closo\text{-}1\text{-}Sn\text{-}2\text{-}(SiMe_3)\text{-}3\text{-}(R)\text{-}2,3\text{-}C_2B_4H_4] + Os_3(CO)_{12} \xrightarrow{\text{150°C, No solvent}}$$

$$3\,[closo\text{-}1\text{-}Os(CO)_3\text{-}2\text{-}(SiMe_3)\text{-}3\text{-}(R)\text{-}2,3\text{-}C_2B_4H_4 + 3\,Sn^\circ + 3\,CO \quad (9.2)$$

$$2\,[closo\text{-}1\text{-}Sn\text{-}2\text{-}(SiMe_3)\text{-}3\text{-}(R)\text{-}2,3\text{-}C_2B_4H_4] + GeCl_4 \xrightarrow{\text{150°C, No solvent}}$$

$$commo\text{-}1,1'\text{-}Ge^{IV}\text{-}[2\text{-}(SiMe_3)\text{-}3\text{-}(R)\text{-}2,3\text{-}C_2B_4H_4]_2 + 2\,SnCl_2 \quad (9.3)$$

$$\text{\textit{closo}-1-Sn-2-(SiMe}_3)\text{-3-(R)-2,3-C}_2\text{B}_4\text{H}_4 + \text{excess GeCl}_4 \xrightarrow{\substack{150°C, \text{ No solvent}}}$$

$$\text{\textit{closo}-1-Ge}^{\text{II}}\text{-2-(SiMe}_3)\text{-3-(R)-5-(Ge}^{\text{IV}}\text{Cl}_3)\text{-2,3-C}_2\text{B}_4\text{H}_3$$

$$+ \text{\textit{closo}-1-Ge}^{\text{II}}\text{-2-(SiMe}_3)\text{-3-(R)-2,3-C}_2\text{B}_4\text{H}_4$$

$$+ \text{\textit{nido}-2,3-(Me}_3\text{Si})_2\text{-2,3-C}_2\text{B}_4\text{H}_6$$

$$\quad (44\% \text{ yield})$$

$$+ \text{\textit{commo}-1,1'-Ge}^{\text{IV}}\text{-[2-(SiMe}_3)\text{-3-(R)-2,3-C}_2\text{B}_4\text{H}_4]_2 + \text{SnCl}_2 \qquad (9.4)$$

where

$$R = SiMe_3, Me, H.$$

The structures of the germacarboranes have been determined by X-ray diffraction and are shown in Figures 9.5 and 9.6. The X-ray crystal structure of the mixed valence germacarborane, $\text{\textit{closo}-Ge}^{\text{II}}\text{-2,3-(Me}_3\text{Si})_2\text{-5-(Ge}^{\text{IV}}\text{Cl}_3)\text{-2,3-}$ $\text{C}_2\text{B}_4\text{H}_3$ shows that the Ge^{II} is η^5 bonded to the open pentagonal face of the carborane and symmetrically situated above this face [bond distances are: Ge– $\text{C}_{\text{(cage)}} = 2.251(4)$ and $2.244(4)$; $\text{Ge–B} = 2.265(6)$, $2.243(6)$, and $2.250(6)$ Å] [39]. The other germanium, in a $+4$ oxidation state, is involved in an exopolyhedral

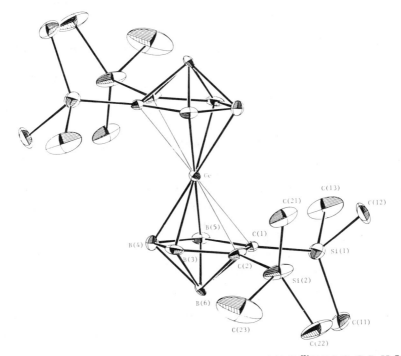

Figure 9.5. Crystal structure of $2,2',3,3'-(\text{SiMe}_3)_4\text{-\textit{commo}-1,1'-Ge}^{\text{IV}}[1,2,3\text{-GeC}_2\text{B}_4\text{H}_4]_2$.

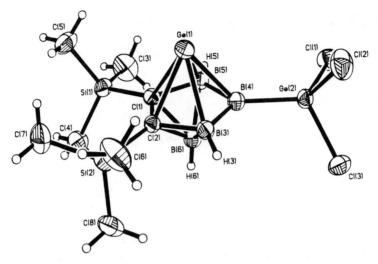

Figure 9.6. Crystal structure of *closo*-Ge^{II}-2,3-(Me$_3$Si)$_2$-5-(Ge^{IV}Cl$_3$)-2,3-C$_2$B$_4$H$_3$.

GeCl$_3$ group bonded to the unique boron of the cage via a Ge–B sigma bond. This is one of the few cases of a main group *closo*-metallacarborane that is not slip distorted (see Figure 9.6) [39, 40]. Theoretical studies on the stannacarboranes [35] indicate that electron withdrawing groups on the unique boron should favor a more centroidal location of the capping heteroatom; this is the case for the mixed valence germacarborane.

The discovery of the stable Na$^+$Li$^+$[2-(SiMe$_3$)-3-(R)-2,3-C$_2$B$_4$H$_4$]$^{2-}$ double salts [16, 17] led to the production of a number of sila- and germacarboranes that could not be produced from the corresponding mono-sodium salts (see Scheme V) [16, 17, 41, 42].

The crystal structures of Si(IV) [16, 17] and Ge(IV) [38, 41] sandwiched carboranes reveal that the central heteroatom adopts an essentially η^5-bonding posture with respect to each of the C$_2$B$_3$ faces (see Figures 9.5, 9.7, and 9.8). However, the M^{IV}-C (M = Si or Ge) distances are substantially longer than the M^{IV}–B distances and thus showing a slippage of the M^{IV}. In these compounds, the cage carbons of the opposing ligands reside in staggered confirmation and the *commo*-group 14 atom is ten-coordinated. Recent MNDO molecular orbital calculations show that the *commo*-heteroatom's d-orbital participation is not necessary to explain its bonding to the two C$_2$B$_4$ cages [43]. It is of interest to note that with the trimethylsilyl-substituted carborane dianion, in addition to M^{IV} insertion, a reductive insertion yielding a liquid [(SiMe$_3$)(R)C$_2$B$_4$H$_4$]M^{II} takes place simultaneously in the reaction vessel (see Scheme V) [16, 17, 42]. If this combination of reductive and nonreductive insertions turns out to be a general phenomenon, it may be possible to routinely carry out "one pot" syntheses of both the M^{II}- and M^{IV}-complexes. Since M^{II}-complex is liquid and

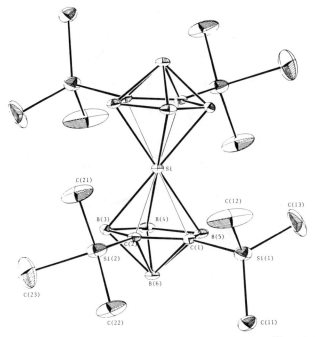

Figure 9.7. Crystal structure of 2,2',3,3'-(SiMe$_3$)$_4$-*commo*-1,1'-Si^{IV}[1,2,3-SiC$_2$B$_4$H$_4$]$_2$.

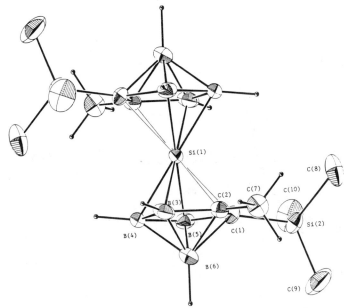

Figure 9.8. Crystal structure of 2,2'-(SiMe$_3$)$_2$-3,3'-(Me)$_2$-*commo*-1,1'-Si^{IV}[1,2,3-SiC$_2$B$_4$H$_4$]$_2$.

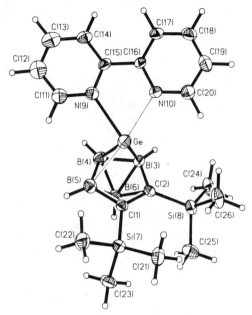

Figure 9.9. Crystal structure of $1\text{-}Ge^{II}(C_{10}H_8N_2)\text{-}2,3\text{-}(SiMe_3)_2\text{-}2,3\text{-}C_2B_4H_4$.

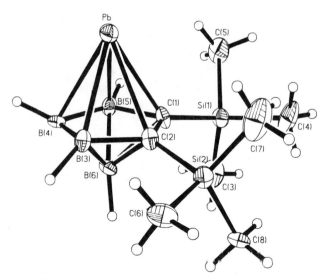

Figure 9.10. Crystal structure of $closo\text{-}1\text{-}Pb^{II}\text{-}2,3\text{-}(SiMe_3)_2\text{-}2,3\text{-}C_2B_4H_4$.

the M^{IV}-complex a solid (at least in the case of M = Si and Ge, and when R = SiMe$_3$), the two compounds can be easily separated. From the above discussion, it is evident that the presence of C-trimethylsilyl groups on carboranes leads to a more systematic, and wide ranging, system of reactions than is found for the C-alkyl or aryl-substituted C$_2$B$_4$-carboranes.

The *closo*-germacarboranes also form adducts with Lewis bases such as 2,2′-bipyridine [42, 44]. The structure of 1-Ge^{II}(C$_{10}$H$_8$N$_2$)-2,3-(Me$_3$Si)$_2$-2,3-C$_2$B$_4$H$_4$, shown in Figure 9.9, is similar to those of the bipyridine-stannacarborane adducts in that the germanium is slipped away from the cage carbons and the bipyridine is situated above the ring borons. However, the structure differs in that the apical germanium is twisted away from the carborane mirror plane so that the Ge-B(3) and Ge-B(5) bond distances are unequal (see Figure 9.9). The germanium can be considered to be η^2 bonded to the unique boron and one basal boron. The two Ge–N bonds are also nonequivalent with one bond distance being 0.153 Å longer than the other. Since the 2,2′-bipyridine nitrogens are equivalent and the *closo*-germacarborane is, presumably, symmetric, there is no ready explanation for these distortions. The solution behavior of the bipyridine-germacarborane complexes is also unusual in that the room temperature proton-decoupled ^{11}B-NMR spectrum shows single boron resonance, indicating fluxional behavior. Thus, it may be that the structure shown in Figure 9.9 represents only one of several structures that exist in solution [42, 44].

Of all the group 14 carboranes the least studied have been the plumbacarboranes. The first example of such compounds was reported by Rudolph et al. in 1970 as a member of the series MC$_2$B$_9$H$_{11}$ (M = group 14 metal) [45]. The smaller cage pentagonal bipyramidal plumbacarboranes of the type *closo*-1-Pb-2,3-(R)$_2$-2,3-C$_2$B$_4$H$_4$ (R = CH$_3$, H) were reported somewhat later by Grimes and co-workers [27]. Structural data were not reported for either of these compounds, but ^{11}B-NMR spectra indicated *closo*-structures. However, it was not clear whether the plumbacarboranes exhibit a slip distortion as has been found in many metallacarboranes. The splitting of the basal BH resonances in the ^{11}B-NMR spectrum of PbC$_2$B$_9$H$_{11}$ was taken by Rudolph as indicating a possible slippage, but such an interpretation has been questioned by Grimes [27]. The X-ray crystal structure of 1-Pb-2,3-(SiMe$_3$)$_2$-2,3-C$_2$B$_4$H$_4$, shown in Figure 9.10, confirms the *closo*-geometry in that it shows the Pb to be essentially symmetrically bonded above the C$_2$B$_3$ face; the Pb-cage atom distances are 2.582(17), 2.634(14), 2.601(16), 2.579(17), and 2.520(20) Å for atoms C(1) to B(5), respectively [46]. The bond distances indicate that, if any distortion exists, it is one in which the Pb is slipped toward B(5), a basal boron that is bonded to a cage carbon. The crystal packing diagram (Figure 9.11) shows that the solid consists of closely associated [Pb(SiMe$_3$)$_2$C$_2$B$_4$H$_4$]$_2$ dimers with a crystallographic center of symmetry halfway between the two Pb atoms. This dimeric structure could give rise to the distortions found in the plumbacarborane cage and, if the dimer exists in solution, to the splitting of the basal BH resonances in its ^{11}B-NMR spectra [46].

As in the cases of *closo*-stanna- and germa-carboranes, the *closo*

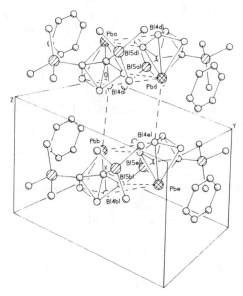

Figure 9.11. The crystal packing diagram showing the extended network of [*closo*-1-Pb^{II}-2,3-(Me$_3$Si)$_2$-2,3-C$_2$B$_4$H$_4$]$_2$ molecular dimers with benzene molecules of crystallization.

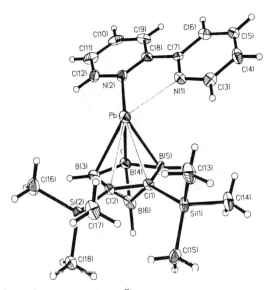

Figure 9.12. Crystal structure of 1-Pb^{II}(C$_{10}$H$_8$N$_2$)-2,3-(SiMe$_3$)$_2$-2,3-C$_2$B$_4$H$_4$.

plumbacarboranes also behave as Lewis acids to form donor-acceptor complexes. The crystal structure of $1\text{-Pb}(C_{10}H_8N_2)\text{-}2,3\text{-}(SiMe_3)_2\text{-}2,3\text{-}C_2B_4H_4$, shown in Figure 9.12 [47], exhibits a less slippage of the apical Pb toward the borons of the C_2B_3 face than is found in the stannacarborane system.

ACKNOWLEDGMENTS

The perseverance of numerous undergraduate students, postdoctoral associates, and other co-workers in many of these studies is gratefully acknowledged. Portions of this work were supported by the National Science Foundation, the Robert A. Welch Foundation, and the donors of the Petroleum Research Fund, administered by the American Chemical Society.

REFERENCES

1. (a) Onak, T., in: *Boron Hydride Chemistry* (E. L. Muetterties, Ed.), Chap. 10, Academic Press, New York, 1975; (b) Grimes, R. N., in: *Carboranes*, Academic Press, New York 1970; Grimes, R. N., in: *Comprehensive Organometallic Chemistry* (G. Wilkinson, F. G. A. Stone and E. W. Abel, Eds.), Vol. 1, Pergamon Press, Oxford, 1982, and references therein.

2. Grimes, R. N., in: *Advances in Boron and the Boranes, Molecular Structure and Energetics* (J. F. Liebman, A. Greenberg, R. E. Williams, Eds.) Vol. 5, Chap. 11, p.235, VCH, New York, 1988.

3. (a) Calhorda, M. J. and Mingos, D. M. P., *J. Organomet. Chem.*, **229**, 229 (1982); (b) Calhorda, M. J., Mingos, D. M. P. and Welch, A. J., *J. Organomet. Chem.*, **228**, 309 (1982).

4. Hosmane, N. S. and Maguire, J. A., *Adv. Organomet. Chem.*, **30**, 99 (1989).

5. Hosmane, N. S. and Grimes, R. N., *Inorg. Chem.*, **18**, 3294 (1979).

6. Maynard, R. B., Borodinsky, L. and Grimes, R. N., *Inorg. Synth.*, **22**, 211 (1983).

7. (a) Onak, T., Williams, R. E. and Weiss, H. G., *J. Am. Chem. Soc.*, **84**, 2830 (1962); (b) Onak, T., Gerhart, F. J. and Williams, R. E., *J. Am. Chem. Soc.*, **85**, 3378 (1963); (c) Onak, T., Drake, R. P. and Dunks, G. B., *Inorg. Chem.*, **3**, 1686 (1964).

8. (a) Hosmane, N. S., Sirmokadam, N. N. and Mollenhauer, M. N., *J. Organomet. Chem.*, **279**, 359 (1985); (b) Hosmane, N. S., Mollenhauer, M. N., Cowley, A. H. and Norman, N. C., *Organometallics*, **4**, 1194 (1985).

9. Hosmane, N. S., Maldar, N. N., Potts, S. B., Rankin, D. W. H. and Robertson, H. E., *Inorg. Chem.*, **25**, 1561 (1986).

10. Ledoux, W. A. and Grimes, R. N., *J. Organomet. Chem.*, **28**, 37 (1971).

11. Hosmane, N. S., Islam, M. S. and Burns, E. G., *Inorg. Chem.*, **26**, 3236 (1987).

12. Hosmane, N. S., Dehghan, M. and Davies, S., *J. Am. Chem. Soc.*, **106**, 6435 (1984).

13. Hosmane, N. S., Barreto, R. D., Tolle, M. A., Alexander, J. J., Quintana, W., Siriwardane, U., Shore, S. G. and Williams, R. E., *Inorg. Chem.*, **29**, 2698 (1990).

14. Onak, T. and Dunks, G. B., *Inorg. Chem.*, **5**, 439 (1966).

15. Hosmane, N. S., Siriwardane, U., Zhang, G., Zhu, H. and Maguire, J. A., *J. Chem. Soc., Chem. Commun.*, 1128 (1989).

16. Hosmane, N. S., de Meester, P., Siriwardane, U., Islam, M. S. and Chu, S. S. C., *J. Chem. Soc., Chem. Commun.*, 1421 (1986).

17. Siriwardane, U., Islam, M. S., West, T. A., Hosmane, N. S., Maguire, J. A. and Cowley, A. H., *J. Am. Chem. Soc.*, **109**, 4600 (1987).

18. Davis, J. H., Jr., Sinn, E. and Grimes, R. N., Papers presented at the Third Chemical Congress of North America, Toronto, Canada, June 1988, *Abstract*: INOR 73, and the First Boron-USA Workshop, Southern Methodist University, Dallas, TX, April 1988, *Abstracts*: TM25.

19. Grimes, R. N., Ed., *Metal Interactions with Boron Clusters*, Plenum Press, New York, 1982.

20. Todd, L. J., in: *Comprehensive Organometallic Chemistry* (G. Wilkinson, F. G. A. Stone and E. W. Abel, Eds.), Vol. 1, Chap. 5, 6, Pergamon Press, Oxford, 1982.

21. Grimes, R. N., *Rev. Silicon, Germanium, Tin, Lead Compd.*, **2**, 223 (1977).

22. Hosmane, N. S. and Maguire, J. A., in: *Advances in Boron and the Boranes, Molecular Structure and Energetics* (J. F. Liebman, A. Greenberg and R. E. Williams, Eds.), Vol. 5, Chap. 14, p. 297, VCH, New York, 1988.

23. Hosmane, N. S., Sirmokadam, N. N. and Herber, R. H., *Organometallics*, **3**, 1665 (1984).

24. Hosmane, N. S., de Meester, P., Maldar, N. N., Potts, S. B., Chu, S. S. C. and Herber, R. H., *Organometallics*, **5**, 772 (1986).

25. Cowley, A. H., Galow, P., Hosmane, N. S., Jutzi, P. and Norman, N. C., *J. Chem. Soc., Chem. Commun.*, 1564 (1984).

26. See reference 13.

27. Wong, K.-S. and Grimes, R. N. *Inorg. Chem.*, **16**, 2053 (1977).

28. Hosmane, N. S., Fagner, J. S., Zhu, H., Siriwardane, U., Maguire, J. A., Zhang, G. and Pinkston, B. S., *Organometallics*, **8**, 1769 (1989).

29. Siriwardane, U., Maguire, J. A., Banewicz, J. J. and Hosmane, N. S., *Organometallics*, **8**, 2792 (1989).

30. Siriwardane, U., Hosmane, N. S., Chu and S. S. C., *Acta Crystallogr., Cryst. Struct. Commun., Sect. C.*, **C43**, 1067 (1987).

31. Hosmane, N. S., Islam, M. S., Siriwardane, U., Maguire, J. A. and Campana, C. F., *Organometallics*, **6**, 2447 (1987).

32. Siriwardane, U. and Hosmane, N. S., *Acta Crystallogr., Cryst. Struct. Commun., Sect. C.*, **C44**, 1572 (1988).

33. Jutzi, P., Kohl, F., Krüger, C., Wolmershäuser, G., Hofmann, P. and Stauffert, P., *Angew. Chem., Int. Ed. Engl.*, **21**, 70 (1982); Kohl, F. X., Schlüter, E., Jutzi, P., Krüger, C., Wolmershäuser, G., Hofmann, P. and Stauffert, P., *Chem. Ber.*, **117**, 1178 (1984).

34. Barreto, R. D., Fehlner, T. P. and Hosmane, N. S., *Inorg. Chem.*, **27**, 453 (1988).

35. Maguire, J. A., Ford, G. P. and Hosmane, N. S., *Inorg. Chem.*, **27**, 3354 (1988).

36. Hosmane, N. S. and Sirmokadam, N. N., *Organometallics*, **3**, 1119 (1984).

37. Hosmane, N. S. and Fagner, J. S., unpublished results

38. Islam, M. S., Siriwardane, U., Hosmane, N. S., Maguire, J. A., de Meester, P. and Chu, S. S. C., *Organometallics*, **6**, 1936 (1987).

39. Siriwardane, U., Islam, M. S., Maguire, J. A. and Hosmane, N. S., *Organometallics*, 7, 1893 (1988).

40. Hosmane, N. S., Lu, K.-J. and Siriwardane, U., manuscript in preparation.

41. Hosmane, N. S., de Meester, P., Siriwardane, U., Islam, M. S. and Chu, S. S. C., *J. Am. Chem. Soc.*, **108**, 6050 (1986).

42. Hosmane, N. S., Islam, M. S., Pinkston, B. S., Siriwardane, U., Banewicz, J. J. and Maguire, J. A., *Organometallics*, 7, 2340 (1988).

43. Maguire, J. A. and Hosmane, N. S., unpublished results.

44. Hosmane, N. S., Siriwardane, U., Islam, M. S., Maguire, J. A. and Chu, S. S. C., *Inorg. Chem.*, **26**, 3428 (1987).

45. Rudolph, R. W., Voorhees, R. L. and Cochoy, R. E., *J. Am. Chem. Soc.*, **92**, 3351 (1970); Voorhees, R. L. and Rudolph, R. W., *J. Am. Chem. Soc.*, **91**, 2173 (1969).

46. Hosmane, N. S., Siriwardane, U., Zhu, H., Zhang, G. and Maguire, J. A., *Organometallics*, **8**, 566 (1989).

47. Hosmane, N. S., Lu, K.-J., Zhu, H., Siriwardane, U., Shet, M. S. and Maguire, J. A., *Organometallics*, **9**, 000 (1990).

CHAPTER 10

Borane Transition Metal Cluster Chemistry: Hydroboration of the Unsaturated Cluster $(\mu H)_2 Os_3 (CO)_{10}$

D. P. WORKMAN and S. G. SHORE

10.1. INTRODUCTION

The triosmium cluster $(\mu\text{-}H)_2Os_3(CO)_{10}$, I, is a triangular molecule in which two osmiums are doubly hydrogen bridged, with a bridge above and a bridge below the Os_3 plane (Figure 10.1) (Orpen et al. 1979, Broach and Williams 1979). This molecule is electronically unsaturated, containing 46 valence electrons instead of the usual 48 electrons associated with a triangular cluster of the iron sub-group (Wade 1975, 1976). Electron pair donors, molecular (Shapley et al. 1975, Deeming and Hasso 1975), and ionic (Kennedy et al. 1981) add to this cluster to

Electron Deficient Boron and Carbon Clusters, Edited by George A. Olah, Kenneth Wade, and Robert E. Williams.
ISBN 0-471-52795-5 © 1991 John Wiley & Sons, Inc.

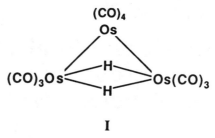

I

Figure 10.1. The structure of $(\mu\text{-H})_2Os_3(CO)_{10}$, **I**.

displace a bridging hydrogen to a terminal site to form adducts such as those represented in Figure 10.2.

The ease of electron addition to **I** encouraged us to consider the possibility of the B–H bond functioning as an electron pair donor to this cluster, since borane complexes such as $[BH_4]^-$ can serve as electron pair donors to transition metal atoms through the formation of metal–H–B 3-center bonds (Gilbert et al. 1982). We found that the reaction of $[BH_4]^-$ with $(\mu\text{-H})_2Os_3(CO)_{10}$ results in hydride transfer (Jan 1985) to give $[H_2(\mu\text{-H})Os_3(CO)_{10}]^-$. On the other hand, reactions of the cluster with BH_3, formed from B_2H_6 in the presence of NEt_3, Me_2O, or THF effectively resulted in the hydroboration of the cluster (Shore et al. 1983, 1984). Two different types of hydroboration products were obtained: a triosmium borylidyne carbonyl, $(\mu\text{-H})_3Os_3(CO)_9(\mu_3\text{-BCO})$, **II**, and a boroxin-supported methylidyne carbonyl, $[(\mu\text{-H})_3Os_3(CO)_9(\mu_3\text{-C-})]_3[O_3B_3O_3]$, **III**; the particular product formed depended upon the reaction conditions employed.

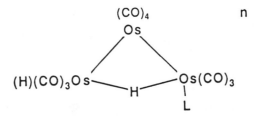

$$L = CO,\ PR_3,\ CNR;\quad n = 0$$
$$L = Cl^-,\ Br^-,\ I^-,\ Mn(CO)_5^-,\ CpMo(CO)_3^-,$$
$$CpFe(CO)_2^-;\quad n = -1$$
$$L = Fe(CO)_4^{2-},\ CpV(CO)_3^{2-};\quad n = -2$$

Figure 10.2. Structures of adducts of nucleophiles to $(\mu\text{-H})_2Os_3(CO)_{10}$.

In this report we summarize our results on the hydroboration of (μ-H)$_2$Os$_3$(CO)$_{10}$ and describe the subsequent reactions and structures of the products derived from these hydroboration studies.

10.2. HYDROBORATION REACTIONS

10.2.1. Hydroboration of (μ-H)$_2$Os$_3$(CO)$_{10}$ in the Presence of NEt$_3$ or OMe$_2$

The cluster (μ-H)$_3$Os$_3$(CO)$_9$(μ_3-BCO), **II**, is prepared according to equation (10.1). A Lewis base is required to cleave the diborane.

$$(\mu\text{-H})_2\text{Os}_3(\text{CO})_{10} + \text{B}_2\text{H}_6 \xrightarrow[\text{CH}_2\text{Cl}_2]{\text{L}} (\mu\text{-H})_3\text{Os}_3(\text{CO})_9(\mu_3\text{-BCO}) + \text{H}_2$$
$$\quad\text{I}\qquad\qquad\qquad\qquad\qquad\qquad\qquad\qquad\qquad\quad\text{II}$$

$$(10.1)$$

L = OMe$_2$, NEt$_3$. It serves as a catalyst in the reaction.

The molecular structure of **II** is shown in Figure 10.3 (Shore et al. 1983). This cluster contains a tetranuclear Os$_3$B core. It is an analog of the ketenylidene cluster (μ-H)$_2$Os$_3$(CO)$_9$(μ_3-CCO) (Sievert et al. 1982; Shapley et al. 1983) in which the α-carbon atom of the CCO fragment is replaced by a boron atom and

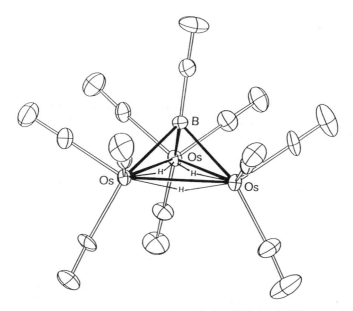

Figure 10.3. The molecular structure of (μ-H)$_3$Os$_3$(CO)$_9$(μ_3-BCO). Reprinted with permission from *Inorg. Chem.*, **29** (1990). Copyright 1990 American Chemical Society.

hydrogen atom. The BCO unit in **II** is nearly linear ($\angle$ B–C–O $= 178(2)°$); it is tilted only $6.4°$ from being perpendicular to the triosmium plane. The boron–carbon bond distance of $1.469(15)$ Å is shorter than observed B–C single bond distances (approx. 1.6 Å) (Saturnino et al. 1975, Hsu et al. 1987) and the B–C distances in BH_3CO, $B_2H_4(CO)_2$, and B_3H_7CO (1.52–1.57 Å) (Rathke and Schaeffer 1974, Gordy et al. 1950, Glore et al. 1973) suggesting partial double bond character.

Cluster **II** is a robust molecule. This bright yellow, air-stable compound sublimes at $60°C$ under 10^{-4} Torr. It decomposes in vacuum at about $90°C$. The carbonyl ligands in this complex do not appear to exchange with added ^{13}CO even under forcing conditions (1000 psi, $80°C$), nor, based upon ^{13}C-NMR spectroscopy, do they appear to undergo any intramolecular exchange processes up to the decomposition temperature.

10.2.2. Hydroboration of $(\mu$-H$)_2$Os$_3$(CO)$_{10}$ in the Presence of THF

When the hydroboration of $(\mu$-H$)_2$Os$_3$(CO)$_{10}$ is conducted in dilute solutions using a quantitative amount of THF and excess diborane, the borylidyne cluster **II** is formed as the major product as observed in the reaction described above (equation (10.1)). In sharp contrast, however, when the reaction is conducted with an excess of THF, another cluster system $[(\mu$-H$)_3$Os$_3$(CO)$_9(\mu_3$-C-$)]_3[O_3B_3O_3]$, **III**, is formed in 80% yield

$$(\mu\text{-H})_2\text{Os}_3(\text{CO})_{10} + B_2H_6 + \text{THF(xs)} \xrightarrow{\text{CH}_2\text{Cl}_2}$$

$$\mathbf{I}$$

$$[(\mu\text{-H})_3\text{Os}_3(\text{CO})_9(\mu_3\text{-C-})]_3[O_3B_3O_3] + C_4H_{10}$$

$$\mathbf{III} \tag{10.2}$$

The molecular structure of **III** is shown in Figure 10.4 (Shore et al. 1984). In the center of the molecule there is a B_3O_3 ring, an approximate regular, hexagonal ring to which three triosmium methylidyne units are bound. Each methylidyne unit consists of a tetranuclear Os_3C core which contains three hydrogen atoms bridging the metal framework and nine terminal carbonyl ligands. This compound is a cream, white solid which is stable at room temperature in the absence of air and moisture.

10.2.3. Proposed Reaction Pathways for the Formation of $(\mu$-H$)_3$Os$_3$(CO)$_9(\mu_3$-BCO) and $[(\mu$-H$)_3$Os$_3$(CO)$_9(\mu_3$-C-$)]_3[O_3B_3O_3]$

Reaction pathways can be proposed which are consistent with the conditions necessary for the formation of $(\mu$-H$)_3$Os$_3$(CO)$_9(\mu_3$-BCO), **II**, and $[(\mu$-H$)_3$Os$_3$(CO)$_9(\mu_3$-C-$)]_3[O_3B_3O_3]$, **III** and are also consistent with model reactions.

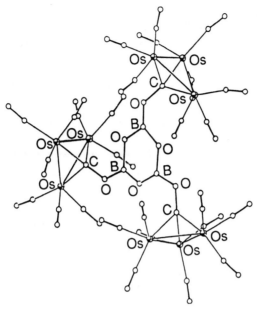

Figure 10.4. The molecular structure of $[(\mu\text{-}H)_3Os_3(CO)_9(\mu_3\text{-}C\text{-})]_3[O_3B_3O_3]$, **III**.

In the proposed pathway for the formation of **II** (Scheme I), an initial adduct **IIa** of BH_3L to **I** forms through an Os–H–B bond. Abstraction of L from **IIa** by the free B_2H_6 in the system allows the incorporation of BH_3 into (μ-$H)_2Os_3(CO)_{10}$ to form **IIb**. Reductive elimination of H_2 from the boron atom could form **II** directly or it could proceed through **IIc** which could then irreversibly isomerize to **II**. However, evidence could not be obtained for **IIc**. An iron analog of **IIc** has been prepared in Fehlner's laboratory (Vites al. 1984). The iron analog does not isomerize to an analog of **II**.

To test the assumption that BH_3L is regenerated as the reaction in Scheme I proceeds, the reaction between **I** and BH_3NEt_3 in the presence of B_2H_6 was followed by NMR spectroscopy. No apparent diminution in BH_3NEt_3 concentration was observed even as the reaction approached completion.

The pathway proposed for the formation of **III** (Scheme II) involves the same first step as in Scheme I, formation of an Os–H–B bond through the addition of BH_3THF to **I**. However, in this case the release of THF from the BH_3 group coordinated to the cluster is inhibited by the presence of excess THF, thus favoring an alternative reaction pathway which involves rupture of the B–H–Os bond and transfer of the resulting $THFBH_2$ fragment to the oxygen atom of a carbonyl group on the unique osmium of the cluster to form the intermediate **IIIb**. The coordinated $THFBH_2$ fragment withdraws electron density from the C–O bond thereby inducing the carbonyl ligand to shift from a terminal

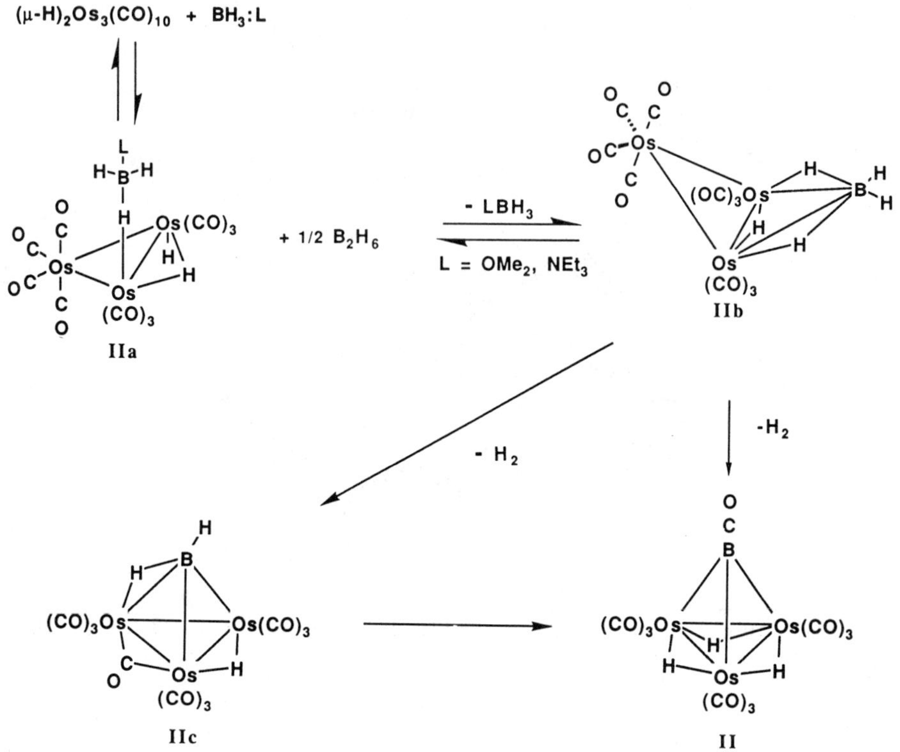

Scheme I.

position to a face capping site, **IIIc**. The two remaining hydrogen atoms on the boron are subsequently transferred to the THF ring and butane is eliminated, identified by GC-mass spectroscopy, leaving an OBO fragment bound to the μ_3-carbon of the triosmium methylidyne cluster, **IIIe**. Trimerization of OB gives the boroxin-supported methylidyne cluster product, **III**. In effect, **III** can be considered to be an analog of trimethoxyboroxine, $(MeO)_3B_3O_3$, with each H atom on the Me group replaced by an $Os(H)(CO)_3$ unit of the triosmium methylidyne cluster. Precedents exist for the key steps in the proposed reaction pathway. Shriver and co-workers were the first to observe the ability of a Lewis acid to induce a shift of a terminal carbonyl ligand to a bridging site (Horwitz and Shriver 1984, Kristoff and Shriver 1974, Alich et al. 1971, 1972. In model reactions described below we have observed the Lewis acid induced transfer of a bridging carbonyl to a μ_3 capping site of three Os atoms (equations (10.4) and (10.5)). The transfer of BH hydrogen to the THF ring and the opening of the ring is analogous to well-known oligomerization reactions of THF in the presence of Lewis acids (Dreyfuss and Dreyfuss 1967, Fried and Kleene 1941, Stone and Shechter 1950, Burwell 1954).

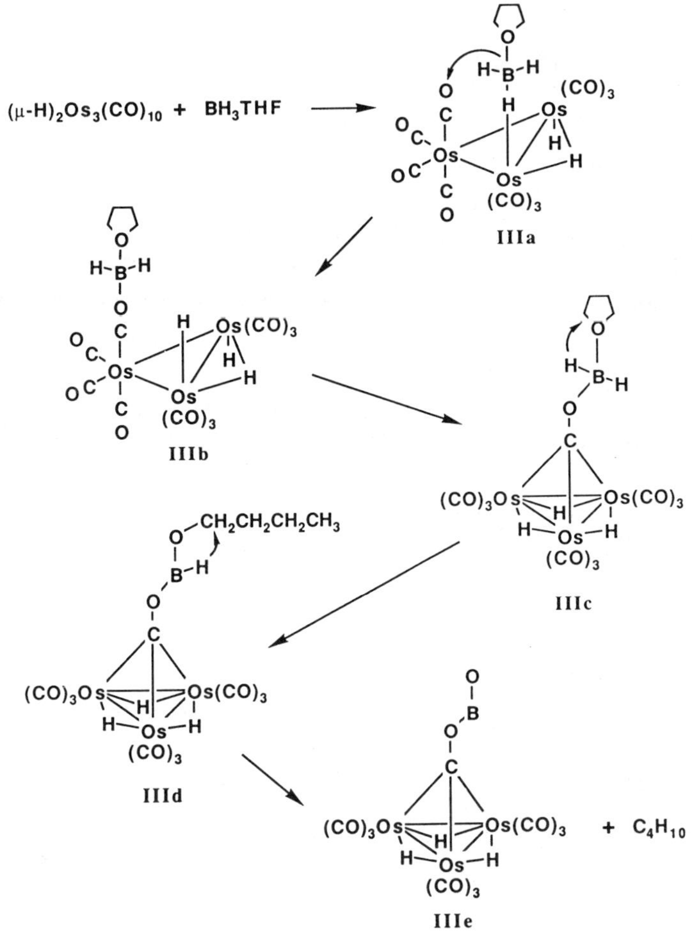

$(\mu\text{-H})_2Os_3(CO)_{10}$ + BH$_3$THF ⟶

IIIa

IIIb

IIIc

IIId

IIIe   + C$_4$H$_{10}$

Scheme II.

Transfer of a terminal carbonyl ligand to a face capping site was modeled in the sequence of reactions given by equations (10.3) and (10.4)

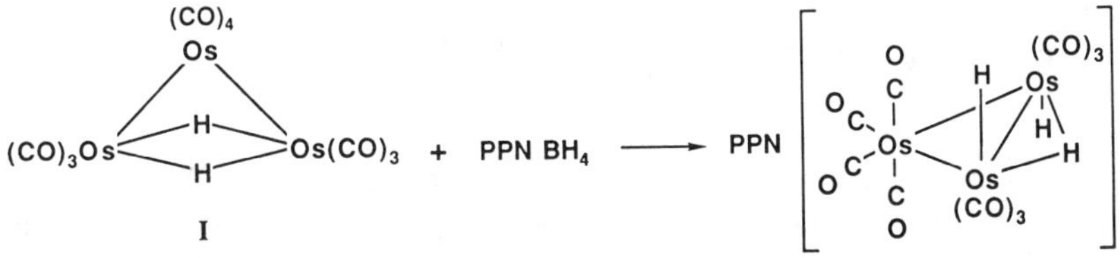

I   + PPN BH$_4$ ⟶   PPN

(10.3)

$$\text{(10.4)}$$

Reaction of $(\mu\text{-H})_2Os_3(CO)_{10}$ with [PPN][BH$_4$] generates $[(\mu\text{-H})_3Os_3(CO)_{10}]^-$, which reacts with Me$_3OBF_4$ to produce $(\mu\text{-H})_3Os_3(CO)_9(\mu_3\text{-COMe})$ (Jan 1985) in 28% yield. This result demonstrates that a strong electrophile can induce a terminal carbonyl to move to a triply bridging site (Sailor et al. 1987).

10.2.4. Hydroboration of $(\mu\text{-H})_2Os_3(CO)_{10}$ by Catecholborane

The hydroboration of **I** by 1,3,2-benzodioxaborole (catecholborane), $C_6H_4O_2BH$ (equation (10.5)) produces $(\mu\text{-H})_3Os_3(CO)_9(\mu_3\text{-COBO}_2C_6H_4)$, **IV** a monomeric analog of the boroxin supported cluster system **III**.

$$\text{(10.5)}$$

The formation of **IV** is believed to occur in a manner similar to the formation of **III**, a Lewis acid induced shift of a terminal carbonyl ligand to a face capping site, and serves as a model for the conversion of $(\mu\text{-H})_3Os_3(CO)_{10}$ to a triosmium methylidyne cluster under hydroboration conditions.

10.3. REACTIONS OF $(\mu\text{-H})_3Os_3(CO)_9(\mu_3\text{-BCO})$ WITH LEWIS ACIDS

Shriver and co-workers (Sailor et al. 1987, Sailor and Shriver 1985, Hriljac and Shriver 1985, 1987, Went et al. 1987, Shriver and Sailor 1988) have demonstrated that the Group VIII dianionic ketenylidenes, $[M_3(CO)_9(\mu_3\text{-CCO})]^{2-}$ (M = Fe, Ru, Os), react with electrophiles. Electrophilic attack can occur at the α-carbon or the oxygen atom of the CCO unit or at the metal atom, depending upon the specific metal ketenylidene and electrophile used. Attack at the CCO unit results in the generation of acetylide and acetylene clusters. Additionally, the acetylene complexes undergo scission of the carbon-carbon bond to form methylidyne and dialkylidyne clusters.

UV-photoelectron spectroscopic and Fenske–Hall molecular orbital investigation of $(\mu\text{-H})_3Os_3(CO)_9(\mu_3\text{-BCO})$ (Barretto et al. 1986), **II**, reveal that the electronic structure and bonding of **II** are similar to the isoelectronic ketenylidene cluster $(\mu\text{-H})_2Os_3(CO)_9(\mu_3\text{-CCO})$.

Both the PES results and Fenske–Hall calculations suggest that when bound to a transition metal cluster, boron can act as a pseudo metal atom. The calculations indicate that a synergistic interaction exists between the CO ligand and boron center analogous to that seen in metal carbonyl systems. As a result the unique carbonyl oxygen atom in **I** is the most negative oxygen atom in the molecule and it is a potential site for reactions with electrophilic reagents (Barretto et al. 1986).

Although the acid catalyzed reaction of $(\mu\text{-H})_2Os_3(CO)_9(\mu_3\text{-CCO})$ with H_2O produces the cluster carboxylic acid $[(\mu\text{-H})_3Os_3(CO)_9(\mu_3\text{-CCO}_2H)]_2$ (Krause et al. 1987), an analogous reaction involving $(\mu\text{-H})_3Os_3(CO)_9(\mu_3\text{-BCO})$, **II**, does not appear to take place. In general, **II**, does not appear to react with proton sources or methylating agents, $(CH_3OSO_2CF_3$ or $CH_3OSO_2F)$. It does react, however, with molecular Lewis acids (Jan et al. 1987, Jan and Shore 1987, Workman et al. 1990a,b). Electrophilic attack is believed to occur solely at the oxygen atom of the unique carbonyl ligand (Barreto et al. 1986). This carbonyl ligand can be induced to shift from its terminal position to a bridging site (Horwitz and Shriver 1984, Kristoff and Shriver 1974, Alich et al. 1971, 1972); depending upon the strength of the Lewis acid employed as discussed in the following sections.

10.3.1. Formation of Vinylidene Analogs

Boron trihalides react with $(\mu\text{-H})_3Os_3(CO)_9(\mu_3\text{-BCO})$, **II**, to produce the vinylidene analogs, $(\mu\text{-H})_3Os_3(CO)_9(\mu_3\text{-CBX}_2)$ $(X = Cl, Br)$ **V** (Workman et al. 1990a)

$$(\mu\text{-H})_3Os_3(CO)_9(\mu_3\text{-BCO}) + BX_3 \rightarrow (\mu\text{-H})_3Os_3(CO)_9(\mu_3\text{-CBX}_2) + 1/3\ X_3B_3O_3$$

$$\text{II} \qquad\qquad\qquad\qquad\qquad\qquad \text{V}$$

$$(10.6)$$

where $X = Cl$, Br. Removal of excess BX_3 results in disproportionation of the trihaloboroxine, $X_3B_3O_3$, to BX_3 and B_2O_3 (Knowles and Buchanan 1965, Gobeau and Keller 1951a,b). Chemical confirmation of the Os_3C core is provided by quantitative hydrolysis to $(\mu\text{-H})_3Os_3(CO)_9(\mu_3\text{-CH})$ and boric acid.

Figure 10.5 depicts the molecular structure of $(\mu\text{-H})_3Os_3(CO)_9(\mu_3\text{-CBCl}_2)$ (Jan et al. 1987). The Os_3C core is like that observed in the methylidyne clusters $(\mu\text{-H})_3Os_3(CO)_9(\mu_3\text{-CX})$ $(X = H, Cl, Br, Ph)$ (Orpen and Koetzle 1984, Jan 1985).

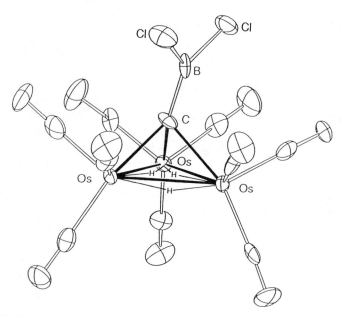

Figure 10.5. The molecular structure of $(\mu\text{-H})_3Os_3(CO)_9(\mu_3\text{-CBCl}_2)$, **V**. Reprinted with permission from *Inorg. Chem.*, **29** (1990). Copyright 1990 American Chemical Society.

The B–C vector is tilted 15° from being perpendicular to the triosmium plane. The two Cl–B–Cl bond angles (123(1)°, 121(1)°, and the Cl–B–Cl bond angle (116(1)°) around the tricoordinate boron atom are consistent with it being sp^2 hybridized. Partial double bond character of the B–C bond is reflected in the B–C bond distance, 1.47(2) Å, which is intermediate between the observed boron-carbon single bond distances (approx. 1.6 Å) (Saturnino et al. 1975, Hsu et al. 1987) and the observed boron-carbon double bond distance, 1.361(5) Å (Boese et al. 1989). The "short" B–C distance might reflect π-interaction between the empty p orbital on the boron and the filled e set of cluster orbitals centered on the capping carbon atom (Sherwood and Hall 1982, Chseky and Hall 1981, DeKock et al. 1982). A similar bonding interaction between the carbido carbon and the tricoordinate boron in $HFe_4(CO)_{12}CBH_2$ has been proposed by Fehlner and co-workers (Meng et al. 1989, Fehlner et al. 1989).

Formation of **V** results in an interchange of the boron and carbon atom positions. This transformation was studied by NMR spectroscopy in the reaction of $(\mu\text{-H})_3Os_3(CO)_9(\mu_3\text{-BCO})$ with $^{10}BCl_3$ and the reaction of $(\mu\text{-H})_3Os_3(CO)_9(\mu_3\text{-}^{10}BCO)$ with BCl_3. Boron from the boron trichloride does not appear to be incorporated in the cluster product **V**. These observations indicate that there is no interchange between the boron in BCl_3 and the boron of $(\mu\text{-H})_3Os_3(CO)_9(\mu_3\text{-BCO})$ in the formation of $(\mu\text{-H})_3Os_3(CO)_9(\mu_3\text{-CBCl}_2)$. Thus the formation of **V** appears to occur through intramolecular exchange of the boron and carbon atoms of **II**. However, on a much slower time scale than the formation of **V** (10 min vs 2 days), ^{10}BCl$_3$ reacts with **V** to noticeably enrich **V** with the ^{10}B isotope.

A proposed reaction pathway for the formation of **V** is given in Scheme III which is initiated through electrophilic attack of BX$_3$ at the unique carbonyl oxygen. The resulting reduction of the CO bond order induces a shift from the terminal position to a bridging site (Horwitz and Shriver 1984, Kristoff and Shriver 1974, Alich et al. 1971, 1972). Movement of the carbonyl ligand into the μ_3-site exposes the boron vertex and results in successive halogen atom transfer from the reagent boron to the cluster boron. With the elimination of X–B–O as trihaloboroxine, X$_3$B$_3$O$_3$, cluster **V** is produced.

A triphenylphosphine derivative of **II** has been prepared (Jan 1985): (μ-H)$_3$Os$_3$(CO)$_8$(PPh$_3$)(μ_3-BCO). Its structure, determined from a single crystal X-ray analysis (Krause 1989), is like that of **II** (Figure 10.3) except that an axial

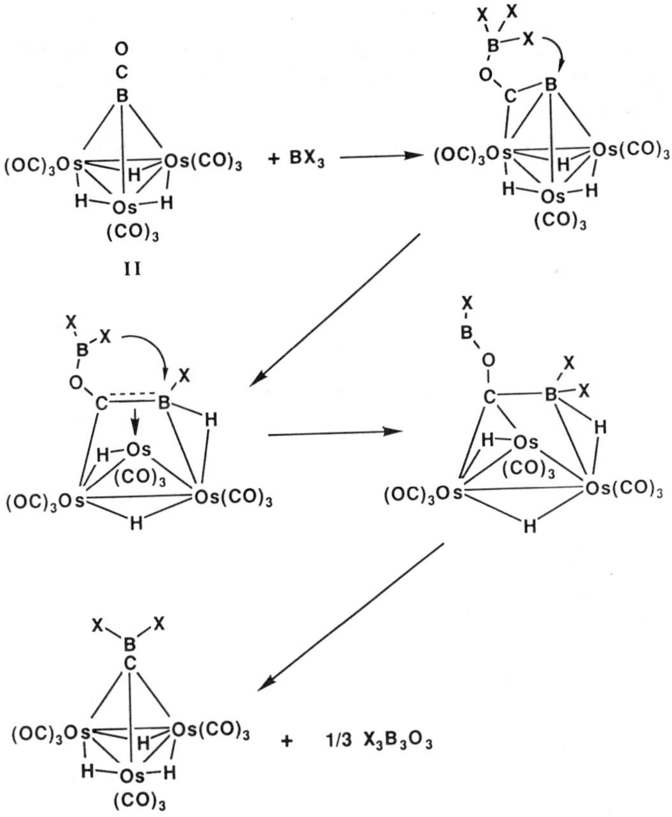

Scheme III.

carbonyl group bound to an osmium atom is replaced by an axial PPh_3 group. The reaction of this cluster with BCl_3 was studied in order to determine if an analog of a vinylidene cluster would be formed or whether the BCl_3 would abstract the PPh_3 group from the cluster. The observed reaction, produces exclusively a vinylidene analog $(\mu\text{-}H)_3Os_3(CO)_8(PPh_3)(\mu_3\text{-}CBCl_2)$, **VI**, similar to **V**

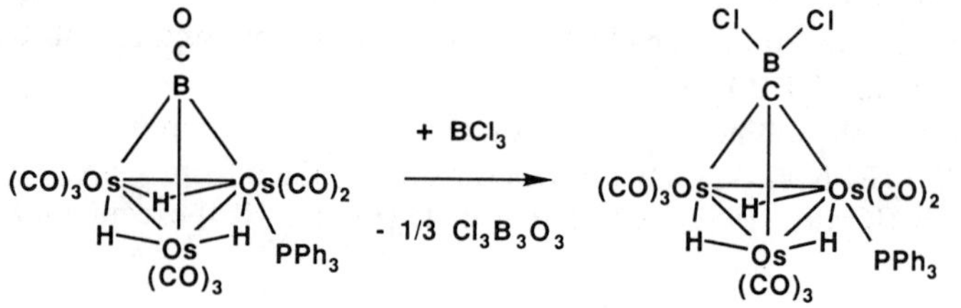

$$(10.7)$$

VI

There is no indication of abstraction of PPh_3 to form PPh_3BCl_3.

At low temperature ($< -40°C$) the Lewis bases NMe_3, PMe_3, and PPh_3 add to the tricoordinate boron of $(\mu\text{-}H)_3Os_3(CO)_9(\mu_3\text{-}CBCl_2)$, **V** (equation (10.8)). Above $-10°C$, the NMe_3 adduct is converted to the salt $[NMe_3H][(\mu\text{-}H)_2Os_3(CO)_9(\mu_3\text{-}CBCl_2)]$ through deprotonation of **V** by the amine (equation (10.8b))

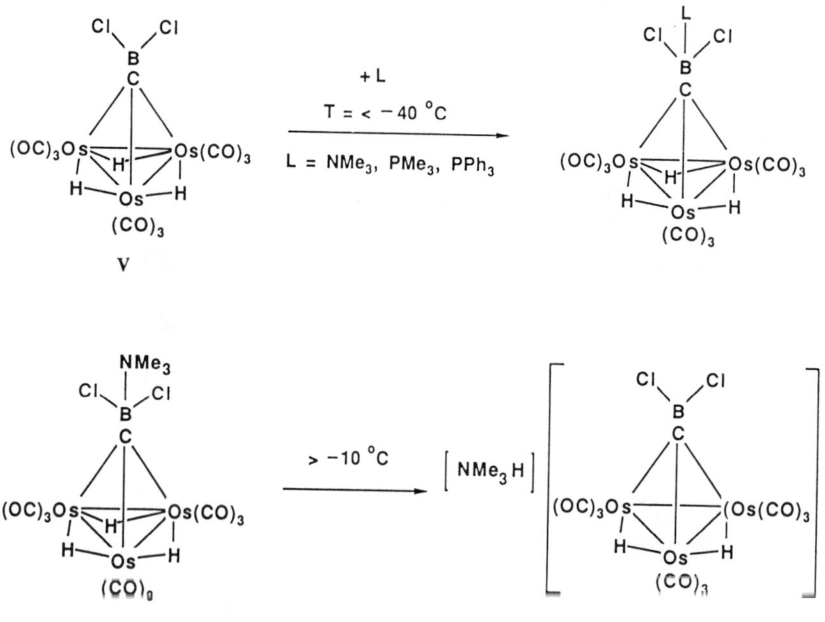

V

$$(10.8)$$

Similarly $HFe_4(CO)_{12}CBH_2$, a cluster substituted monoborane, was shown by Fehlner and co-workers to be deprotonated by NM_3 to form $[NMe_3H][Fe_4(CO)_{12}CBH_2]$ (Fehlner et al. 1989).

Above $-10°C$, the PMe_3 adduct is unstable. The PPh_3 adduct is sufficiently stable at $30°C$ for its NMR spectra to be obtained.

10.3.2. Formation of Alkyne Analogs

In an attempt to isolate compounds which would relate to proposed intermediates (Scheme III) in the reaction of $(\mu-H)_3Os_3(CO)_9(\mu_3-BCO)$, **II**, with BX_3 to produce $(\mu-H)_3Os_3(CO)_9(\mu_3-CBX_2)$, **V** (equation (10.5)), reactions of **II** with the mono- and dichloroborane reagents B-chloro-9-borobicyclo(3.3.1)nonane (B-chloro-9-BBN) (Kramer and Brown 1974), and phenylboron dichloride were studied (Workman et al. 1990a,b).

Reaction of **II** with B-chloro-9-BBN produces $(\mu-H)_3Os_3(CO)_9[\mu_3,\eta^2-C(OBC_8H_{14})B(Cl)]$, **VII**

$$(10.9)$$

The molecular structure of the product **VII** is shown in Figure 10.6. In the formation of **VII**, the unique carbonyl of **II** is shifted to a μ_3 site capping two osmiums and the boron atom while the chlorine atom of the B-chloro-9-BBN is transferred to the μ_3 boron of the cluster. This compound provides the basis for a proposed intermediate containing μ_3-BCl in the formation of **V** (Scheme III) from the reaction of **II** with BX_3.

Compound **VII** can be considered to be an alkyne analog having a carbon atom replaced by a BH group. The B–C unit along with the hydrogen atom bridging Os–B donate four electrons to the cluster through a π-interaction with an Os atoms, an Os–C σ-bond, and a hydrogen bridged Os–H–B bond. The B–C unit adopts the μ_3, η^2 bonding mode and is oriented nearly parallel, within $10°$, to an Os–Os bond (Sappa et al. 1983, Raithby and Rosales 1985).

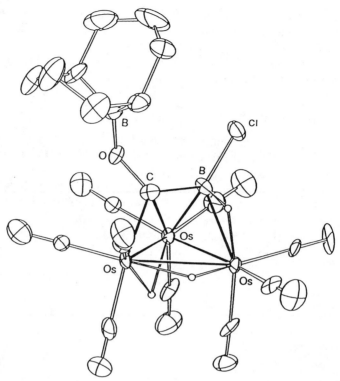

Figure 10.6. The molecular structure of $(\mu\text{-H})_3Os_3(CO)_9[\mu_3,\eta^2\text{-C}(OB_8C_8H_{14})B(Cl)]$, **VII**. Reprinted with permission from *Inorg. Chem.*, **29** (1990). Copyright 1990 American Chemical Society.

Analogously, alkyne products resulting from electrophilic attack at the oxygen atom of a CCO ligand have been observed for the iron and osmium ketenylidene dianions (Hriljac and Shriver 1985, 1987) and a mononuclear tungsten complex (Kriessl et al. 1983a,b).

The B–C distance in **VII**, 1.46(2) Å, is comparable to the B–C distances observed in $(\mu\text{-H})_3Os_3(CO)_9(\mu_3\text{-BCO})$, **II** (Shore et al. 1983), $(\mu\text{-H})_3Os_3 \times (CO)_9(\mu_3\text{-CBCl}_2)$, **V** (Jan et al. 1987), and the ditungsten alkyne analog $W_2[\mu\text{-MeCB(H)Et}](CO)_4(\eta^5\text{-C}_5H_5)_2$ (Carriendo et al. 1984, Barratt et al. 1987) [1.469(15), 1.47(2), and 1.46(1) Å, respectively]. These bonds are all considered to have boron-carbon partial double bond character and have bond distances that are between the boron-carbon double bond distance of 1.361(5) Å observed by Paetzold and co-workers (Boese et al. 1989) and observed B–C single bond distances (approx. 1.6 Å) (Saturnino et al. 1975, Hsu et al. 1987).

Phenylboron dichloride, $PhBCl_2$, reacts with **II** to produce the alkyne analog $(\mu\text{-H})_3Os_3(CO)_9\{\mu_3, \eta^2\text{-C}[OB(Ph)Cl]B(Cl)]\}$, **VIII**,

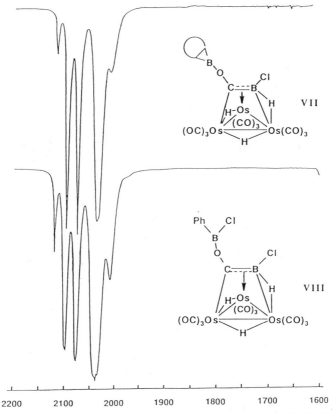

$$(10.10)$$

The proposed structure of **VIII** is analogous to that of **VII**. Infrared spectra of **VII** and **VIII** (Figure 10.7) show strong similarities. NMR data from the cluster components of **VII** and **VIII** are also very similar.

Figure 10.7. Infrared spectra in CH_2Cl_2 (30°C). Reprinted with permission from *Inorg. Chem.*, **29** (1990). Copyright 1990 American Chemical Society.

The reaction of **VII**, $(\mu\text{-H})_3\text{Os}_3(\text{CO})_9[\mu_3,\eta^2\text{-C(OBC}_8\text{H}_{14})\text{B(Cl)}]$ with BCl_3 produces $(\mu\text{-H})_3\text{Os}_3(\text{CO})_9(\mu_3\text{-CBCl}_2)$, **V**, while the reaction with BBr_3 gives evidence for the formation of the mixed halogen complex $(\mu\text{-H})_3\text{Os}_3(\text{CO})_9(\mu_3\text{-}$ CBClBr). The formation of these vinylidene analogs supports the contention that $(\mu\text{-H})_3\text{Os}_3(\text{CO})_9[\mu_3,\eta^2\text{-C(OBC}_8\text{H}_{14})\text{B(Cl)}]$ is a reasonable model for the intermediate formed in the reaction of **II** with a boron trihalide to form **V**. $^{11}\text{B-}$ NMR studies of reaction of **VII** with BX_3 show rapid formation of B-halo-9-BBN. Presumably, this involves initial coordination of BX_3 to the oxygen atom of the C–O–B unit (Scheme IV). Halogen atom transfer to the 9-BBN boron ruptures the B–O bond and eliminates B-halo-9-BBN to produce the proposed intermediate for the reaction of $(\mu\text{-H})_3\text{Os}_3(\text{CO})_9(\mu_3\text{-BCO})$ with BX_3 (Scheme III). A similar step has been proposed for the reaction of trihaloboranes with the boroxin-supported methylidyne cluster $[(\mu\text{-H})_3\text{Os}_3(\text{CO})_9(\mu_3\text{-C-})]_3[\text{O}_3\text{B}_3\text{O}_3]$ (Shore et al. 1984, Jan 1985) and for the reaction of BX_3 (X = Cl, Br) with the ruthenium and osmium methylidyne clusters, $(\mu\text{-H})_3\text{M}_3(\text{CO})_9(\mu_3\text{-COMe})$

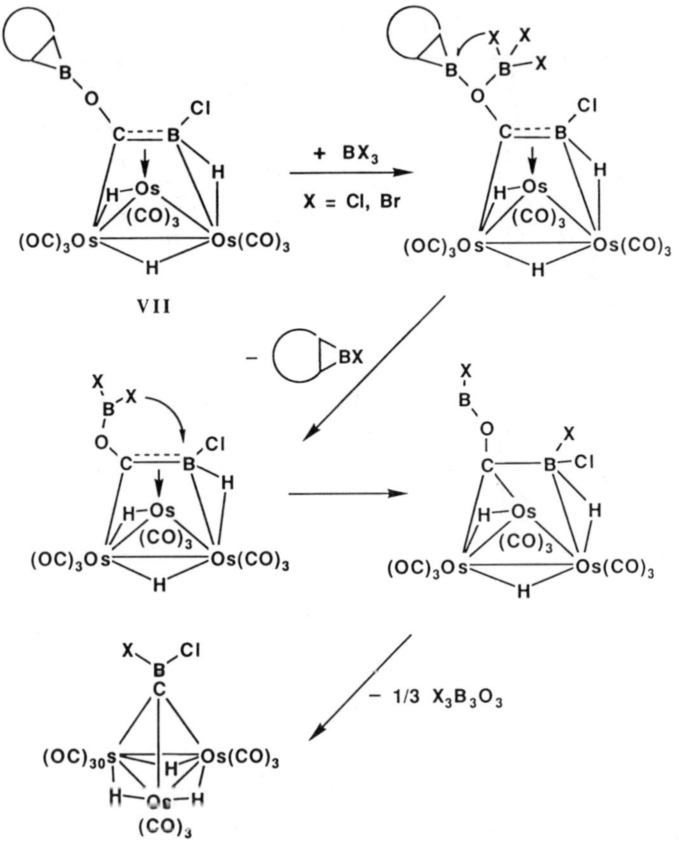

Scheme IV.

(M = Ru, Os) (Keister 1979, Keister and Horling 1980, Keister et al. 1983). In a reaction analogous to that given by equation (10.10), **VIII** reacts with BCl_3 to form **III**, $PhBCl_2$, and $Cl_3B_3O_3$.

The methylidyne complex $(\mu\text{-H})_3Os_3(CO)_9(\mu_3\text{-CH})$ is formed quantitatively in the reaction of **VII** with HCl

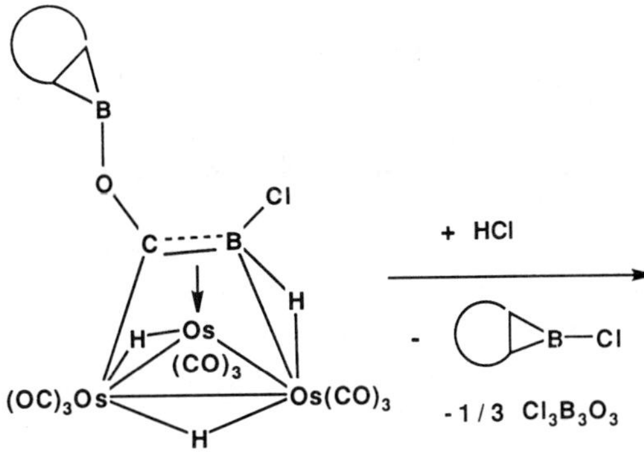

VII

$$(10.11)$$

The reaction of **VIII** with HCl also produces $(\mu\text{-H})_3Os_3(CO)_9(\mu_3\text{-CH})$ but in reduced yields (20–25%).

10.3.3. Reduction of the Unique Carbonyl in $(\mu\text{-H})_3Os_3(CO)_9(\mu_3\text{-BCO})$ with BH_3THF

The unique carbonyl of **II** is reduced to a methylene group by $THFBH_3$ to produce the vinylidene analog, $(\mu\text{-H})_3Os_3(CO)_9(\mu_3,\eta^2\text{-BCH}_2)$, **IX** (Jan and Shore 1987)

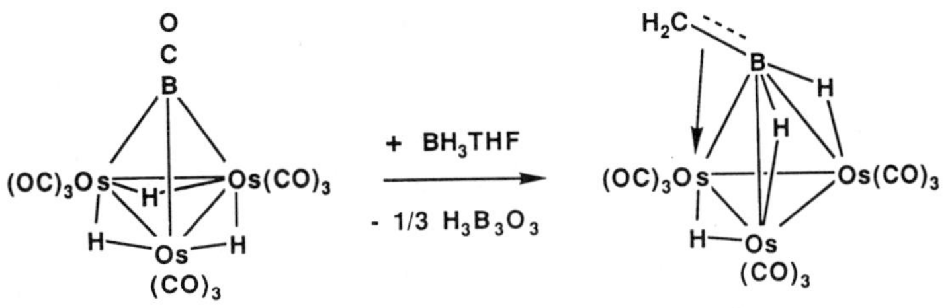

II

IX

$$(10.12)$$

The reduction reaction is believed to occur through initial coordination of BH_3 to the oxygen atom of the unique carbonyl to give $(\mu\text{-H})_3(CO)_9Os_3(\mu_3\text{-BCOBH}_3)$ followed by transfer of two BH hydrogens to the carbon atom. Elimination of H–B–O as boroxine, $H_3B_3O_3$, would then result in the formation of **IX**. Deuterium labeling experiments indicate that reduction occurs with no scrambling of B–H and Os–H–Os hydrogen atoms. Reaction of (μ-H)$_3$Os$_3$(CO)$_9$(μ_3-BCO) with BD$_3$THF gives (μ-H)$_3$Os$_3$(CO)$_9$(μ_3,η^2-BCD$_2$) while reaction of (μ-D)$_3$Os$_3$(CO)$_9$(μ_3-BCO) with BH$_3$THF gives (μ-D)$_3$Os$_3$(CO)$_9$(μ_3,η^2-BCH$_2$). The electrophile BH$_3$ is much weaker Lewis acid toward oxygen bases than the trihaloboranes, BX$_3$ (X = F, Cl, Br) (Fratiello et al. 1968). It probably does not induce a shift of the unique carbonyl ligand from its terminal position to a bridging site, which would account for the fact that interchange of boron and carbon atom positions does not occur.

The molecular structure of **IX** is shown in Figure 10.8 (Jan and Shore 1987). The arrangement of the nonhydrogen atoms in **IX** closely resembles that observed for the vinylidene cluster (μ-H)$_2$Os$_3$(CO)$_9$(μ_3,η^2-CCH$_2$) (Deeming and Underhill 1973, 1974). The B–C distance of 1.498(15) Å is approximately 0.1 Å shorter than observed B–C single bond distances (Saturnino et al. 1975, Hsu et al. 1987). The BCH$_2$ fragment is bound to the triosmium framework by two hydrogen bridged bonds, Os and B atoms and through interaction of the BCH$_2$ π-system with an Os atom. This results in a tilting of the BCH$_2$ unit by 60° from perpendicular toward the Os atom. This tilt is significantly larger than observed in the structurally characterized vinylidene clusters such as Co$_2$Ru(CO)$_9$(μ_3,η^2-CCHR) (R = t-Bu (43°) and Ph (40°) (Roland et al. 1985). The two B–H–Os bridges in the structure **IX** probably force the μ_3-BCH$_2$ to an extreme tilt angle compared to the vinylidene complexes. Although the extreme tilt angle of the BCH$_2$ unit in **IX** implies that the compound can also be described as a methylene bridged complex, the "short" B–C distance and the relatively long Os–C distance 2.325(17) Å favor the vinylidene analogy.

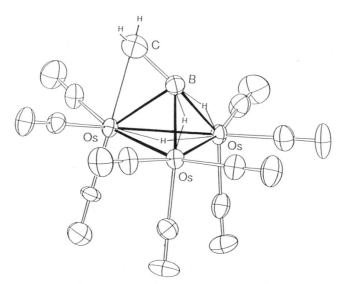

Figure 10.8. The molecular structure of (μ-H)$_3$Os$_3$(CO)$_9$(μ_3,η^2-BCH$_2$), **IX.** Reprinted with permission from *Inorg. Chem.*, **29** (1990). Copyright 1990 American Chemical Society.

Proton and carbon-13 NMR spectra of **IX** indicate that the molecule is asymmetric and nonfluxional up to its decomposition temperature, 90°C, in toluene (Workman et al. 1990a,b). On the other hand, the vinylidene analog $(\mu\text{-H})_2Os_3(CO)_9(\mu_3,\eta^2\text{-CCH}_2)$ is fluxional above room temperature (approx. 72°C) (Deeming and Underhill 1974).

10.4. REACTIONS OF $[(\mu\text{-H})_3Os_3(CO)_9(\mu_3\text{-C-})]_3[O_3B_3O_3]$

10.4.1. Reactions with Boron Trihalides

The boroxin-supported triosmium methylidyne cluster, **III** (Figure 10.4) proves to be a useful reagent in the preparation of triosmium methylidyne cluster derivatives. When a suspension of **III** in CH_2Cl_2 is treated with excess BCl_3 or BBr_3, the chloro- and bromo-methylidyne clusters, $(\mu\text{-H})_3Os_3(CO)_9(\mu_3\text{-CX})$ ($X = Cl, Br$), are produced and isolated in yields of 44% and 60%, respectively (Scheme V) (Shore et al. 1984). These reactions are believed to occur through a sequence that involves initial coordination of the boron halide to the oxygen atom of the C–O–B unit followed by transfer of a halogen atom to the carbon atom with rupture of the carbon-oxygen bond (Keister 1979, Keister and Horling 1980, Keister et al. 1983, Lappert 1958, Dolby and Robinson 1972). Compound **IV**, $(\mu\text{-H})_3Os_3(CO)_9(\mu_3\text{-COBO}_2C_6H_4)$, contains a C–O–B similar linkage to that in **III**. Reaction of **IV** with BBr_3 produces the expected bromomethylidyne cluster $(\mu\text{-H})_3Os_3(CO)_9(\mu_3\text{-CX})$, **X**.

When BF_3 reacts with **III** in CH_2Cl_2 a highly air and moisture-sensitive, and as yet uncharacterized, orange product is obtained. However, when the reaction is undertaken with benzene as the solvent, the phenyl-substituted methylidyne cluster, $(\mu\text{-H})_3Os_3(CO)_9(\mu_3\text{-CC}_6H_5)$, **XI** is isolated in 70% yield (Scheme V) (Shore et al. 1984). Initial coordination of BF_3 to the methylidyne oxygen,

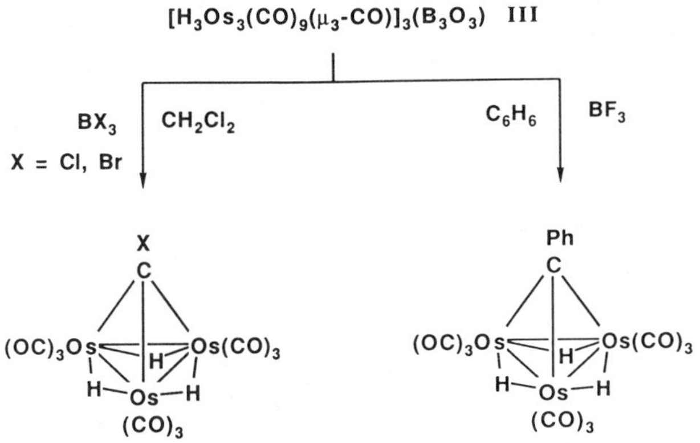

Scheme V.

similar to BCl_3 and BBr_3 is envisioned, however, the accompanying fluorine atom transfer does not appear to be favored, possibly due to the stronger B–F bond. Instead, addition of BF_3 results in either heterolytic cleavage of the C–O bond to produce a triosmium methylidyne carbonium ion or perhaps by causing extreme polarization of the C–O bond to form an intermediate capable of electrophilic attack of the benzene solvent and facilitate a Friedel–Crafts type of reaction (Dolby and Robinson 1972, Seyferth et al. 1974).

10.4.2. "Alkylation" of Pentaborane(9) and 1,2-$C_2B_{10}H_{12}$ Carborane

Selected boron hydrides (Blay et al. 1960, Ryschkewitsch et al. 1963, Lipscomb 1965, Dupont and Hawthorne 1962, Gaines and Martens 1968) and carboranes (Grimes 1970, Beall 1975, Zakharkin et al. 1968) undergo Friedel–Crafts type reactions resulting in electrophilic displacement of a B–H hydrogen. This ability prompted attempts to link a borane or carborane cage to a triosmium methylidyne unit, $(\mu\text{-H})_3Os_3(CO)_9(\mu_3\text{-C-})$, through the formation of a carbon-boron bond (Werner et al. 1988). In the presence of BF_3, pentaborane(9), B_5H_9, is "alkylated" by the $(\mu\text{-H})_3Os_3(CO)_9(\mu_3\text{-C-})$ unit of $[(\mu\text{-H})_3Os_3(CO)_9(\mu_3\text{-C-})]_3 \times [O_3B_3O_3]$, **III**, to form $(\mu\text{-H})_3Os_3(CO)_9(\mu_3\text{-CB}_5H_8)$, **XII**

$$[\mu\text{-H})_3Os_3(CO)_9(\mu_3\text{-C})]_3(O_3B_3O_3) + B_5H_9 \xrightarrow{\ BF_3\ } (\mu\text{-H})_3Os_3(CO)_9(\mu_3\text{-C})(B_5H_8)$$

III   **XII**

$$(10.13)$$

Compound **XII** can also be prepared by the reaction of the triosmium chloromethylidyne cluster, $(\mu\text{-H})_3Os_3(CO)_9(\mu_3\text{-CCl})$, with B_5H_9 in the presence of $AlCl_3$. Compound **XIII** is very moisture sensitive; it is readily hydrolyzed in air to produce $(\mu\text{-H})_3Os_3(CO)_9(\mu_3\text{-CH})$. The proposed structure of **XII** (Figure 10.9) is readily deduced from its NMR and mass spectra (Wermer et al. 1988). It

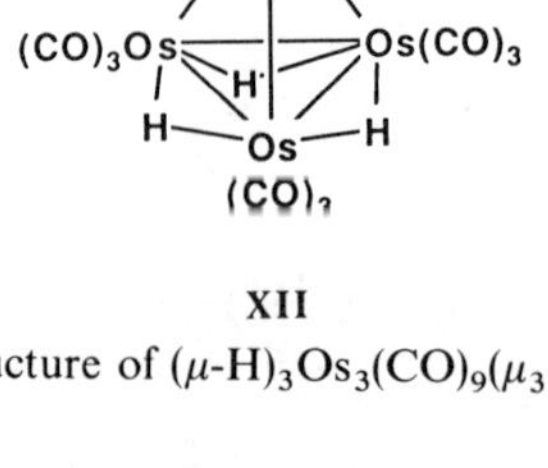

XII

Figure 10.9. Structure of $(\mu\text{-H})_3Os_3(CO)_9(\mu_3\text{-C})(B_5H_8)$, **XII**.

contains an apically substituted pentaborane(9) unit. Formation of the carbon-boron bond at the apical atom (Blay et al. 1960, Ryschkewitsch et al. 1963) of the B_5 cluster supports a Friedel–Crafts type reaction where electrophilic attack is expected at the most negative boron atom in B_5H_9 (Lipscomb 1965), the apical boron.

The carborane $1,2\text{-}C_2B_{10}H_{12}$ also reacts with **III** in the presence of BF_3 to form the boron substituted complex $(\mu\text{-}H)_3Os_3(CO)_9(\mu_3\text{-}C)(C_2B_{10}H_{11})$, **XIII** (Wermer et al. 1988)

$$[\mu\text{-}H)_3Os_3(CO)_9(\mu_3\text{-}C)]_3(O_3B_3O_3) + 1,2\text{-}C_2B_{10}H_{12} \xrightarrow{BF_3}$$

III

$(\mu\text{-}H)_3Os_3(CO)_9(\mu_3\text{-}C)(C_2B_{10}H_{11})$

XIII   (10.14)

On the other hand, no reaction is observed between **III** and the 1,7 isomer of $C_2B_{10}H_{12}$

$$[\mu\text{-}H)_3Os_3(CO)_9(\mu_3\text{-}C)]_3(O_3B_3O_3) + 1,7\text{-}C_2B_{10}H_{12} \xrightarrow{BF_3} \text{ No Reaction}$$

III

(10.15)

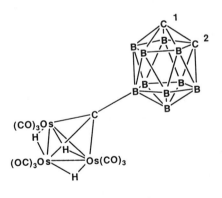

XIII

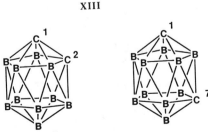

Figure 10.10. Structures of $(\mu\text{-}H)_3Os_3(CO)_9(\mu_3\text{-}C)(C_2B_{10}H_{11})$, **XIII**, $1,2\text{-}C_2B_{10}H_{11}$, and $1,7\text{-}C_2B_{10}H_{11}$.

The structure of **XIII** could not be deduced from its NMR spectra because of overlapping resonances that could not be resolved. However, it was proposed that the most likely structure is the 9-B substituted isomer shown in Figure 10.10. The favored site for electrophilic substitution in carboranes is the boron farthest away from the carbon atoms (Grimes 1970, Beall 1975, Zakharkin et al. 1968). Electrophilic monohalogenation (Cl_2, Br_2, or I_2 in the presence of aluminum halide), substitution occurs predominately at boron atom 9(12) in 1,2-$C_2B_{10}H_{12}$ and at boron 9(10) in 1,7-$C_2B_{10}H_{12}$, the boron atoms farthest from both carbon atoms (Zakharkin et al. 1968). Since only one isomer for **XIII** is obtained, it is assumed to be the 9-B substituted isomer.

ACKNOWLEDGMENT

This research was supported by the National Science Foundation.

REFERENCES

Alich, A., Nelson, N. J. and Shriver, D. F., *J. Chem. Soc. D*, 254 (1971).

Alich, A., Nelson, N. J., Strope, D. and Shriver, D. F., Inorg. Chem., **11**, 2976 (1972).

Barratt, D., Davies, S. J., Elliott, G. P., Howard, J. A. K., Lewis, D. B. and Stone, F. G. A., *J. Organomet. Chem.*, **325**, 185 (1987).

Barreto, R. D., Fehlner, T. P., Hsu, L.-Y., Jan, D.-Y. and Shore, S. G., *Inorg. Chem.*, **25**, 3572 (1986).

Beall, H., in: *Boron Hydride Chemistry* (E. L. Muetterties, Ed.), p. 319, Academic Press, New York, 1975.

Blay, N. J., Dunstan, I. and Williams, R. L., *J. Chem. Soc.*, 430 (1960).

Boese, R., Paetzold, P., Tapper, A. and Ziembinski, R., *Chem. Ber.*, **122**, 1057 (1989).

Broach, R. W. and Williams, J. M., *Inorg. Chem.*, **18**, 314 (1979).

Burwell, R. L., *Chem. Rev.*, **54**, 615 (1954).

Carriendo, G. A., Elliott, G. P., Howard, J. A. K., Lewis, D. B. and Stone, F. G. A., *J. Chem. Soc., Chem. Commun.*, 1585 (1984).

Chesky, P. T. and Hall, M. B., *Inorg. Chem.*, **20**, 4419 (1981).

Deeming, A. J. and Hasso, S., *J. Organomet. Chem.*, **88**, C21 (1975).

Deeming, A. J. and Underhill, M., *J. Chem. Soc., Chem. Commun.*, 277 (1973).

Deeming, A. J. and Underhill, M., *J. Chem. Soc., Dalton Trans.*, 1415 (1974).

DeKock, R. L., Wong, K. S. and Fehlner, T. P., *Inorg. Chem.*, **21**, 3203 (1982).

Dolby, R. and Robinson, B. H., *J. Chem. Soc., Dalton Trans.*, 2046 (1972).

Dreyfuss, P. and Dreyfuss, M. P., *Frotschr. Hochpolym.-Forsch.*, **4**, 528 (1967).

Dupont, J. A. and Hawthorne, M. F., *J. Am. Chem. Soc.*, **84**, 1804 (1962).

Fehlner, T. P., Meng, X., Rath, N. P. and Rheingold, A. L., 198th National Meeting of American Chemical Society, Miami Beach, Florida, September 10–15, Abstract #275, 1989.

Fratiello, A., Onak, T. P. and Schuster, R. E., *J. Am. Chem. Soc.*, **90**, 1194 (1968).

Fried, S. and Kleene, R. D., *J. Am. Chem. Soc.*, **63**, 2691 (1941).

Gaines, D. and Martens, J. A., *Inorg. Chem.*, **7**, 704 (1968).

Gilbert, K. B., Boocock, S. K. and Shore, S. G., in: *Comprehensive Organometallic Chemistry* (G. Wilkinson, F. G. A. Stone and E. W. Abel, Eds.), Vol. IV, pp. 897–914, Pergamon, Oxford, U.K., 1982.

Glore, W., Rathke, S. W. and Schaeffer, R., *Inorg. Chem.*, **12**, 2175 (1973).

Gobeau, J. and Keller, H. Z., *Z. Anorg. Allg. Chem.*, **265**, 73 (1951a).

Gobeau, J. and Keller, H. Z., *Z. Anorg. Allg. Chem.*, **267**, 1 (1951b).

Gordy, W., Ring, H. and Burg, A. B., *Phys. Rev.*, **78**, 512 (1950).

Grimes, R. N., *Carboranes*, p. 172, Academic Press, New York, 1970.

Horwitz, C. P. and Shriver, D. F., *Adv. Organomet. Chem.*, **23**, 219 (1984).

Hriljac, J. A. and Shriver, D. F., *Organometallics*, **4**, 2225 (1985).

Hriljac, J. A. and Shriver, D. F., *J. Am. Chem. Soc.*, **109**, 6010 (1987).

Hsu, L.-Y., Mariategui, J. L., Niendenzu, K. and Shore, S. G., *Inorg. Chem.*, **26**, 143 (1987).

Jan, D.-Y., Ph.D. Dissertation Ohio State University, 1985.

Jan, D.-Y. and Shore, S. G., *Organometallics*, **6**, 428 (1987).

Jan, D.-Y., Hsu, L.-Y., Workman, D. P. and Shore, S. G., *Organometallics*, **6**, 1984 (1987).

Johnson, B. F. G. and Benfield, R. F., in: *Transition Metal Clusters* (B. F. G. Johnson, Ed.), Chap. VII, Wiley, New York, 1980.

Keister, J. B., *J. Chem. Soc., Chem. Commun.*, 214 (1979).

Keister, J. B. and Horling, T., *Inorg. Chem.*, **19**, 2304 (1980).

Keister, J. B., Payne, M. W. and Muscatella, M. J., *Organometallics*, **2**, 219 (1983).

Kennedy, S., Alexander, J. J. and Shore, S. G., *J. Organomet. Chem.*, **291**, 385 (1981).

Knowles, D. J. and Buchanan, A. S., *Inorg. Chem.*, **4**, 1799 (1965).

Kramer, G. W. and Brown, H. C., *J. Organomet. Chem.*, **73**, 1 (1974).

Kristoff, J. S. and Shriver, D. F., *Inorg. Chem.*, **13**, 499 (1974).

Krause, J. A. K., Ph.D. Dissertation, The Ohio State University, 1989.

Krause, J. K., Jan, D.-Y. and Shore, S. G., *J. Am. Chem. Soc.*, **109**, 4416 (1987).

Kriessl, F. R., Sieber, W. and Wolfgruber, M., *Angew. Chem. Int. Ed. Engl.*, **22**, 493 (1983a).

Kriessl, F. R., Sieber, W. and Wolfgruber, M., *Z. Naturforsch.*, **38B**, 1419 (1983b).

Lappert, M. F., *J. Chem. Soc.*, A, 3256 (1983).

Lipscomb, W. N., *Boron Hydrides*, pp. 171–180, Benjamin, New York, 1965.

Meng, X., Rath, N. P. and Fehlner, T. P., *J. Am. Chem. Soc.*, **111**, 3422 (1989).

Orpen, A. G. and Koetzle, T. F., *Acta Crystallogr.*, **40b**, 606 (1984).

Orpen, A. G., Rivera, A. V., Bryan, D. Pippard, G. M. Sheldrick and Rouse, K. D., *J. Chem. Soc., Chem. Commun.*, 723 (1978).

Raithby, P. R. and Rosales, M. J., *Adv. Inorg. Chem. Radiochem.*, **29**, 169 (1985).

Rathke, J. and Schaeffer, R., *Inorg. Chem.*, **13**, 760 (1974).

Roland, E., Wolfgang, B. and Vahrenkamp, H., *Chem. Ber.*, **118**, 2858 (1985).

Ryschkewitsch, G. F., Harris, S. W., Mezey, E. J., Sisler, H. H., Weilmuenster, E. A. and Garrett, A. B., *Inorg. Chem.*, **2**, 890 (1963).

Sailor, M. J., Brock, C. P. and Shriver, D. F., *J. Am. Chem. Soc.*, **109**, 6015 (1987).

Sailor, M. J. and Shriver, D. F., *Organometallics*, **4**, 1476 (1985).

Sappa, E., Tiripicchio, A. and Braunstein, P., *Chem. Rev.*, **83**, 203 (1983) and references therein.

Saturnino, D. J., Yamauchi, M., Clayton, W. R., Nelson, W. R. and Shore, S. G., *J. Am. Chem. Soc.*, **97**, 6063 (1975).

Seyferth, D., Hallgren, J. E. and Eschbach, C. S., *J. Am. Chem. Soc.*, **96**, 1730 (1974).

Shapley, J. R., Keister, J. B., Churchill, M. R. and DeBoer, B. G., *J. Am. Chem. Soc.*, **97**, 4145 (1975).

Shapley, J. R., Strickland, D. S., St. George, G. M., Churchill, M. R. and Bueno, C., *Organometallics*, **2**, 185 (1983).

Sherwood, D. E. Jr. and Hall, M. B., *Organometallics*, **1**, 1519 (1982).

Shore, S. G., Jan, D.-Y., Hsu, L.-Y., Hsu, W.-L., *J. Am. Chem. Soc.*, **105**, 5923 (1983).

Shore, S. G., Jan, D.-Y., Hsu, W.-L., Hsu, L.-Y., Kennedy, S., Hoffman, J. C., Lin Wang, T.-C. and Marshall, A. G., *J. Chem. Soc. Chem. Commun.*, 392 (1984).

Shriver, D. F. and Sailor, M. J., *Acc. Chem. Res.*, **21**, 374 (1988).

Sievert, A. C., Strickland, D. S., Shapley, J. R., Steinmetz, G. R. and Geoffroy, G. L., *Organometallics*, **1**, 214 (1982).

Stone, H. and Shechter, H., *J. Org. Chem.*, **15**, 491 (1950).

Vites, J. C., Housecroft, C. E., Jacobsen, G. B. and T. P., *Organometallics*, **3**, 1591 (1984).

Wade, K., *Chem. Ber.*, **11**, 177 (1975).

Wade, K., *Adv. Inorg. Chem. Radiochem.*, 1 (1976).

Went, M. J., Sailor, M. J., Bogdan, P. L., Brock, C. P. and Shriver, D. F., *J. Am. Chem. Soc.*, **109**, 6023 (1987).

Wermer, J. R., Jan, D.-Y., Getman, T. D. Moher, E. and Shore, S. G., *Inorg. Chem.*, **27**, 4274 (1988).

Workman, D. P., Deng, H.-B. and Shore, S. G., *Angew. Chem. Int. Ed. Engl.* **29**, 309 (1990a).

Workman, D. P., Jan, D.-Y. and Shore, S. G., *Inorg. Chem.* **29**, (1990).

Zakharkin, L. I., Kalinin, V. N. and Lozovskaya, V. S., *Bull. Acad. Sci. USSR, Div. Chem. Sci. (Engl. Transl.)*, 1683 (1968).

CHAPTER 11

Cyclocarborane-Stabilized Multidecker/Multicluster Sandwich Compounds and Linked Molecular Systems

R. N. GRIMES

11.1. INTRODUCTION

Metal-boron cluster chemistry is an area of enormous scope and variety, which continues to expand with the discovery of new cluster geometries, bonding modes, and reaction pathways [1]. With growing awareness of common ground between boron and other clusters (a theme of this volume), an aspect of this field to which transition metal-organometallic chemists can easily relate is the utilization of carboranes as ligands in the construction of metal sandwich

Electron Deficient Boron and Carbon Clusters, Edited by George A. Olah,
Kenneth Wade, and Robert E. Williams.
ISBN 0-471-52795-5 © 1991 John Wiley & Sons, Inc.

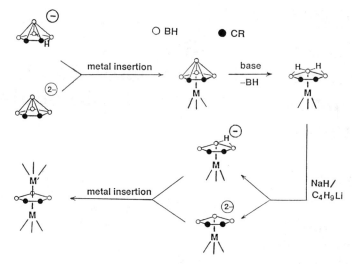

Figure 11.1. Synthesis of mono- and bimetallic sandwich complexes from *nido*-$R_2C_2B_4H_5^-$ and $R_2C_2B_4H_4^{2-}$ anions.

complexes. While many different carborane ligands are represented in known compounds, those which have found extensive use in synthesis are the 11-vertex $R_2C_2B_9H_9^{2-}$ "dicarbollide" ions, in which insertion of a metal ion into the open face completes a 12-vertex icosahedral cage [2], and the 6-vertex $R_2C_2B_4H_4^{2-}$ dianions which generate 7-vertex pentagonal bipyramidal MC_2B_4 clusters upon incorporation of a metal [3]. As is schematically illustrated in Figure 11.1, removal of the apex BH unit from a metal-bound $R_2C_2B_4H_4^{2-}$ ligand affords a planar $R_2C_2B_3H_5^{2-}$ "cyclocarborane" group, which in turn can be bridge-deprotonated to give the corresponding metal complex containing a formal $R_2C_2B_3H_3^{4-}$ ligand; addition of a second metal to the open face is then possible [3].

The C_2B_9 and C_2B_4 *nido*-carborane ligands have pentagonal C_2B_3 open faces and are isoelectronic and isolobal with $C_5H_5^-$ and arenes, all being 6-electron donors in metal sandwich binding (Figure 11.2); moreover, it is apparent that carborane ligands in general bind metals more tightly than do cyclopentadieneide and its derivatives [4, 5]. In consequence of this, there are numerous examples of metal-carborane complexes whose cyclopentadienyl-metal analogs are unstable or nonexistent.

In comparison to $C_5H_5^-$, the favorable metal-bonding properties of the anionic carborane ligands can be attributed to (1) the lower electronegativity of boron vs. carbon, which leads to greater covalent metal-ligand character, (2) the dinegative charge, which helps to stabilize higher metal oxidation states, and (3) the inward-directed facial bonding orbitals, which achieve better overlap with metal orbitals than do the corresponding cyclopentadienyl orbitals which are constrained to be perpendicular to the plane of the C_5 ring (Figure 11.2) [5].

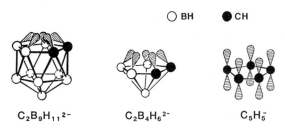

Figure 11.2. Facial orbitals available for bonding to metals in carborane and cyclopentadienide ligands.

The C_2B_4 and C_2B_3 ligands, while similar in many respects to their large C_2B_9 homologs, nevertheless exhibit unique chemical behavior which has no counterpart in the larger carboranes. A prime example is their metal-promoted room temperature oxidative fusion to create a single large cage, illustrated by the high-yield formation of $R_4C_4B_8H_8$ (Figure 11.3). This surprising reaction, first noted in 1974 [6], occurs in a variety of small boron clusters and has been extensively investigated [7]. While oxidative fusion is exhibited by small metallacarboranes, boranes, and metallaboranes as well as small carboranes, it has not been observed in large icosahedral-fragment clusters such as $C_2B_9H_{11}^{2-}$.

Our work in recent years has exploited certain inherent advantages afforded by the pyramidal $R_2C_2B_4H_4^{2-}$ and cyclic planar $R_2C_2B_3H_3^{4-}$ units in organometallic synthesis. In addition to their isoelectronic and isolobal relationship to arenes and cyclopentadienide, mentioned above, these small carboranes are also effective *steric* surrogates for aromatic hydrocarbons, in that the C_2B_3 ring is about the same size as $C_5H_5^-$. Consequently, these carboranes can serve as versatile building-block units for the construction of transition-metal sandwich complexes. The effectiveness of $R_2C_2B_4H_4^{2-}$ and $R_2C_2B_3H_3^{4-}$ ligands in stabilizing organometallic systems is illustrated by the robustness in air of species such as $(\eta^8\text{-}C_8H_8)M(Et_2C_2B_4H_4)$ where M = Ti or V [8] and by the synthesis of a wide variety of $(arene)M^{II}(R_2C_2B_4H_4)$ compounds (M = iron [9] or ruthenium [10]) and their reversible, facile electrochemical oxidation to the

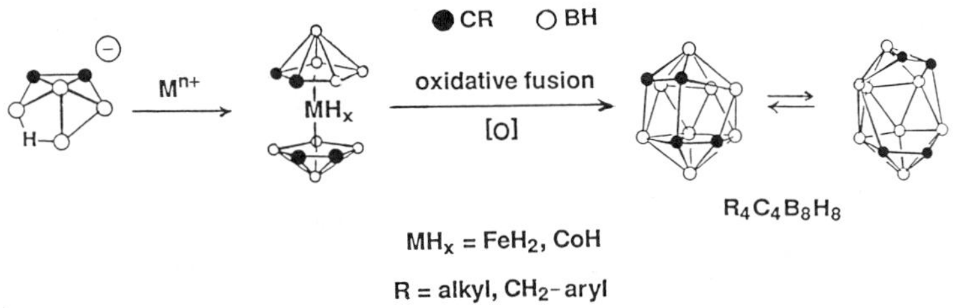

Figure 11.3. Formation of C_4B_8 cages via metal-promoted fusion of C_2B_4 units.

M^{III} species (Figure 11.4) [11]. This latter property is particularly noteworthy and stands in sharp contrast to previously studied iron- and ruthenium-arene complexes.

Also important in designed synthesis is the "decapitation" reaction wherein the apex BH unit of a metal-coordinated $R_2C_2B_4H_4^{2-}$ is removed to create a planar C_2B_3 open face, which can be deprotonated and then coordinated to a second metal (Figure 11.1). This approach was employed some time ago [12] to prepare the first electrically neutral triple-decker sandwich complexes, green and red isomers of $(C_5H_5)Co(R_2C_2B_3H_3)Co(C_5H_5)$ (Figure 11.5).

Crystal structure analyses of these compounds confirmed the structures proposed from NMR data [12a, 13]. Unfortunately, attempts to prepare other triple-deckers by analogous reactions have been mostly unsuccessful; moreover, the cyclopentadienyl capping ligands in these complexes are not displaceable, and hence these triple-decker complexes do not lend themselves readily to further stacking or to utilization as construction units for extended sandwich systems. In addition, until recently methods have not been available for attaching substituents to small carboranes and metallacarboranes in a regio-specific, controlled manner.

In recent years, new approaches to these problems have allowed us to develop systematic routes to carborane-stabilized organometallic sandwich compounds, including not only new double- and triple-decker complexes but also multimetallic electron-delocalized linked systems in which the metals are

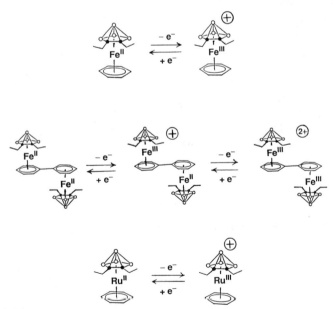

Figure 11.4. Electrochemical interconversions of small arene-ferracarborane and arene-ruthenacarborane sandwich complexes.

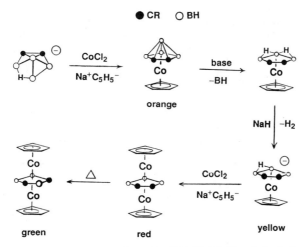

Figure 11.5. Synthetic route to CpCo(R$_2$C$_2$B$_3$H$_3$)CoCp triple-decker complexes.

first, second- or third-row transition elements or main-group elements. These advances, in turn, have opened the way to the synthesis of even larger systems including, potentially, electrically conducting or semiconducting polymers of unusual stability. Conceivably, such materials could be tailored (via appropriate choice of metals and ligands) to exhibit specific desired electronic properties. This chapter outlines some of the recent developments in our laboratory which are directed toward these goals, and is organized to illustrate the design and stepwise construction of successively larger and more complex systems from simple building-block carborane and hydrocarbon units.

11.2. SYNTHESIS OF BUILDING-BLOCK *nido*-CARBORANES

11.2.1. Carbon-Substituted Derivatives

The preparation of *nido*-RR′C$_2$B$_4$H$_6$ carboranes via the base-promoted addition of alkynes to pentaborane(9) is illustrated in Figure 11.6. This method was developed in our laboratory [14] as a bench-scale improvement on the original gas-phase and lutidine-promoted solution syntheses of Onak et al. [15], and exploits the large U.S. Government stockpile of B$_5$H$_9$. The reaction is remarkably general and, in our hands, has worked with almost all alkynes tried including di-, tri-, and cyclic acetylenes [16, 17]. (An exception is the synthesis of the C,C′-diphenylcarborane from diphenylacetylene, which is conducted at 180°C in hexane in the absence of base [14c].) In most cases the alkynes employed are commercially available, but the bis(indenylmethyl) and bis(fluorenylmethyl) acetylenes (all but one of which were previously unknown) required new syntheses [17].

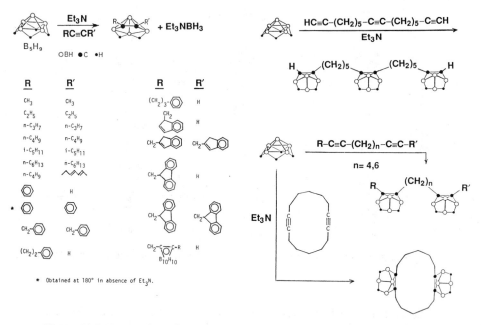

Figure 11.6. Conversion of B_5H_9 to *nido*-$RR'_2C_2B_4H_6$ carborane derivatives.

The carborane products are generally volatile liquids (e.g., R = methyl, ethyl) or nonvolatile oils (R = phenyl, arylalkyl, or large alkyl groups); the latter compounds are kinetically air-stable and can be used as benchtop reagents. Important recent additions to this family are the C-trimethylsilyl and C,C'-bis(trimethylsilyl) derivatives prepared by Hosmane and co-workers [18] and a viable large-scale synthesis of the parent $C_2B_4H_8$ by the same group [19]. In all of these carboranes, one [20] or both [21] bridge protons are removed on treatment with alkali metal hydrides or *n*-butyllithium to form the conjugate base mono- or di-anion. The mono-deprotonation has been examined kinetically and was found to be pseudo-first order, with the rate decreasing with increasing size of the alkyl substituents [22].

The principal feature of interest in these carboranes for our purposes is, of course, the C_2B_3 face which can be deprotonated and η^5-coordinated to metal ions. However, the versatility of these reagents is further increased in the aryl derivatives, which present additional sites for metal attack as shown in Figure 11.7 [23].

11.2.2. Boron-Substituted Derivatives

The preparative route to $RR'C_2B_4H_6$ carboranes described in Figure 11.6 is dependent on the accessibility of acetylenes having the desired R and R' groups. For our synthetic purposes, it is also important to be able to make boron-substituted derivatives, by controlled, regiospecific routes affording single pure

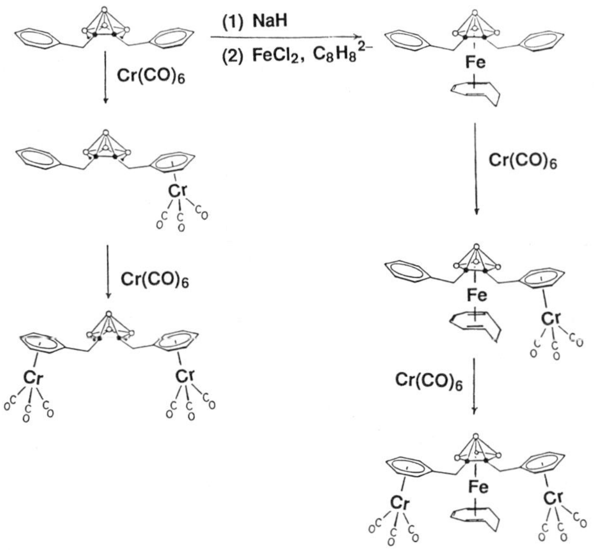

Figure 11.7. Synthesis of *nido*-C,C'dibenzylcarborane transition metal complexes.

isomers [24]. This has been accomplished by the attack of organic halides on carborane monoanions (Figure 11.8), which occurs with high specificity (99% or better) at the equivalent basal B(4,6) locations via proposed bridged intermediates [27]. Only mono-B-alkylation is observed under the conditions employed, even when dianionic carborane substrates are used; while the mono-B-substituted derivatives readily undergo bridge-deprotonation, treatment of the anions with alkyl halides regenerates the neutral carboranes via bridge-protonation.

The clean regiospecific B(4) [B(6)]-monosubstitution observed in *nido*-$R_2C_2B_4H_5^-$ anions is nicely complemented by the observed B-alkyl substitution on the structurally analogous *nido*-(arene)$M(R_2C_2B_3H_4)^-$ metallacarborane anions, discussed below, which occurs exclusively at the middle boron (B(5)) location [28]. The synthetic advantages afforded by these combined observations are outlined in a later section.

11.3. SYNTHESIS OF ARENE-METAL-CARBORANE DOUBLE-DECKER COMPLEXES

11.3.1. Iron Complexes

Our overall synthetic goals, which focus on extended multimetal centered organometallic oligomers and polymers, require controlled, "rational" synthetic routes to monomeric arene-metal-carborane sandwich complexes that can be assembled to form larger systems. Although (η^6-arene)$M(C_2B_9H_9)$ metal-

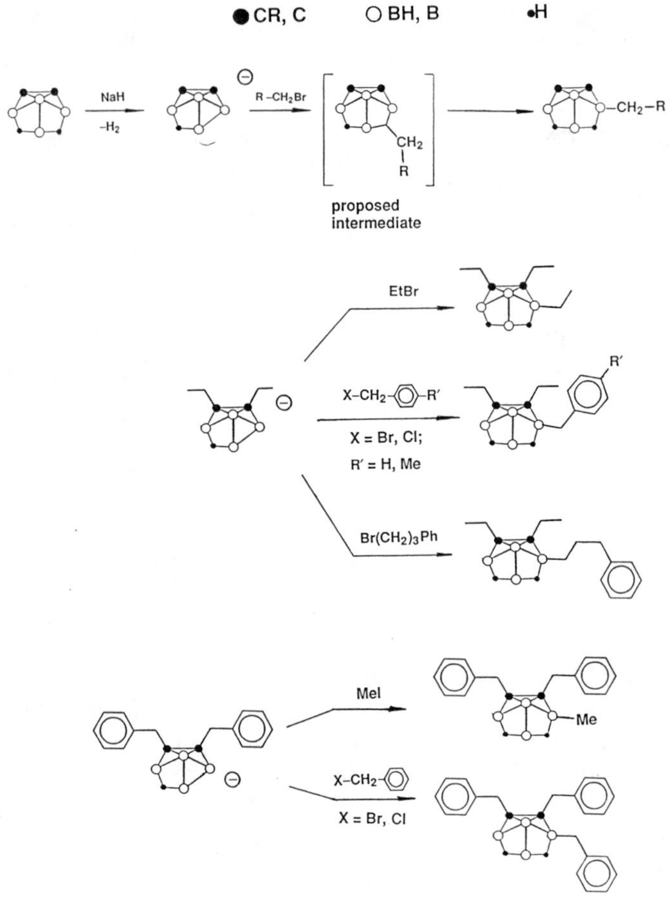

Figure 11.8. Regiospecific synthesis of $4(6)$-R-$Et_2C_2B_4H_5$ boron-substituted *nido*-carboranes.

lacarboranes ($M = Fe$ or Ru) have been reported [29] and (η^6-toluene)Fe(Et$_2$C$_2$B$_4$H$_4$) was produced in an iron vapor reaction [30], a general method for preparing small arene-metallacarboranes was lacking until we found that the η^6-cyclooctatriene iron complex [31] **1** readily undergoes displacement of the C$_8$H$_{10}$ ligand by arenes [3, 9, 32] (Figure 11.9). The reaction is conducted over AlCl$_3$, or in some cases via thermolysis without solvent, but is quite general and has been used to prepare a wide variety of (η^6-arene)Fe(C$_2$B$_4$H$_4$) complexes (**2**), usually obtained as yellow to red-orange air-stable crystalline solids. The C$_8$H$_{10}$ displacement also occurs intramolecularly to give "carboranophane" complexes such as **4** (Figure 11.9, bottom) [32b].

In some instances, when the arene can be reduced to an anion (e.g. naphthalene or fluorene), reaction of the areneide ion with the carborane ion

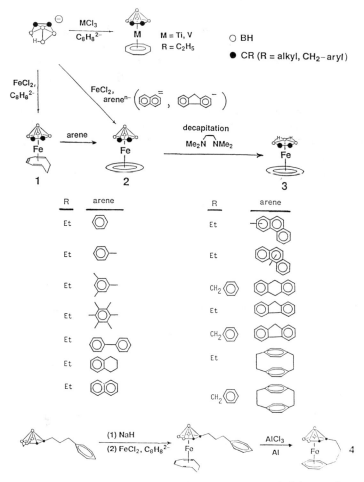

Figure 11.9. Synthetic routes to arene-metallacarborane sandwich complexes [3].

and $FeCl_2$ affords the desired complex directly [9c], eliminating the need for the $(\eta^6\text{-}C_8H_{10})Fe(R_2C_2B_4H_4)$ intermediate as depicted in Figure 11.9.

All of these MC_2B_4 clusters are easily decapitated via base attack in ethereal solvents (Figure 11.9) to give the corresponding $(arene)M(R_2C_2B_3H_5)$ species (**3**), which are the primary construction modules for larger sandwich systems as will be described.

11.3.2. Cobalt, Ruthenium, and Osmium Complexes

The routes to arene-iron-carborane compounds just outlined have not, in general, proved useful for other transition metals, so alternative approaches are required. The three-component reaction of $CoCl_2$, $C_5H_5^-$, and a carborane substrate, illustrated by the formation of the dicobalt triple-decker complex in Figure 11.5, similarly is not readily extended to other metals or cyclic hydrocarbon ligands (however, it has been utilized in the preparation of triple-decker sandwiches containing cobalt, as described below). In order to prepare (arene)M($R_2C_2B_4H_4$) complexes of second- and third-row transition metals, of which there were no prior examples (although the related clusters $(Et_3P)_2Pt(R_2C_2B_4H_4)$ (R = H, Me) [33] and $(CO)_3Os[(Me_3Si)_2C_2B_4H_4]$ [34] have been synthesized), we have employed reactions of $R_2C_2B_4H_4^{2-}$ dianions with (arene)ruthenium dichloride reagents [10], as shown in Figure 11.10. This method exploits the recent discovery by Hosmane et al. [21] that the *nido*-carborane dianions can be stabilized as the Na$^+$, Li$^+$ double salts. It should also be noted that a similar approach has been used by Todd to prepare 12-vertex (arene)Fe($C_2B_9H_{11}$) ferracarboranes [29c]. The corresponding reaction of (cymene)osmium dichloride (cymene = *p*-isopropyltoluene) does not yield the desired osmacarborane, but this reagent is useful in triple-decker synthesis, as will be demonstrated. Decapitation of the (arene)Ru($R_2C_2B_4H_4$) complexes proceeds readily, as shown [10].

11.3.3. Regiospecific Synthesis of B-Substituted Complexes

The open-faced *nido*-($R_2C_2B_3H_5$)M(L) complexes are structural and electronic analogs of the *nido*-$R_2C_2B_4H_6$ carboranes, in which the apex BH is replaced by an isolobal M(L) unit. On this basis, the *nido*-complexes were expected to exhibit a substitution chemistry similar to that of the carboranes, described

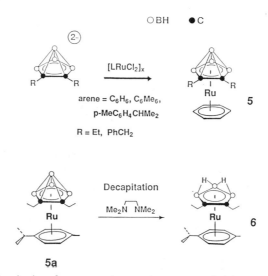

Figure 11.10. Synthesis of arene-ruthenacarborane sandwich complexes from the $Et_2C_2B_4H_4^{2-}$ dianion [10a].

above; however, we have found that the alkylation of $(R_2C_2B_3H_4)M(arene)^-$ anions via attack by alkyl halides occurs exclusively at B(5), the central boron in the C_2B_3 ring (Figure 11.11) [28]. This stereospecificity is completely opposite that of the $R_2C_2B_4H_5^-$ carboranes (Figure 11.8), a fortuitous circumstance for designed synthesis since it is now possible to prepare metal-$R_2C_2B_3H_5$ complexes having alkyl or arylalkyl groups at any designated boron or carbon sites on the C_2B_3 ring. Moreover, in contrast to the $R_2C_2B_4H_6$ carboranes, the $(R_2C_2B_3H_5)M(arene)$ complexes can be repeatedly deprotonated and alkylated, giving ultimately peralkylated products (Figure 11.11b).

Reactions of $(R_2C_2B_3H_4)M(arene)^-$ anions with bifunctional halides produce B,B'-arylalkyl-linked metallacarborane systems such as 7 (Figure 11.11c) [28].

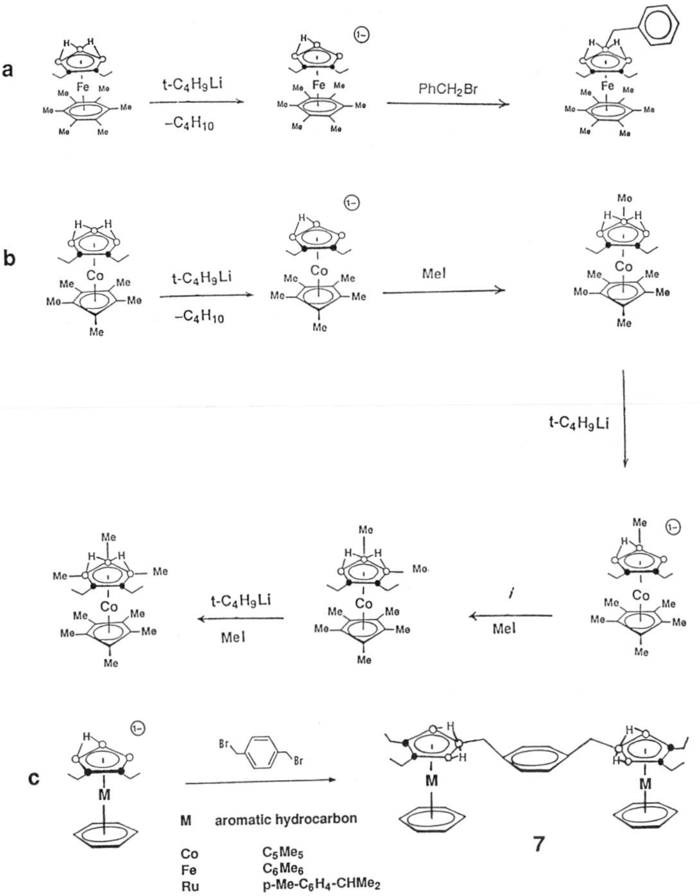

Figure 11.11. Substitution at boron in (L)metal(Et$_2$C$_2$B$_3$H$_5$) complexes (L = arene or Cp*) [28].

11.4. SYNTHESIS OF ARENE-METAL-CARBORANE TRIPLE-DECKER COMPLEXES

11.4.1. Monomeric Triple-Deckers

As was noted earlier, the preparation of the original $Cp_2Co_2RR'C_2B_3H_3$ carborane triple-decker compounds (Figure 11.5) did not provide a general route to such complexes (although related triple- and multidecker sandwiches having C_4B or C_3B_2 bridging rings are known, primarily through the work of Herberich [35] and Siebert [36] and their associates). However, the large repository of organotransition-metal carboranes now at our disposal has opened several avenues for the designed synthesis of homo- and heterobimetallic triple-decker sandwich complexes. Three approaches, shown in Figure 11.12, have been demonstrated thus far, the first two of which will be discussed here with the third to be described in Section 11.6.

Methods 1 and 2 (Figure 11.12) both involve the coordination of a second

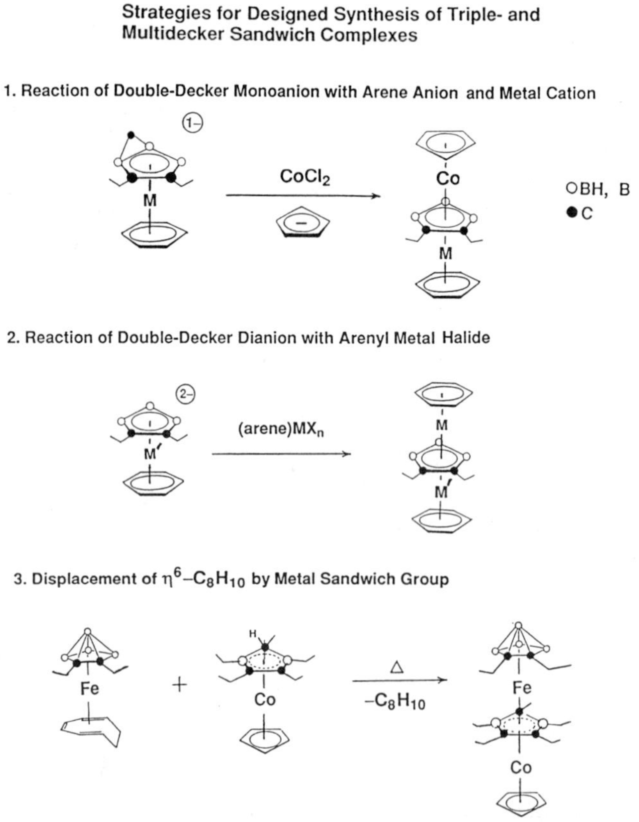

Strategies for Designed Synthesis of Triple- and Multidecker Sandwich Complexes

1. Reaction of Double-Decker Monoanion with Arene Anion and Metal Cation

2. Reaction of Double-Decker Dianion with Arenyl Metal Halide

3. Displacement of η^6–C_8H_{10} by Metal Sandwich Group

Figure 11.12. Synthetic routes to multidecker sandwich complexes.

metal to the C_2B_3 face of a *nido*-MC_2B_3 cluster, but differ significantly in that the metallacarborane *mono*-anion is required for insertion of a CpCo unit, whereas the insertion of (arene)Ru or (arene)Os is observed only with the *di*-anion [10a]. Figure 11.13 outlines the pathways we have employed to prepare a series of air-stable crystalline triple-decker sandwiches 8–11 in 30–50% yield. Of these, the structures of the diruthenium and cobalt-ruthenium compounds 8 and 10 have been crystallographically established [10a]. All of the complexes shown are diamagnetic and were characterized from their ^1H- and ^{11}B-NMR, IR, UV-visible, and mass spectra; the NMR spectra in particular are quite diagnostic of triple-decker (MC_2B_3M' cluster) geometry. As a rule the ruthenium-cobalt and osmium-cobalt complexes are dark green while the others are orange or yellow-orange. At this writing, electrochemical and ESR studies in collaboration with other laboratories are under way on selected species, with results supporting extensive electronic delocalization between the metal centers [37].

Different strategies, which have not yet been demonstrated to have general applicability in synthesis, were employed to prepare the iron-cobalt complex 12, a black-brown paramagnetic solid (Figure 11.14a), and the emerald diamagnetic "pseudo-triple-decker" ruthenium-cobalt species 13 (Figure 11.14b) [10a]. In the first reaction shown, the cyclooctatriene ligand was unexpectedly displaced

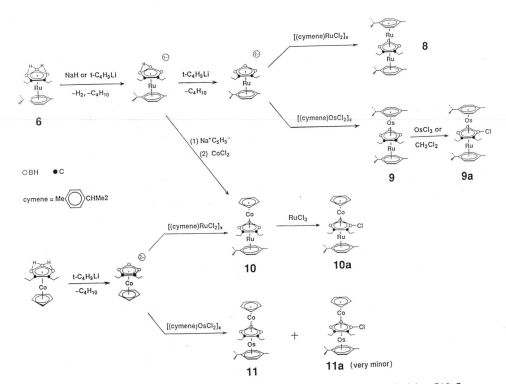

Figure 11.13. Synthesis of homo- and heterobimetallic triple-decker sandwiches [10a].

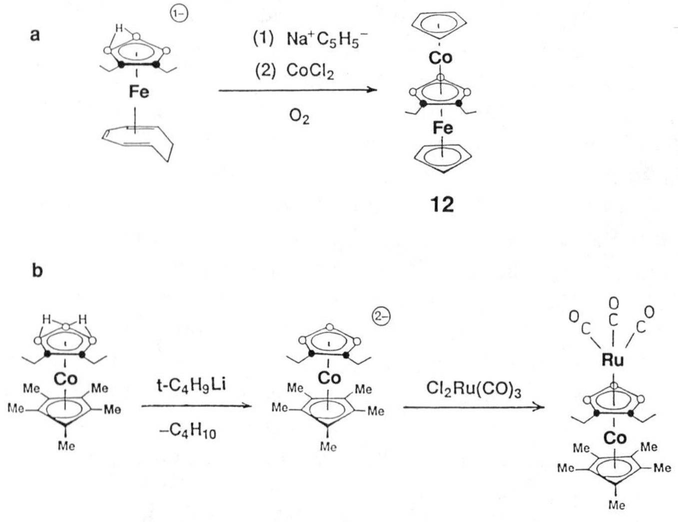

Figure 11.14. Syntheses of (a) $CpCo(Et_2C_2B_3H_3)FeCp$ and (b) $Cp^*Co(Et_2C_2B_3H_3)$ $\times Ru(CO)_3$ [10a].

by $C_5H_5^-$; while C_8H_{10} replacement by *arenes* is common in our systems (see Section 11.3), this was our first observation of such a process involving cyclopentadienide ion.

11.4.2. Linked Double- and Triple-Decker Sandwich Complexes

As a possible route to extended electron-delocalized sandwich systems, we explored reactions of the fulvalenide dianion $(C_5H_4)_2^{2-}$ with suitable metals and carborane anions [10b]. Figure 11.15 depicts the reaction with $CoCl_2$ and $Et_2C_2B_4H_5^-$, which affords the orange dimer $[(Et_2C_2B_4H_4)Co(C_5H_4)]_2$ (**14**) together with its yellow mono-decapitated derivative (not shown) and other side products. "Double-decapitation" of **14** occurs readily, generating yellow **15** almost quantitatively.

Compounds **14** and **15** are novel, as the first characterized fulvalene-metallacarborane complexes. The structure of **14** has been confirmed by an X-ray structure determination [10b] which revealed a centrosymmetric molecule (constraining the fulvalene ligand to be planar) with a central C–C distance of 1.474(6) Å, typical of fulvalene-metal π-complexes. Since the formation of **14** is inefficient owing to the presence of $C_5H_5^-$ in solutions of fulvaleneide dianion [38], we turned to more tractable polycyclic aromatic hydrocarbon ions as linking agents. The bis(tetramethylcyclopentadienyl)phenyleneide dianions have proved useful for this purpose, giving the desired phenylene- and biphenylene-linked orange complexes **16** and **17** (Figure 11.16). As shown, both

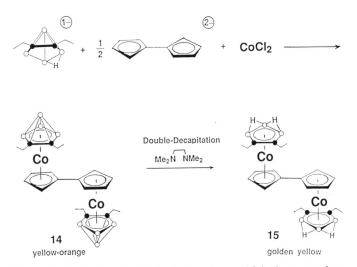

Figure 11.15. Synthesis of bis(cobaltacarboranyl)fulvalene complexes.

products undergo double-decapitation to give the yellow species **18** and **19**, respectively. A crystallographic study of **18** established the structure, in which the central C_6H_4 plane is twisted by 47° with respect to each end cyclopentadienyl ring (Figure 11.17) [10b].

We had anticipated that the "double-open-ended" complexes **15**, **18**, and **19** could serve as templates for construction of larger sandwiches via metal complexation at the open faces, in a manner analogous to the formation of triple-deckers from $CpCo(Et_2C_2B_3H_5)$. These complexes are indeed apparently tetradeprotonated on treatment with n-butyllithium, but the resulting tetra-anions give no characterizable products on reaction with cymeneruthenium dichloride (however, the outlook for polymerization of these species is more

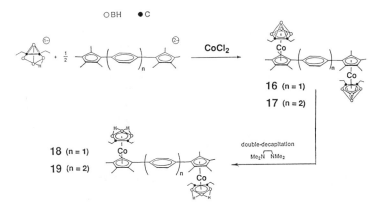

Figure 11.16. Synthesis of p-phenylene-linked bis(cobaltacarborane) complexes.

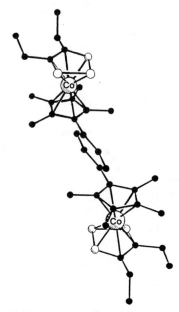

Figure 11.17. Molecular structure of $[(Et_2C_2B_3H_5)Co(C_5Me_4)]_2C_6H_4$ (**18** in Figure 11.16), with hydrogens omitted [10b].

positive, as suggested below). As an alternative approach, we reacted polycyclic hydrocarbon dianions with $CoCl_2$ to form dimetallic intermediates presumed to be structurally analogous to the complex [39] $[CoCl_2(C_5Me_5)]_2$. Reaction of the intermediate with the cymeneruthenium sandwich $\mathbf{6}^-$ generates the phenylene-linked triple-deckers **20** and **21**, as shown in Figure 11.18.

The crystallographically determined structure of emerald, air-stable crystalline **20**, which contains CH_2Cl_2 in the unit cell, consists of ruthenium-cobalt triple-decker units like those of the monomeric species **10**, joined by a phenylene ring whose plane is rotated by 51° relative to the attached cyclopentadienyl planes (Figure 11.19) [10b].

A directly linked bis(triple-decker) complex (the first reported example of this class) was obtained via the reaction of $\mathbf{6}^-$ with fulvaleneide dianion and $CoCl_2$ to give **22**, another emerald green air-stable solid (Figure 11.18). This compound, which incorporates a central fulvalene ligand, is expected to exhibit substantial electron-delocalization among the four metal centers and possibly highly interesting mixed-valence behavior; these properties are currently under investigation.

11.4.3. Oligomeric and Polymeric Triple-Deckers

At this writing, our efforts in this area are focused primarily on the designed synthesis of monomeric sandwich species and studies of their molecular and electronic structures. However, we have begun to extend the work to larger

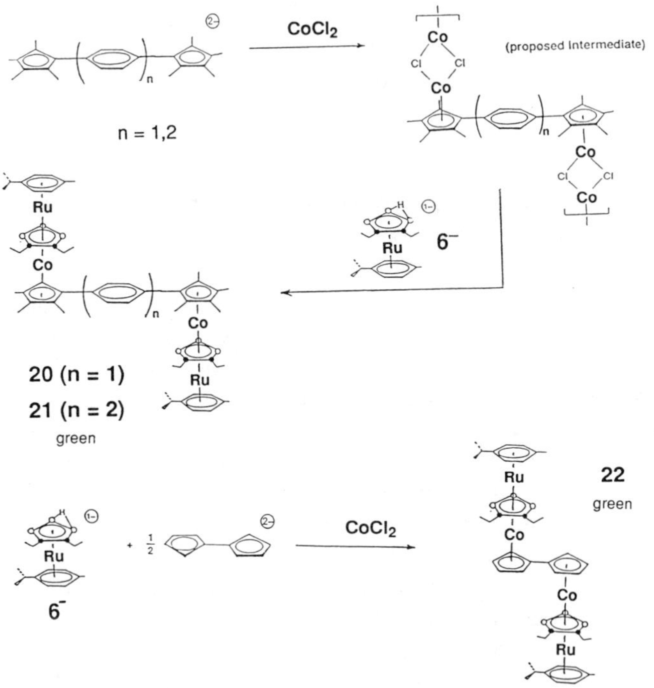

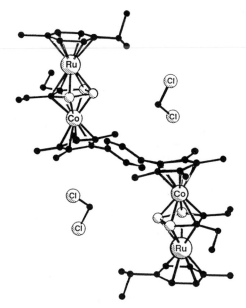

Figure 11.18. Synthesis of linked triple-decker sandwich complexes [10b].

Figure 11.19. Molecular structure of $[(\eta^6\text{-}MeC_6H_4CHMe_2)Ru(Et_2C_2B_3H_3)Co(\eta^5\text{-}Me_4C_5)]_2C_6H_4 \cdot 2CH_2Cl_2$ (**20** in Figure 11.18), with hydrogens omitted [10b].

oligomers and polymers as a step toward the eventual goal of constructing new types of electron-delocalized networks. Figure 11.20 outlines one synthetic strategy we are currently investigating, and on which preliminary data indicates that oligomerization/polymerization has occurred as indicated [40].

11.5. SYNTHESIS OF HETEROCYCLE-METAL-CARBORANE TRIPLE-DECKER COMPLEXES

The versatility of small carborane ligands in forming stable organometallic sandwiches, demonstrated in the chemistry described above, led us to explore still other possibilities including complexes of heterocyclic analogues of cyclo-pentadienyl. Nitrogen-containing heterocycles such as pyrrolyl ($C_4H_4N^-$) tend to coordinate to metals via N-metal σ bonds [41], and consequently there are relatively few well-characterized η^5-bonded metal-pyrrolyl sandwich species. In contrast to this general picture, we have found that both double- and triple-decker carboranyl-metal-pyrrolyl complexes can be obtained as air-stable and thermally stable products (Figure 11.21) [42]. More recently, this chemistry has been extended to analogous pholpholyl and pyrrolyl-phospholyl C_2B_3-bridged triple-decker complexes, [42b].

The pyrrolyl nitrogen in **25** is, as anticipated, nucleophilic and readily combines with MeI to form the N-methyl pyrrole iodide salt. This property is potentially exploitable in synthesis, as a means of effecting direct stereospecific ring-ring intermolecular linkage; when conducted at both ends of bis(pyrrolyl)

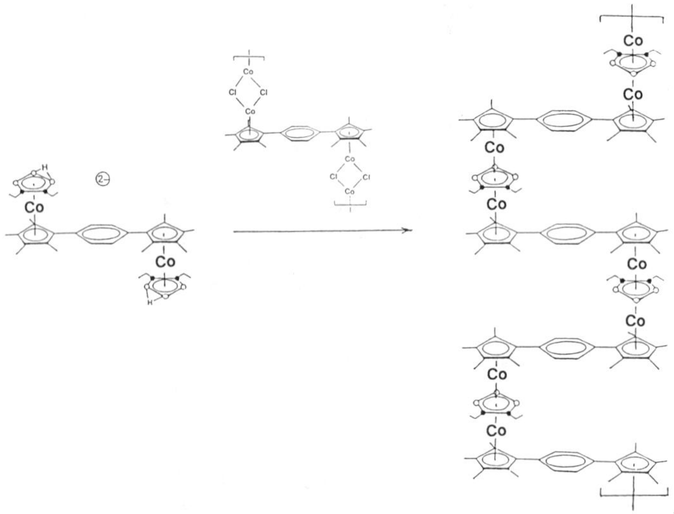

Figure 11.20. An approach to a phenylene-linked triple-decker dicobaltacarborane sandwich polymeric system [40].

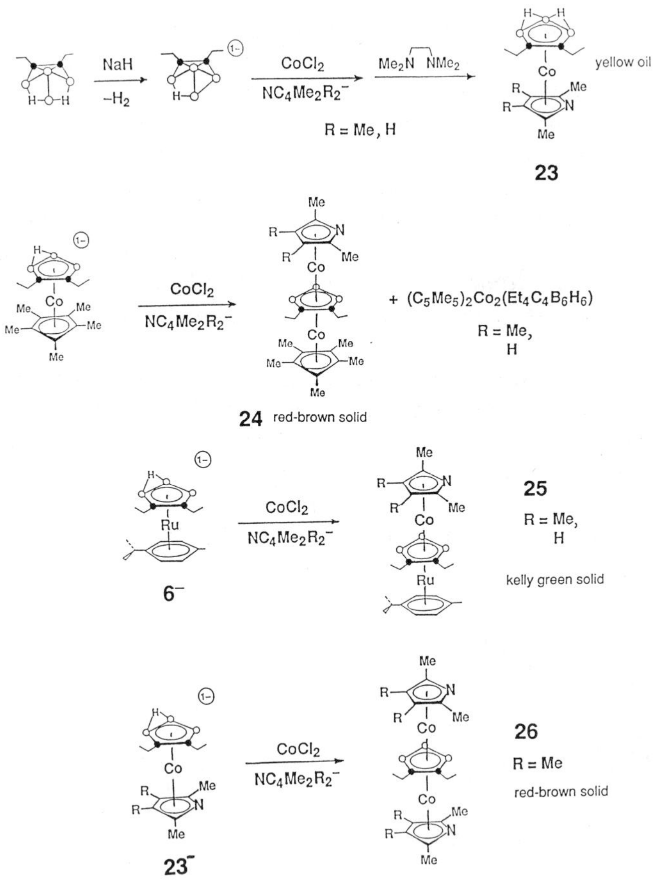

Figure 11.21. Synthesis of pyrrolyl-metallacarborane triple-decker sandwich complexes [42].

species such as **26**, for example, such linkage could lead to controlled polymerization. The triple-decker geometry proposed for these species is supported by spectroscopic data in comparison with their cyclopentadienyl counterparts [42], and by X-ray diffraction studies on several complexes [42c].

11.6. SYNTHESIS OF DIBOROLYL-METAL-CARBORANE TRIPLE-DECKER COMPLEXES

The extensive development of metallacarborane chemistry based on C_2B_n carborane ligands has been paralleled by a similarly productive effort in studies of metal complexes of boron heterocycles, especially those having C_4B (borolene), C_3B_2 (diborolene), C_5B (borabenzene), and C_4B_2 (diborabenzene) rings

[35, 36, 43]. Numerous multidecker complexes of borolene or diborolene have been characterized, including even a polydecker which is a semiconductor [44]. This chemistry, while closely related to the metallacarboranes in terms of structure and bonding, has developed entirely independently based on organoboron (as opposed to borane) starting materials. It appeared potentially useful to utilize the properties of both classes of ligands together as building-block units in metal sandwich synthesis. Accordingly, in a joint effort of our group and that of Professor W. Siebert, we have prepared and characterized a family of heteroligand triple-decker compounds which contain both alkylated 1,3-dihydrodiborolyl ($Et_2MeC_3B_2Et_2^{3-}$) and carboranyl ($R_2C_2B_4H_4^{2-}$) ligands [45, 46].

As depicted in Figure 11.22, the synthesis of the heterobimetallic-heteroligand sandwiches **28** entails displacement of the η^8-C_8H_{10} ligand from the cyclooctatriene-iron-carborane complex **1** by the dihydrodiborolyl cobalt reagent **27**; this approach is analogous to the subrogation of cyclooctatriene by arenes (see Figure 11.9). Alternatively, combination of the deprotonated anion **27⁻** with the diethylcarborane monoanion and $CoCl_2$ affords the red, dia-

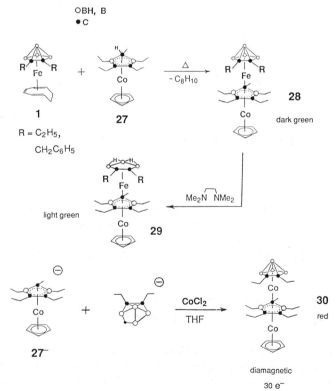

Figure 11.22. Synthesis of carboranyl-diborolyl "hybrid" triple-decker complexes.

magnetic dicobalt complexes [46] $(R_2C_2B_4H_4)Co(Et_2MeC_3B_2Et_2)Co(C_5H_5)$ (**30**, R = Et, PhCH$_2$) which contain formal Co^{III}. A possibly extensive chemistry for these carboranyl/diborolyl complexes is presaged by the observation [45] that **28** can be decapitated to give the open-ended sandwiches **29**.

The iron-cobalt complexes are greenish-brown crystalline paramagnetic solids containing formal Fe^{III} and Co^{III}; the ESR spectrum of the C,C'-diethyl compound is shown in Figure 11.23 (middle) [45]. Potassium reduction of the neutral complex (and its C,C'-dibenzyl counterpart) affords the diamagnetic monoanion **28$^-$** which can be observed via its ^{11}B-NMR spectrum. Further

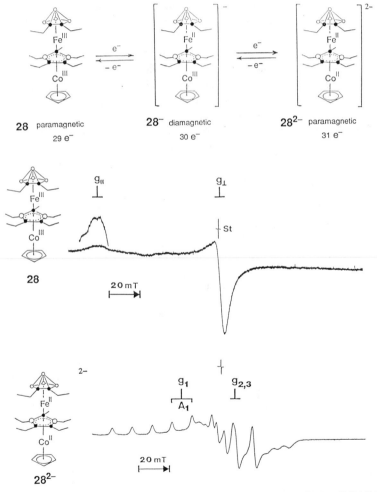

Figure 11.23. (Top) Redox behavior of iron-cobalt hybrid complexes. (Middle) EPR spectrum of $(Et_2C_2B_4H_4)Fe^{III}(Et_2MeC_3B_2Et_2)Co^{III}Cp$ [45]. (Bottom) EPR spectrum of $(Et_2C_2B_4H_4)Fe^{II}(Et_2MeC_3B_2Et_2)Co^{II}Cp$ [45].

reduction produces the dianion 28^{2-} containing formal Fe^{II} and Co^{II}, whose ESR spectrum (Figure 11.23, bottom) reveals cobalt hyperfine structure. In both spectra the observed g values are indicative of one unpaired electron. These observations have been augmented by cyclic voltammetry studies which indicate reversible 1- and 2-electron reductions as well as a reversible 1-electron oxidation for the neutral species [45].

Very recent work has generated tetradecker complexes incorporating C_2B_3 and C_3B_2 bridging ligands with three transition metal clusters [47].

11.7. CURRENT AND FUTURE DIRECTIONS

The synthetic tools are now in place to enable the planned construction of a truly broad array of arene-metal-cyclocarborane linked-multidecker complexes, from monomers to polymers and possibly solid-state or liquid crystal networks. The general air and thermal stability of these compounds, and their ability typically to undergo chemical and electrochemical modification without decomposition, increases enormously their potential utility as electronically active materials. One can envision, for example, electron-delocalized polymers designed to exhibit variable anisotropic conductivity which is controlled via reversible changes in metal oxidation states. However, such applications will require a much deeper understanding than we currently have, of the electronic nature of the monomers and the smaller linked oligomers. It will be interesting to see how the electron-transport properties of these systems change in going from the monomers to progressively larger linked complexes and finally to true polymers.

Our present efforts in this area are centered on (1) structural, spectroscopic, electrochemical, and related studies aimed at elucidating the nature of metal-ligand bonding, metal-metal communication, and electron delocalization in these systems, and (2) the further development of synthetic strategies which are required in order to assemble more complex stacked and linked sandwich systems.

ACKNOWLEDGMENT

Our current work described in this review is generously supported by the National Science Foundation, the U.S. Army Research Office, and a NATO International Collaborative Research Grant.

REFERENCES

1. (a) Liebman, J. F., Greenberg, A. and Williams, R. E. (Eds.), *Advances in Boron and the Boranes* (*Mol. Struct. Energ.*, Vol. 5), VCH Publishers, New York, 1988; (b) Grimes, R. N. (Ed), *Metal Interactions with Boron Clusters*, Plenum Press, New York, 1982.

2. (a) Callahan, K. P. and Hawthorne, M. F., *Adv. Organomet. Chem.*, **14**, 175 (1976); (b) Schubert, D. M., Rees, W. S. Jr, Knobler, C. B. and Hawthorne, M. F., *Pure Appl. Chem.*, **59**, 869 (1987).

3. Grimes, R. N., *Pure Appl. Chem.*, **59**, 847 (1987) and references therein.

4. Grimes, R. N., in: *Comprehensive Organometallic Chemistry* (G. Wilkinson, F. G. A. Stone and E. Abel, Eds.), Chap. 5.5, Pergamon Press, Oxford, U.K., 1982.

5. Calhorda, M. J. and Mingos, D. M. P., *J. Organomet. Chem.*, **229**, 229 (1982).

6. (a) Maxwell, W. M., Miller, V. R. and Grimes, R. N., *Inorg. Chem.*, **15**, 1343 (1976); (b) Maxwell, W. M., Miller, V. R. and Grimes, R. N., *J. Am. Chem. Soc.*, **96**, 7116 (1974).

7. (a) Grimes, R. N., Synthetic strategies in boron cage chemistry, in: *Advances in Boron and the Boranes* (*Mol. Struct. Energ.*, Vol. 5) (J. F. Liebman, A. Greenberg and R. E. Williams, Eds), Chap. 11, pp. 235–263, VCH Publishers, New York, 1988; (b) Grimes, R. N., *Adv. Inorg. Chem. Radiochem.* **26**, 55 (1983); (c) Grimes, R. N., *Acc. Chem. Res.*, **16**, 22 (1983).

8. Swisher, R. G., Sinn, E. and Grimes, R. N., *Organometallics*, **3**, 599 (1984).

9. (a) Swisher, R. G., Butcher, R. J., Sinn, E. and Grimes, R. N., *Organometallics*, **4**, 882 (1985) and references therein; (b) Swisher, R. G., Sinn, E. and Grimes, R. N., *Organometallics*, **4**, 896 (1985); (c) Spencer, J. T. and Grimes, R. N., *Organometallics*, **6**, 323 (1987); (d) Spencer, J. T. and Grimes, R. N., *Organometallics*, **6**, 328 (1987).

10. (a) Davis, J. H., Jr., Sinn, E. and Grimes, R. N., *J. Am. Chem. Soc.*, **111**, 4776 (1989); (b) Davis, J. H., Jr., Sinn, E. and Grimes, R. N., *J. Am. Chem. Soc.*, **111**, 4784 (1989).

11. Merkert, J. M., Geiger, W. E., Jr., Davis, J. H., Jr., Attwood, M. D. and Grimes, R. N., *Organometallics*, **8**, 1580 (1989).

12. (a) Beer, D. C., Miller, V. R., Sneddon, L. G., Grimes, R. N., Mathew, M. and Palenik, G. J., *J. Am. Chem. Soc.*, **95**, 3046 (1973); (b) Grimes, R. N., Beer, D. C., Sneddon, L. G., Miller, V. R. and Weiss, R., *Inorg. Chem.*, **13**, 1138 (1974).

13. (a) Robinson, W. T. and Grimes, R. N., *Inorg. Chem.*, **14**, 3056 (1975); (b) Pipal, J. R. and Grimes, R. N., *Inorg. Chem.*, **17**, 10 (1978).

14. (a) Hosmane, N. S. and Grimes, R. N., *Inorg. Chem.*, **18**, 3294 (1979); (b) Maynard, R. B., Borodinsky, L. and Grimes, R. N., *Inorg. Synth.*, **22**, 211 (1983); (c) Modified procedure: Boyter, H. A., Jr. and Grimes, R. N., *Inorg. Chem.*, **27**, 3075 (1988).

15. (a) Onak, T. P., Williams, R. E. and Weiss, H. G., *J. Am. Chem. Soc.*, **84**, 2830 (1962); (b) Onak, T. P., Gerhart, F. J. and Williams, R. E., *J. Am. Chem. Soc.*, **85**, 3378 (1963); (c) Williams, R. E. and Onak, T. P., *J. Am. Chem. Soc.*, **86**, 3159 (1964); (d) Onak, T. P., Drake, R. P. and Dunks, G. B., *Inorg. Chem.*, **3**, 1686 (1964).

16. Boyter, H. A., Jr. and Grimes, R. N., *Inorg. Chem.*, **27**, 3080 (1988).

17. Fessler, M. E., Spencer, J. T., Lomax, J. F. and Grimes, R. N., *Inorg. Chem.*, **27**, 3069 (1988).

18. (a) Hosmane, N. S., Sirmokadam, N. N. and Mollenhauer, M. N., *J. Organomet. Chem.*, **279**, 359 (1985); (b) Hosmane, N. S., Maldar, N. M., Potts, S. B., Rankin, D. W. H. and Robertson, H. E., *Inorg. Chem.*, **25**, 1561 (1986).

19. Hosmane, N. S., Islam, M. S. and Burns, E. G., *Inorg. Chem.*, **26**, 3236 (1987).

20. Onak, T. P., Drake, R. P. and Dunks, G. B., *Inorg. Chem.*, **5**, 439 (1966).

21. Siriwardane, U., Islam, M. S., West, T. A., Hosmane, N. S., Maguire, J. A. and Cowley, A. H., *J. Am. Chem. Soc.*, **109**, 4600 (1987).

22. Fessler, M. E., Whelan, T., Spencer, J. T. and Grimes, R. N., *J. Am. Chem. Soc.*, **109**, 7416 (1987).

23. (a) Spencer, J. T., Pourian, M. R., Butcher, R. J., Sinn, E. and Grimes, R. N., *Organometallics*, **6**, 335 (1987); (b) Whelan, T., Spencer, J. T., Pourian, M. R. and Grimes, R. N., *Inorg. Chem.*, **26**, 3116 (1987).

24. Previously reported thermal [25] and metal-catalyzed [26] processes generate mixtures of B-organosubstituted $C_2B_4H_8$ derivatives.

25. Onak, T., Marynick, D., Mattschei, P. and Dunks, G., *Inorg. Chem.*, **7**, 1754 (1968).

26. Wilczynski, R. and Sneddon, L. G., *Inorg. Chem.*, **21**, 506 (1982).

27. Davis, J. H., Jr. and Grimes, R. N., *Inorg. Chem.*, **27**, 4213 (1988).

28. (a) Davis, J. H., Jr., Attwood, M. D. and Grimes, R. N., *Organometallics*, **9**, 1171 (1990); (b) Allwood, M. D., Davis, J. H., Jr. and Grimes, R. N., *Organometallics*, **9**, 1177 (1990).

29. (a) Garcia, M. P., Green, M., Stone, F. G. A., Somerville, R. G., Welch, A. J., Briant, C. E., Cox, D. N. and Mingos, D. M. P., *J. Chem. Soc., Dalton Trans.*, 2343 (1985); (b) Garcia, M. P., Green, M., Stone, F. G. A., Somerville, R. G. and Welch, A. J., *J. Chem. Soc., Chem. Commun.*, 871 (1981); (c) Hanusa, T. P., Huffman, J. C. and Todd, L. J., *Polyhedron*, **1**, 77 (1982).

30. Micciche, R. P. and Sneddon, L. G., *Organometallics*, **2**, 674 (1983).

31. Maynard, R. B., Swisher, R. G. and Grimes, R. N., *Organometallics*, **2**, 500 (1983).

32. (a) Swisher, R. G., Sinn, E. and Grimes, R. N., *Organometallics*, **2**, 506 (1983); (b) Swisher, R. G., Sinn, E. and Grimes, R. N., *Organometallics*, **2**, 890 (1983).

33. Barker, G. K., Green, M., Stone, F. G. A. and Welch, A. J., *J. Chem. Soc., Dalton Trans.*, 1186 (1980).

34. Hosmane, N. S. and Sirmokadam, N. N., *Organometallics*, **3**, 1119 (1984).

35. Herberich, G. E., in: *Comprehensive Organometallic Chemistry* (G. Wilkinson, F. G. A. Stone and E. Abel, Eds.), Chap. 5.3, Pergamon Press, Oxford, U.K., 1982 and references therein.

36. Siebert, W., *Angew. Chem. Int. Ed. Engl.*, **24**, 943 (1985); (b) *Pure Appl. Chem.*, **59**, 947 (1987) and references therein.

37. Davis, J. H., Jr., Attwood, M. D., Grimes, R. N., Merkert, J. M. and Geiger, W. E., Jr., Abstracts of Papers, American Chemical Society National Meeting, Boston, MA, April 1990, Abstract INOR 92.

38. Pittman, C. U. and Surynarayanan, B., *J. Am. Chem. Soc.*, **96**, 7916 (1976).

39. Koelle, U., Fuss, B., Belting, M. and Raabe, E., *Organometallics*, **5**, 980 (1986).

40. Davis, J. H., Jr. and Grimes, R. N., to be submitted for publication.

41. Pannell, K. H., Kalsotra, B. L. and Parkanyi, C., *J. Heterocyclic Chem.*, **15**, 1057 (1978).

42. (a) Chase, K. J., and Grimes, R. N., *Organometallics* **8**, 2492 (1989). (b) Chase, K. J., Ph.D. Thesis, Univ. of Virginia, 1990. (c) Chase, K. J., Grimes, R. N., Bryan, R. F. and Sinn, E., to be submitted for publication.

43. Grimes, R. N., *Coord. Chem. Rev.*, **28**, 47 (1979).

44. (a) Kuhlmann, T., Roth, S., Roziere, J. and Siebert, W., *Angew. Chem. Int. Ed. Engl.*, **25**, 105 (1986); (b) Kuhlmann, T., Roth, S., Roziere, J., Siebert, W. and Zenneck, U., *Synth. Metals*, **19**, 757 (1987).

45. Attwood, M. A., Fonda, K. K., Grimes, R. N., Brodt, G., Hu, D., Zenneck, U. and Siebert, W., *Organometallics*, **8**, 1300 (1989).
46. Fessenbecker, A., Attwood, M. A., Bryan, R. F., Woode, M. K., Grimes, R. N., Stephan, M., Zenneck, U. and Siebert, W., *Inorg. Chem.*, in press (1991).
47. Fessenbecker, A., Attwood, M. D., Grimes, R. N., Stephan, M., Pritzkow, H., Zenneck, U. and Siebert W., *Inorg. Chem.*, in press (1991).

CHAPTER 12

Boron-Transition Metal Compounds Containing Direct Metal-Boron Bonds. Recent Progress

T. P. FEHLNER

12.1. INTRODUCTION

The characterization of direct bonding between transition and main group elements in three-dimensional structures has led to advances on a number of fronts. First, the main group element is placed in a new chemical environment leading to new chemistry. For example, the number of transition element-boron "bonds" is not limited to four and clusters of transition metals present special properties when acting as ligands to a main group element. Likewise the transition element is perturbed by the new main group element "ligands" in unusual ways. Second, the new and unusual structures serve as tests of present theory ranging from electron counting rules to the estimation of physical and spectroscopic properties and constitute springboards for the extension of bonding descriptions in new directions. Finally, when synthesized in adequate quantities, the compounds resulting from the combination of transition and

Electron Deficient Boron and Carbon Clusters, Edited by George A. Olah,
Kenneth Wade, and Robert E. Williams.
ISBN 0-471-52795-5 © 1991 John Wiley & Sons, Inc.

main group elements in novel ways constitute potential precursors of materials containing the same elements. In some cases materials synthesis via these compounds possesses advantages over conventional routes in terms of control of stoichiometry, mildness of preparative conditions, and purity of materials.

For these reasons we have sought to draw an analogy between the historical development of organometallic chemistry and transition-main group element chemistry by coining the term "inorganometallic chemistry" (Fehlner 1988a). By doing so the basic similarity between the ways in which transition metals interact with carbon and with other main group elements, for example, boron, is made evident. Often direct, isoelectronic comparisons can be made as, for example, B⁻ is isoelectronic with C. These comparisons can yield significant insights into the nature of the bonding in both systems. As importantly, the differences between the organometallic and "inorganometallic" systems are highlighted.

Substituent effects external to a reaction center play a large role in the chemist's armamentum. Changing the reaction center itself while keeping the total number of valence electrons constant is a method of control that is only now beginning to be exploited. In a real sense nuclear charge can become an empirical variable in the hands of the inorganic chemist. A beautiful example is the synthesis of pharmacologically active boron analogs of amino acids that act as antimetabolites (Spielvogel 1988). However, much more knowledge of synthesis, structure, and reactivity of compounds containing direct transition metal-main group atom interactions must be gained before similar control in these interesting systems will be achieved.

Our work has been focused mainly on the preparation and characterization of compounds in which there is at least one direct interaction of boron with a transition metal but work emanating from many laboratories has served to define the scope of "inorganometallic chemistry" both for boron (Grimes 1982, Kennedy 1984, 1986, Housecroft and Fehlner 1982, Housecroft 1987, Gilbert et al. 1982, Greenwood 1984, Paetzold 1987, Fehlner 1988b) as well as many other main group elements (Cowley 1984, Herrmann 1986, Huttner and Knoll 1987, Schmid 1978, Vahrenkamp 1983, Whitmire 1988). The sweep of synthesis, structure, and reactivity involving compounds of this type promises a rich future full of new compound types, molecular architecture, and chemical reactions.

12.2. NEW COMPOUND TYPES

One of the principal themes of our research has been the synthesis and characterization of compounds containing boron in unusual bonding environments. Our characterization includes not only geometric structure but also electronic structure and reactivity, that is, we attempt to determine how these new bonding environments are expressed in observed properties and reactions. In the following we briefly summarize selected recent results from our laboratory.

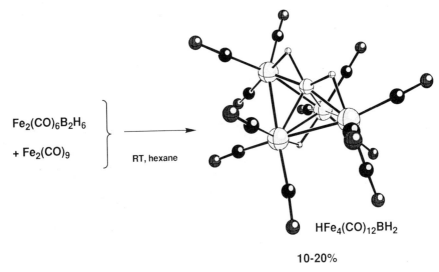

$Fe_2(CO)_6B_2H_6$

$+ Fe_2(CO)_9$

RT, hexane

$HFe_4(CO)_{12}BH_2$

10-20%

Figure 12.1. The original preparation and the structure of $HFe_4(CO)_{12}BH_2$.

12.2.1. Discrete Transition Metal Borides

Discrete molecular clusters containing interstitital carbon atoms are now well known (Bradley 1983), and the structural types observed mimic some of the solid state carbides. Although clusters with interstitial atoms containing electron precise, e.g., C, or electron rich, for example, N, are well known, those containing boron are not. On the other hand, metal borides are well known in the solid state (Greenwood and Earnshaw 1984) and there is no a priori reason why boron should not form stable interstitial, discrete compounds.

12.2.1.1. A Synthesis and Characterization

Our approach to the preparation of a discrete transition metal boride began with the synthesis of $HFe_4(CO)_{12}BH_2$; a molecule we described as one containing an effectively interstitial boride (Figure 12.1) (Wong et al. 1982, Fehlner et al. 1983). Indeed we demonstrated that the three cluster bound hydrogens could be removed as protons with base yielding a boride with an exposed boron atom (Rath and Fehlner 1987). This has led directly to the preparation of a discrete boride, $[Rh_2Fe_4(CO)_{16}B]^-$, as the sole product of a cluster expansion reaction, that is, reaction of $[HFe_4(CO)_{12}BH]^-$ (Housecroft et al. 1987) with $[Rh(CO)_2Cl]_2$. (Khattar et al. 1989). The structure of this compound, shown in Figure 12.2, contains an octahedral metal core with the rhodium atoms trans to each other and a boron atom in the octahedral hole. The new boride is closely related to the carbide $[Fe_6(CO)_{16}C]^{2-}$ which was characterized 15 years earlier and is the first example of an interstitial boride in an octahedral transition metal environment to be reported (Churchill and

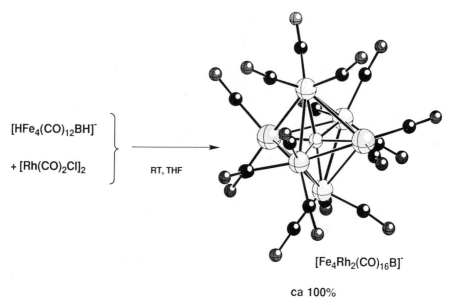

[HFe₄(CO)₁₂BH]⁻

+ [Rh(CO)₂Cl]₂

RT, THF

$[Fe_4Rh_2(CO)_{16}B]^-$

ca 100%

Figure 12.2. The preparation and the structure of $[Fe_4Rh_2(CO)_{12}B]^-$.

Wormald 1974). There are some interesting structural differences. Although $[Rh_2Fe_4(CO)_{16}B]^-$ has D_{2d} symmetry, the carbide has a single plane of symmetry. In the boride there are four short Rh–Fe bonds and four longer Rh–Fe bonds with the latter being bridged by semibridging carbonyls. This causes the Fe_4 square to be folded slightly. On the other hand, in the carbide while there are again two groups of Fe–Fe distances similar in length, it is the shorter Fe–Fe bonds that are found to be more symmetrically bridged with μ-CO ligands.

Once the composition was known by FAB mass spectrometry, the interstitial nature of the boron in $[Rh_2Fe_4(CO)_{16}B]^-$ was clear from the characteristics of the ^{11}B-NMR resonance. The very sharp triplet observed at extremely low field showed that the boron was bound to a large number of metal atoms (low field shift), was in a highly symmetric environment (increased relaxation time due to the quadrupolar nature of boron leading to narrow line widths) and was bound to two rhodium atoms (two nuclei of spin 1/2).

We have recently shown by explicitly evaluating the paramagnetic contribution to the shielding in the Ramsey SOS approach utilizing Fenske–Hall wave functions that the high deshielding of the boron atoms in these metal-rich metallaboranes is due to the large B 2p character in high-lying filled MOs (Fehlner et al. 1990). The greater the number of direct metal-boron interactions, the greater the B 2p character, the greater the paramagnetic contribution and the more deshielded the boron atom. Expressions in the literature concerning interstitial carbides to the contrary (Chini 1980), the low field shifts are not due to any positive character on the boron atom. On the contrary the calculated Mulliken charges on boron in the metal rich clusters is always rather negative. The boron of $[Rh_2Fe_4(CO)_{16}B]^-$ has the most negative charge of any molecule we have ever investigated with Fenske–Hall MO calculations.

A characteristic feature of metal clusters containing main group atoms in general is the vibrations associated with the main group atom rattling around in the cage formed by the much more massive transition metal atoms. This characteristic spectroscopic feature is observed for $[Rh_2Fe_4(CO)_{16}B]^-$ as well. Comparing the observed vibrational frequency with that reported for $[Fe_6(CO)_{16}C]^{2-}$ using a simple mass-on-a-spring model leads to the conclusion that the force constant for the boride is $\approx 80\%$ of that for the carbide. This is consistent with the somewhat longer Fe–B distances in the boride compared to the Fe–C distances in the carbide.

An example of another type of boride, $Fe_4(CO)_{12}(AuPPh_3)_3B$, has also been reported (Harpp et al. 1988). In fact, one, two, and three protons of $HFe_4(CO)_{12}BH_2$ have been replaced by the isolobal fragment $[AuPR_3]^+$ leading to $HFe_4(CO)_{12}(AuPPh_3)BH$, $Fe_4(CO)_{12}(AuPPh_3)_2BH$, and $Fe_4(CO)_{12}(AuPPh_3)_3B$ (Housecroft and Rheingold 1986, 1987). The solid state structure of the second shows one gold atom bridging an edge not bridged in the parent compound $HFe_4(CO)_{12}BH_2$ by a hydrogen (see above). In the third case, $Fe_4(CO)_{12}B(AuPPh_3)_3$ contains a cluster core with an encapsulated boron atom. The butterfly array of iron atoms of the starting material is retained with the gold atoms asymmetrically arranged about the open face totally enclosing the single boron with metal atoms. In solution, the phosphine ligands on the gold atoms are equivalent and the CO ligands on the "wing-tip" and "hinge" iron atoms are pairwise equivalent. Hence the solid state structure, in so far as the gold atoms are concerned, must not be a rigid one.

Recently, another example of a boride, $HRu_6(CO)_{17}B$, which is closely related to $[Rh_2Fe_4(CO)_{16}B]^-$, has been reported (Hong et al. 1989a). It is isolated from the reaction of $Ru_3(CO)_{12}$ with diborane and the cluster core is analogous to that of $Ru_6(CO)_{17}C$. This compound is neutral and contains a hydride bound to the metal framework. As is often the case with metallaboranes, this skeletal hydrogen can be removed as a proton with a base. In the homonuclear metal cluster the boron atom is again found in an octahedral environment of six metal atoms.

12.2.1.2. Rearrangement of the Boride

It was very clear from the spectroscopic information on the reaction of $[HFe_4(CO)_{12}BH]^-$ (Housecroft et al. 1987) with $[Rh(CO)_2Cl]_2$ that the isolated cluster product is preceded by an initial product. The first formed product has the spectroscopic signature of a boride and is rearranged cleanly into the crystallographically characterized material. That is, the large low field chemical shift is consistent with a boron atom directly bonded to six transition metal atoms (Rath and Fehlner 1988, Fehlner et al. 1990). The narrow line width in the

^{11}B-NMR and correspondingly long relaxation time demonstrate a highly symmetrical environment, that is, octahedral (Kidd 1983), and the observed triplet with $J_{BRh} = 23.3\,Hz$ shows direct bonding of the boron atom to two rhodium atoms. Thus, the most probable identification of the initial product is $[1,2\text{-}\{Rh(CO)_2\}_2\text{-}3,4,5,6\text{-}\{Fe(CO)_3\}_4B]^-$ (*cis*-isomer) which rearranges to $[1,6\text{-}\{Rh(CO)_2\}_2\text{-}2,3,4,5\text{-}\{Fe(CO)_3\}_4B]^-$ (*trans*-isomer). As may be seen from a comparison of the Fe atom positions in Figures 12.1 and 12.2, the *cis*-isomer is the expected initial product of reaction in the absence of skeletal rearrangement of $HFe_4(CO)_{12}BH_2$.

The kinetics of this reaction has now been examined (Bandyopadhyay et al. 1989). The reaction proceeds via an associative process exhibiting a kinetic rate law that is first order in the boride anion and first order in CO. The CO independent pathway is negligible within experimental error. Comparison of the effects of NEt_3 (none) and $PPhMe_2$ (enhances rate) on the reaction demonstrates that the role of the Lewis base is coordination to a metal site. The composition of the activated complex for rearrangement, $[Rh_2Fe_4(CO)_{17}B]^-$, suggests reaction proceeding via an open structure.

The intermediates for two of the most frequently proposed pathways for rearrangement of a 6-atom *closo* cluster (see Figure 12.3) (Hoffmann and Lipscomb 1963) have higher skeletal electron counts than an octahedral cluster (Wade 1976, Mingos and Johnston 1987). The one via a trigonal prismatic intermediate has always been appealing from the point of view of simplicity and symmetry; however, it is well known experimentally (Martinengo et al. 1979) as well as theoretically (Wijeyesehera and Hoffman 1984) that a trigonal prismatic cluster requires two more electron pairs than an octahedron. As a second order dependence on CO might well be expected, this pathway is ruled out. The alternate pathway involves a pentagonal pyramidal intermediate which is a *nido* six atom cluster requiring one additional electron pair for stability. It is the structure predicted for $[Rh_2Fe_4(CO)_{17}B]^-$ using the skeletal electron counting rules. Several arrangements of the Rh and Fe atoms are possible for pentagonal pyramidal $[Rh_2Fe_4(CO)_{17}B]^-$ and only one is illustrated in Figure 12.3. The added CO ligand satisfies the electronic requirements of the intermediate and the opening of the cluster provides sufficient room for good binding. This is the simplest mechanism that will explain existing observations.

12.2.1.3. Related Work

Our work in the area of metallaborane chemistry constitutes only one contribution to the chemistry of metal-rich metallaboranes. The recent reviews mentioned in Section 12.1 give a comprehensive account of the state of metal-rich metallaborane chemistry as well as related areas but a brief summary of work preceding our own is appropriate. The first fully characterized metal-laborane with more metal atoms than boron atoms is $HMn_3(CO)_{10}B_2H_6$ (Kaesz et al. 1965). Here two of three metals are separated from the third by an ethane-like B_2H_6 bridging "ligand". Second, the pioneering work of Grimes and

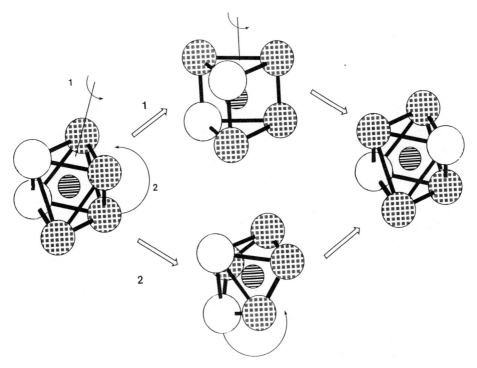

Figure 12.3. Two proposed pathways for the rearrangement of the *cis* isomer of $[Fe_4Rh_2(CO)_{12}B]^-$ to the *trans* isomer. The upper pathway proceeds via a trigonal prismatic geometry while the lower pathway is via a pentagonal bipyramidal cluster geometry.

co-workers (Grimes 1978, 1983) in which a significant number of metallaboranes with more than three metal atoms were characterized must be pointed out. Other compounds, clearly members of the class of metal-rich metallaboranes, have been reported on the basis of spectroscopic and analytical data. For example, $(CO)_9Co_3BNMe_3$ and $(CO)_{18}CO_6B$ have been described; however, no structural evidence was reported (Schmid et al. 1975). The ruthenaboranes formulated as $Ru_4(CO)_{12}BH_3$ and $Ru_3(CO)_9B_2H_6$ were observed in low yields as products of the reaction of BH_4^- with $Ru_3(CO)_{12}$ (Johnson et al. 1977). The former cluster, an analog of $HFe_4(CO)_{12}(H_2B)$, has been recently reported as the product of the reaction of $BH_3.THF$ and $H_4Ru_4(CO)_{12}$ (Hong et al. 1989b). It has also been prepared by the hydride activated reaction of $Ru_3(CO)_{12}$ with $BH_3.THF$ and some of its chemistry explored (Chipperfield et al. 1990). In a similar attack on the ruthenium-boron cluster problem, the $Ru_3(CO)_9BH_5$ cluster system has been shown to exist in solution as a tautomeric equilibrium mixture of two structures $[HRu_3(CO)_9(H_3BH)$ and $H_2Ru_3(CO)_9(H_2BH)]$ (Chipperfield and Housecroft 1988). The osmium rich metallaborane

$H_3Os_3(CO)_9BCO$ has been reported and shown to exhibit some extremely interesting chemistry (Shore et al. 1983, Jan and Shore 1987). We have described $HFe_3(CO)_{10}(HBH)$ which is an analog of $H_3Os_3(CO)_9BCO$ but clearly one with a very different structure (Vites et al. 1984).

12.2.2. Metal Substituted Boranes

The reaction $Co_2(CO)_8 + 2BH_3 \cdot THF \rightarrow 2(CO)_4CoBH_2 \cdot THF + H_2$ has been demonstrated to occur cleanly at $-15°C$ in THF (Basil et al. 1990). The very reactive $(CO)_4CoBH_2 \cdot THF$ molecule has been characterized by low temperature ^{11}B-NMR and infrared spectroscopies as well as classical chemical analysis and the proposed structure is shown in Figure 12.4. The formation of $(CO)_4CoBH_2 \cdot THF$ bears a remarkable similarity to that of $(CO)_4CoSiR_3$ (Chalk and Harrod 1967). Displacement of the bound THF of $(CO)_4CoBH_2 \cdot THF$ occurs with Lewis bases and the Lewis acidity of $(CO)_4CoBH_2 \cdot THF$ relative to that of $BH_3 \cdot THF$ for SMe_2 has been estimated. Apparently $(CO)_4CoBH_2$ is a somewhat harder acid than BH_3 just as BMe_3 is. Displacement of $[Co(CO)_4]^-$ from $(CO)_4CoBH_2 \cdot THF$ occurs very easily, for example, reaction with PhMgBr yields $PhBH_2$. $(CO)_4CoBH_2 \cdot THF$ readily accepts hydride from $[HFe_2(CO)_8]^-$ losing $[Co(CO)_4]^-$ but reduces the CO ligands of hydride free metal carbonylate anions. $(CO)_4CoBH_2.THF$ is a very active reducing agent and above $10°C$ cleaves THF and condenses with hydrocarbyl and metal fragments to yield a mixture of clusters including an unusual tailed cluster $(CO)_9Co_3C(CH_2)_nOH$, $n = 4,5$. These results provide some of the first information on the effect of a direct, unsupported M–B bond on the reactivity of B–H hydrogens.

The preparation and structural characterization of a tetrairon carbido cluster substituted tricoordinate borane, $HFe_4(CO)_{12}CBHX$, X = H, Cl, Br, have been carried out (Meng et al. 1989). The solid state structures of $HFe_4(CO)_{12}CBHX$ were determined by single crystal X-ray diffraction methods (Figure 12.5). The

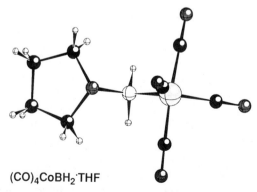

$(CO)_4CoBH_2 \cdot THF$

Figure 12.4. The proposed structure of $(CO)_4CoBH_2.THF$.

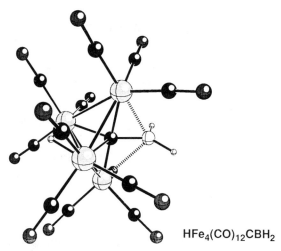

$HFe_4(CO)_{12}CBH_2$

Figure 12.5. The structure of $[HFe_4(CO)_{12}C]BH_2$.

molecule may be described as a $[HFe_4(CO)_{12}C]^-$ 62 electron butterfly cluster structure bound via the carbido carbon to a trigonal planar $[BHX]^+$ fragment. The fact that neither dimerization nor coordination to Lewis bases is observed suggests that the cluster substituent is both sterically demanding as well as a good π donor to the empty boron 2p orbital. The π donation derives from the high-lying filled MO with large carbido carbon-"wing-tip" iron atom character and leads to weak ($d_{FeB} = 2.43$ Å), direct iron-boron bonding. The π back-donation is confirmed by upfield ^{11}B chemical shift ($\delta = 9.6$). This compound constitutes the first example of a metal cluster substituted monoborane and illustrates the properties of a carbido cluster when viewed as a substituent. The steric bulk and strong π back-donating ability of this cluster substituent may lead to further utility in preparing other element compounds in unusual states. It constitutes a rare example of a monosubstituted borane that neither dimerizes nor readily adds a Lewis base in that it either deprotonates (from the cluster framework) or reacts in other ways.

12.2.3. New Metallacarboranes

The reaction of the metallacycle $Cp(PPh_3)CoC_4Ph_4$ with monoboranes containing at least one B–H bond provides a new route to the formation of $CpCoC_4Ph_4BH$. Using $BH_2Cl.SMe_2$ as the source of the BH fragment, yields of 30% are obtained (Hong et al. 1989a). The structure was proved by X-ray methods (Figure 12.6). Although B substituted derivatives of $CpCoC_4Ph_4BH$ were known previous to our work (Herberich et al. 1980), they were prepared by totally different routes and our reaction constitutes an example of a potentially general route to similar compounds containing main group fragments. Note that the BH fragment in $CpCoC_4Ph_4BH$ constitutes a functional group. We

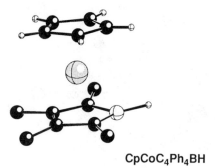

CpCoC$_4$Ph$_4$BH

Figure 12.6. The structure of CpCoC$_4$Ph$_4$BH where the four phenyl groups are represented by the *ipso* carbon atoms.

have shown that the B–H bond can be reversibly halogenated and can be alkylated with alkyllithium reagents. The resulting B–X bond can be hydrolyzed to yield the B–OH derivative or substituted with Grignard reagents.

Most metallacarborane cages contain a predominance of main group atoms and there are few examples where the number of main group and transition metals in the cage are equal. The work of Grimes et al. constitutes a fine example (Bowser et al. 1979). These systems can exhibit unusual behavior. The synthesis and structural characterization of the first example of a boracyclopropene ring coordinated to a trimetal fragment, Fe$_3$(CO)$_9$[η^3-B(H)C(H)C(Me)], has been carried out and the solid state structure of Fe$_3$(CO)$_9$[η^3-B(H)C(H)C(Me)] has been determined with a single crystal X-ray diffraction study (Figure 12.7) (Meng et al. 1990). Conventional analysis of the bonding based on a three metal system leads to a 44 electron unsaturated cluster; however, there is no geometric evidence that the metal fragment is unsaturated. On the other hand, the compound has a distorted octahedral structure which is isolobal with C$_2$B$_4$H$_6$

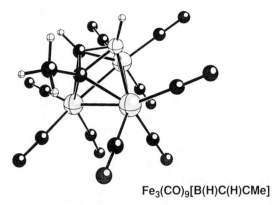

Fe$_3$(CO)$_9$[B(H)C(H)CMe]

Figure 12.7. The structure of Fe$_3$(CO)$_9$[η^3-B(H)C(H)CMe].

and structurally analogous to it as well as to $(C_5H_5)_3Co_3B_3H_5$. Hence, $Fe_3(CO)_9[\eta^3\text{-}B(H)C(H)C(Me)]$ constitutes an unambiguous example of a cluster containing three transition metals whose bonding can only reasonably be considered in terms of a six-atom cluster containing three iron, two carbon and one boron atom. This particular derivative of the Fe_3C_2B cluster core also constitutes an example of a chiral cluster.

12.3. ADVANCES IN SYNTHESIS

12.3.1. New Metallaboranes

Many of the compounds characterized in our laboratories containing direct iron-boron bonds were initially isolated from rather complex reaction mixtures in very low yield. Evidence continues to build supporting a hypothesis of limited thermodynamic control where the borane effectively acts a a dual reagent in "activating" the metal and in condensing with the unsaturated fragments formally produced in situ. That is, reactive fragments are produced by reaction with BH_3 and condense with BH_3 to form those products of sufficient stability to survive the reaction conditions. The distribution of products appears to be controlled by the concentration of reactants among other parameters. Relatively small changes leads to significantly different product distributions. Yields have been improved by optimizing reaction conditions. We have now extended this approach to the synthesis of metal-rich ferraboranes to metals other than iron. For example, the reaction shown in Figure 12.8 summarizes our knowledge on the formation of $H_2Cp_4Co_4B_2H_2$ (Feilong et al. 1987a). This molecule is a member of the 6-atom *closo* series $(CpCo)_nB_{6-n}H_{8-n}$, two members of which $(n = 2, 3)$ have been characterized earlier (Pipal and Grimes 1977, 1979a). The

CHANGE IN METAL

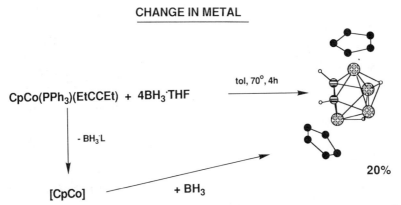

$$CpCo(PPh_3)(EtCCEt) + 4BH_3\cdot THF \xrightarrow{\text{tol, } 70°, \text{ 4h}}$$

$- BH_3\cdot L$

$[CpCo]$ ⟶ $+ BH_3$

20%

Figure 12.8. The preparation and structure of $(\mu^3\text{-}H)_2(CpCo)_4B_2H_2$. The lower pathway is the suggested route via the generation of [CpCo] fragments and subsequent reaction with BH_3. Two Cp ligands have been omitted from $(\mu^3\text{-}H)_2(CpCo)_4B_2H_2$ for clarity.

compound is also a borane mimic of $Co_4(CO)_{10}C_2R_2$. The highly metallic environment of the boron atoms are reflected in a very low field ^{11}B chemical shift, that is, $\delta 114$.

In the above reaction, the borane effectively acts as a dual reagent in removing the ligands from cobalt and in condensing with the unsaturated CpCoL or CpCo fragments formally produced in situ. A very similar approach to large gold clusters has been described by Schmid (1985). On the face of it, this fragment condensation is related to the preparation of Fe-S clusters in which "spontaneous self assembly" is thought to take place (Whitener et al. 1986). In the latter case simple variation of reagent stoichiometry serves to provide selective routes to a range of cluster compositions. Hence, we have varied reaction conditions in this CpCo/BH system and have demonstrated the formation of several new clusters.

By reducing the borane/cobalt ratio and changing the mode of addition of reagents, clusters containing PPh_x fragments $(x < 3)$ along with borane and cobalt fragments have been isolated. Examples of this type of product are $(CpCo)_4B_2H_2PPh$ and $Cp_3Co_3BPhPPh$ (Feilong et al. 1987b, 1988a). The former metal-rich cobaltaborane consists of a pentagonal bipyramid containing four cobalt, one phosphorus, and two boron atoms in the cluster core (right-hand side of Figure 12.9). As such, it constitutes the first example of a metal rich metallaphosphaborane. We have shown that $(CpCo)_4B_2H_2PPh$ is not formed by the direct reaction of $(\mu_3\text{-}H)_2Cp_4Co_4B_2H_2$ with PPh_3 or $BH_3 \cdot PPh_3$ and, therefore, under BH_3 deficient conditions one phosphine is retained in the cluster building process. Benzene is a product of the reaction and is presumably formed by reductive elimination of a metal hydride with a phenyl group. The second compound illustrates the transfer of a phenyl group from phosphorus to

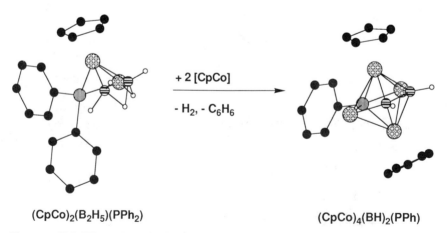

$$(CpCo)_2(B_2H_5)(PPh_2) \qquad +2\,[CpCo] \qquad (CpCo)_4(BH)_2(PPh)$$

$$-H_2, -C_6H_6$$

$(CpCo)_2(B_2H_5)(PPh_2)$ \qquad\qquad\qquad $(CpCo)_4(BH)_2(PPh)$

Figure 12.9. The hypothetical conversion of $(CpCo)_2(B_2H_5)(PPh_2)$ into $(CpCo)_4(B_2H_2)(PPh)$. Two Cp ligands have been omitted from $(CpCo)_4(B_2H_2)(PPh)$ for clarity.

boron; however, it should be noted that Cp_3Co_3BHPPh has been spectroscopically characterized (Feilong and Fehlner 1988).

Increasing the borane to cobalt ratio results in the formation of $Cp_2Co_2(PPh_2)B_2H_5$ in modest yield (Feilong et al. 1988b). This compound has been fully characterized and structurally is a formal analog of a dinuclear transition metal complex containing a C_2H_3 ligand bound in a σ-π mode (left-hand side of Figure 12.9). The same type of bonding of a B_2H_5 ligand has been observed earlier in the more complex metallaborane $Pt_2(PMe_2Ph)_2$ $\times (B_2H_5)(B_6H_9)$ (Ahmad et al. 1982, 1986).

It is significant that all of these products can be simply related by the formal addition or subtraction of CpCo, PPh_x, Ph or H fragments. For example, as illustrated in Figure 12.9, the condensation of $Cp_2Co_2(PPh_2)B_2H_5$ with CpCo fragments can be thought of as generating $(CpCo)_4B_2H_2PPh$. Likewise, as shown in Figure 12.10, the formal condensation of two $Cp_2Co_2(PPh_2)B_2H_5$ leads to the formation of $Cp_4Co_4B_4H_4$, a prominent product in our system originally characterized in low yield by Pipal and Grimes (1979b). There is no evidence that these reactions take place as written but such relationships suggest all products result from the same set of progenitor fragments.

12.3.2. New Routes to Known Compounds

These observations, combined with the fact that mixtures of products are always observed, suggest that the generation of reactive fragments is followed by condensation to form those products of sufficient stability to survive the reaction conditions. The distribution of products appears to be controlled by the concentration of reactants among other parameters. Relatively small changes

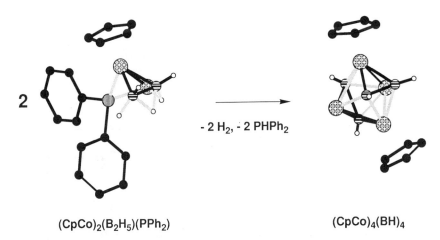

$(CpCo)_2(B_2H_5)(PPh_2)$ $(CpCo)_4(BH)_4$

$- 2 H_2, - 2 PHPh_2$

Figure 12.10. The hypothetical conversion of $(CpCo)_2(B_2H_5)(PPh_2)$ into $(CpCo)_4(BH)_4$. Two Cp ligands have been omitted from $(CpCo)_4(BH)_4$ for clarity.

led to significantly different product distributions. This appears to be most easily seen in the cobalt system where the initiation reaction is the straightforward dissociation of a ligand from the metal. Indeed for $CpCo(CO)_2$ a cluster building process is well established (Vollhardt et al. 1975). As illustrated in Figure 12.11, the well-known dimer, trimer, and tetramer can be thought of as being formed from sequential additions of CpCoCO fragments with or without loss of additional CO ligands. Hence, it is reasonable to postulate that a similar cluster building process takes place with $L = PPh_3$. The observed metallaboranes can be thought of as arising from the trapping of the dimer, trimer, et cetera with the elimination of the appropriate ligand fragments. This scheme explains the observed dependence of product nuclearity distribution on borane to cobalt ratio, that is, at high B/Co ratios metal fragment association is inhibited and the major product contains two cobalt atoms while at low B/Co ratios the cobalt tetramer is the major species leading to the predominance of Co_4 products. Obviously this is a very qualitative scheme that does not really deserve the appellation of mechanism but nonetheless does summarize the essence of what we have learned. Further, the iron carbonyl system is similar except that the initiation reaction leading to metal cluster formation is not obvious although we have suggested some possibilities previously (Fehlner 1988b).

We have continued to extend this thinking in trying to discover improved routes to some of our known ferraboranes and, based on the above reasoning, have sought precursors of iron carbonyl fragments that react under mild conditions. We have sought to increase the reactivity by replacing a CO ligand with a better σ donor and poorer π acceptor ligand. We have prepared $(CO)_4FeSMe_2$, a new and rather labile derivative, that at high B/Fe ratios reacts

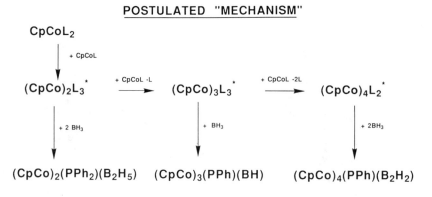

POSTULATED "MECHANISM"

$CpCoL_2$

↓ + CpCoL

$(CpCo)_2L_3^*$ →(+ CpCoL -L)→ $(CpCo)_3L_3^*$ →(+ CpCoL -2L)→ $(CpCo)_4L_2^*$

↓ + 2 BH₃   ↓ + BH₃   ↓ + 2BH₃

$(CpCo)_2(PPh_2)(B_2H_5)$ $(CpCo)_3(PPh)(BH)$ $(CpCo)_4(PPh)(B_2H_2)$

All known for L = CO

Figure 12.11. A working mechanism showing the generation of known cobaltaboranes from the trapping of cobalt clusters of differing nuclearity. The cobalt clusters are all known for $L = CO$ but the cobaltaboranes are prepared from the mononuclear compound with $L = PPh_3$.

with $BH_3 \cdot SMe_2$ to produce $B_2H_6Fe_2(CO)_6$ in 25% yield (Meng 1990). A single ferraborane product is produced simplifying isolation, the reaction can be carried out on a gram scale, and the yield is increased a factor ≈ 10 from our previous method. Additional evidence arises from the use of Grevels' $(CO)_3Fe(cco)_2$, where (cco) = cis-cyclooctene (Fleckner et al. 1984). This compound is an established source of $Fe(CO)_3$ fragments at low temperature and reacts with borane at $-30°C$ to give modest to good yields of a number of our known ferraboranes (Bandyopadhyay and Fehlner 1989). As in the cobalt system discussed above, the B/Fe ratio in the products depends on the ratio of BH_3 to $(CO)_3Fe(cco)_2$ used in the reaction. Hence, we conclude that the qualitative scheme in Figure 12.11 also explains the major features of the synthesis of ferraboranes.

12.4. TRANSITION TO THE SOLID STATE

12.4.1. BORIDES

The metal borides are one of the five major classes of boron compounds (Greenwood and Earnshaw 1984). Because of their technological value as well as promise the metal borides have received considerable attention (Greenwood et al. 1966, 1973, Matkovich 1977, Post 1964). Several hundred binary borides are known and they range in stoichiometry from M_5B to MB_{12} and higher. Nonstoichiometric phases of variable composition and ternary compounds add to the list of interesting systems. The inner penetration of the boron and metal lattices leads to some beautiful, if complex, solid state structures (Kuzma 1983). As is often the case in solid state systems, stoichiometry is less informative than structure simply because there is no immediately evident relationship between the two. The characteristic forms taken up by the boron network have direct connections to the structures of borane cages observed in discrete species in the gas phase or in solution. Indeed, there has been a symbiotic relationship between the advances in the development of structure and bonding in the solid state borides containing boron atom networks and that of discrete borane cages (Lipscomb 1981). As there are a number of solid state borides with isolated boron atoms, it is of interest to compare some of the observations made on the discrete boride described in Section 12.2.1.1.

First, with few exceptions, the distribution of metal atoms around the boron is trigonal prismatic, cubic antiprismatic or tetrahedral rather than a octahedral as found for the discrete boride (Lundström 1969). This observation has often been discussed in the past on the basis of the radius ratio rule, i.e., r_B/r_M exceeds 0.414 the value most favored for the octahedron. Using a radius for boron of 0.88 Å, the calculated ratios for typical borides, for example, Re_3B, range from 0.64 to 0.71 Å thereby substantially exceeding the radius ratios for both octahedral and trigonal prismatic (0.528) arrangements. In the characterized discrete boride, $[Rh_2Fe_4(CO)_{16}B]^-$, the metal-metal distances are typical of those for species with direct bonding. From this structure the calculated r_B is 0.67–0.69 Å and the radius ratio is 0.52–0.51 which is much smaller than previously thought. Hence, in the discrete octahedral case there appear to be no restraints caused by the requirements of metal-metal bonding. Thus, it seems unlikely that it is size alone that lies behind the trigonal prismatic geometry in the solid state.

The differences between the geometries of discrete and solid state borides may be due not so much to the metal-boron interactions but rather to the fact that the exo-cluster connections are very different. In the former, connections are to CO ligands whereas in the latter they are to additional metal atoms. This constitutes a significant difference. First, the good, π-backbonding, σ-donor CO ligands in the discrete clusters induce directional character to the metal-metal bonding. Second, the exo-ligands of the discrete cluster are intimately involved in the cluster electron count, for example, adding two more electron pairs via two additional exo-ligands can result in an octahedral interstitial nitride cluster being converted to a trigonal prismatic cluster (Bordoni et al. 1988). Finally, it is also possible that weak B–B interactions are in fact important in determining the metal environment of the boron atoms in the solid state. For example, even in Rh_5B_4 which contains boron atoms in octahedral holes in a close-packed metal lattice, "strings" of four boron atoms (B–B distance of 2.22 Å) separated by 3.306 Å are observed (Noläng et al. 1981). The B–B bonding is considered "doubtful" but the chains are still considered elements of the structure.

12.4.2. Alloy Thin Films from Metallaboranes

At minimum there is a formal connection between fragments of discrete species and the same fragments found in solid state compounds containing the same elements. An interesting question then is whether the discrete species can serve as precursors for the efficient production of specific solid state materials containing the cluster cores of the discrete compounds serving as building blocks of the solid.

Metal organic chemical vapor deposition (MOCVD) is a well established, practical technique for forming simple as well as complex solid state films (Powell et al. 1966). For binary systems the conventional approach is to use mixtures of the most readily available molecules containing the elements of interest. This approach has been employed to prepare borides of several types. For example, iron-boron alloys have been prepared with 9–13% Fe (Dasseau et al. 1981) and $FeB_{29.5}$ has been made directly from β-rhombohedral boron and iron metal (Werheit et al. 1981). However, the reagents used in this approach are difficult to activate and require rigorous conditions for deposition. This leads to restrictions in the substrates that can be utilized and, in the case of thermal activation, often results only in the formation of the most stable form of the solid.

In recent years a number of groups have developed compounds that serve as single precursors for specific binary materials (Aylett and Colquhoun 1977,

Jefferies and Girolami 1989, Czekaj and Geoffroy 1988, Bochmann et al. 1988, Steigerwald 1989, Boyd et al. 1989, Cowley and Jones 1989, Gross et al. 1989, Interrante et al. 1986, Seyferth and Wiseman 1984). In this approach, the objective is to synthesize precursor molecules that are easy to handle with low toxicity and which intrinsically contain not only the desired stoichiometry of the material but also a low energy decomposition pathway. When successful these rational precursors have the desired properties of convenience and safety but, as importantly, create opportunities for preparing films of presently known stoichiometry with new properties as well as for exploring systems of unknown stoichiometry and properties.

Three groups have reported an MOCVD approach to the preparation of metal boride films. In one, $Ti(BH_4)_3(1,2\text{-}(MeO)_2C_2H_4)$ was used to deposit pure, amorphous, thin films of TiB_2 on pyrex substrates under mild conditions (Jensen et al. 1988). The films were pure showing neither titanium metal or elemental boron and other impurities such as oxygen were low. Films of ZrB_2 and HfB_2 were made in an analogous fashion. In a closely related study films of ZrB_2 and HfB_2 were prepared from the boron hydrides and characterized (Wayda et al. 1988). Recently a variety of approaches for the synthesis of ZrB_2 from $Zr(BH_4)_4$ have been explored (Rice and Woodin 1988). These authors report films formed at higher temperatures contained excess boron while those at lower temperatures were boron deficient.

In conjunction with our synthetic work, research into the utilization of metal clusters for the formation of alloy thin films has been carried out to demonstrate the potential utility of metallaboranes in generating alloy thin films. Preliminary experiments showed that $HFe_4(CO)_{12}BH_2$ decomposes under thermolytic conditions to yield CO and H_2 plus a residual solid containing boron and iron (Fehlner et al. 1989). In more recent studies films were prepared by subliming $HFe_4(CO)_{12}BH_2$ in a low pressure (base pressure 10^{-8} Torr, deposition pressure 10^{-4} Torr) MOCVD reactor of our own design on substrates resistively heated to 180°C. Uniform, contiguous films of 1000 Å in thickness with a metallic luster were grown in approx. 30 min on glass, silicon, and metal substrates. The films adhere well to all substrates examined and were amorphous to X-rays. Analysis by Auger and XPS showed a FeB ratio of $\approx 4:1$ and no impurities other than carbon and oxygen ($< 5\%$). The chemical environment of the boron atom as measured by the 1s binding energy is identical to that of the authentic material. Mössbauer spectroscopy demonstrates a distribution of hyperfine fields typical of the authentic amorphous metal. Resistivity measurements yield values two times larger than those of the bulk materials. These results unambiguously define $HFe_4(CO)_{12}BH_2$ as a useful single precursor for the production of thin films of an authentic metallic glass under mild conditions.

Secondly, the thermal decomposition of $HFe_3(CO)_9BH_4$ on glass or aluminum substrates at 175–200°C at pressures between 10^{-5}–10^{-4} Torr results in the deposition of uniform, amorphous alloy films of approximate composition $Fe_{75}B_{25}$ with individual thicknesses ranging from 1000 to 10,000 Å. The glassy metal films have been characterized by Auger, X-ray, and Mössbauer spectroscopies. The Mössbauer spectra show that the local structure of the film is similar to that observed for films prepared by the rapid quenching technique. In contrast, however, films prepared from $HFe_3(CO)_9BH_4$ behave as an ordered magnetic material with moments having a preferential direction normal to rather than in the film plane (Amini et al. 1990). Although the films are stable in air, oxidation during deposition readily takes place under poor vacuum conditions and leads to films containing oxidized boron. The stoichiometry of the oxide phase has been shown to be B_2O_3. The resistivity of the film increases rapidly with increasing oxygen content.

These studies show that metallaboranes constitute precursors for solid state borides. As it is well known that there is a relationship between the structure of borides and boranes as far as the boron networks are concerned, metallaboranes may serve as sources of presently unknown boride phases. That is, comparison of the literature shows that the present known structural diversity of the boranes is much greater than that of the borides. This may well be due to the rigorous conditions required for boride preparation in that only the most stable boron networks are formed and isolated. Thus, the low temperature routes to Fe_4B and Fe_3B illustrated in the foregoing work suggest an approach to presently unknown compounds with novel boride networks. Hence, as suggested by Lipscomb (1981), the multitude of known metallaboranes constitute a stockpile of potential precursors to metal borides with new and perhaps useful properties.

ACKNOWLEDGMENT

The generous support of our work by the National Science Foundation and Army Research Office is greatly appreciated.

REFERENCES

Ahmad, R., Crook, J. E., Greenwood, N. N., Kennedy, J. D. and McDonald, W. S., *J. Chem. Soc., Chem. Commun.*, 1019 (1982).

Ahmad, R., Crook, J. E., Greenwood, N. N. and Kennedy, J. D., *J. Chem. Soc., Dalton Trans.*, 2433 (1986).

Amini, M. M., Fehlner, T. P., Long, G. J. and Politowski, M., *Chemistry of Materials*, **2**, 432 (1990).

Aylett, B. J. and Colquhoun, H. M., *J. Chem. Soc., Dalton Trans.*, 2058 (1977).

Bandyopadhyay, A. K. and Fehlner, T. P., *J. Organomet. Chem.*

Bandyopadhyay, A. K., Khattar, R. and Fehlner, T. P., *Inorg. Chem.*, **28**, 4434 (1989).

Basil, J. D., Aradi, A. A., Bhattacharyya, N. K., Rath, N. P., Eigenbrot, C. and Fehlner, T. P., *Inorg. Chem.*, **29**, 1260 (1990).

Bochmann, M., Hawkins, I. and Wilson, L. M., *J. Chem. Soc., Chem. Commun.*, 344 (1988).

Bordoni, S., Heaton, B. T., Seregni, C., Strona, L., Goodfellow, R. J., Hursthouse, M. B., Thornton-Pett, M. and Martinengo, S., *J. Chem. Soc., Dalton Trans.*, 2103 (1988).

Bowser, J. R., Bonny, A., Pipal, J. R. and Grimes, R. N., *J. Am. Chem. Soc.*, **101**, 6229 (1979).

Boyd, D. C., Haasch, R. T., Mantell, D. R., Schulze, R. K., Evans, J. F. and Gladfelter, W. L., *Chemistry of Materials*, **1**, 119 (1989).

Bradley, J. S., *Adv. Organomet. Chem.*, **22**, 1 (1983).

Chalk, A. J. and Harrod, J. F., *J. Am. Chem. Soc.*, **89**, 1640 (1967).

Chini, P., *J. Organomet. Chem.*, **200**, 37 (1980).

Chipperfield, A. K. and Housecroft, C. E., *J. Organomet. Chem.*, **349**, C17 (1988).

Chipperfield, A. K., Housecroft, C. E. and Rheingold, A. L., *Organometallics*, **9**, 681 (1990).

Churchill, M. R. and Wormald, J., *J. Chem. Soc., Dalton Trans.*, 2410 (1974).

Cowley, A. H., *Polyhedron*, **3**, 389 (1984).

Cowley, A. H. and Jones, R. A., *Angew. Chem. Int. Ed. Engl.*, **28**, 1208 (1989).

Czekaj, C. L. and Geoffroy, G. L., *Inorg. Chem.*, **27**, 8 (1988).

Dasseau, J. M., Robert, J. L., Armas, B. and Combescure, C., *J. Less Common Met.*, **82**, 137 (1981).

Fehlner, T. P., *Comments on Inorg. Chem.*, **7**, 307 (1988a).

Fehlner, T. P., *New J. Chem.*, **12**, 307 (1988b).

Fehlner, T. P., Housecroft, C. E., Scheidt, W. R. and Wong, K. S., *Organometallics*, **2**, 825 (1983).

Fehlner, T. P., Amini, M. M., Zeller, M. V., Stickle, W. F., Pringle, O. A., Long, G. J. and Fehlner, F. P., in: *Chemical Perpsectives of Microelectronic Material* (M. E. Gross, J. M. Jasinski and J. T. Yates, Jr., Eds.), pp. 413–418. Materials Research Soc., Pittsburgh, PA, 1989.

Fehlner, T. P., Czech, P. T. and Fenske, R. F., *Inorg. Chem.*, in press.

Feilong, J. and Fehlner, T. P., unpublished results (1988).

Feilong, J., Fehlner, T. P. and Rheingold, A. L., *J. Am. Chem. Soc.*, **109**, 1860 (1987a).

Feilong, J., Fehlner, T. P. and Rheingold, A. L., *J. Chem. Soc., Chem. Commun.*, 1395 (1987b).

Feilong, J., Fehlner, T. P. and Rheingold, A. L., *Angew. Chem., Int. Ed. Engl.*, **27**, 424 (1988a).

Feilong, J., Fehlner, T. P. and Rheingold, A. L., *J. Organomet. Chem.*, **348**, C22 (1988b).

Fleckner, H., Grevels, F.-W. and Hess, D., *J. Am. Chem. Soc.*, **106**, 2027 (1984).

Gilbert, K. B., Boocock, S. K. and Shore, S. G., *Comp. Organomet. Chem.*, **6**, 879 (1982).

Greenwood, N. N., *Comprehensive Inorganic Chemistry*, vol. 1, pp. 665–991, Pergamon, Oxford, 1973.

Greenwood, N. N., *Chem. Soc. Rev.*, **13**, 353 (1984).

Greenwood, N. N. and Earnshaw, A., *Chemistry of the Elements*, pp. 162–170, Pergamon Press, New York, 1984.

Greenwood, N. N., Parish, R. V. and Thornton, P., *Q. Rev.*, **20**, 441 (1966).

Grimes, R. N., *Acc. Chem. Res.*, **11**, 420 (1978).

Grimes, R. N. (Ed.), *Metal Interactions with Boron Clusters*, Plenum Press, New York, 1982.

Grimes, R. N., *Acc. Chem. Res.*, **16**, 22 (1983).

Gross, M. E., Jasinski, J. M., Yates, J. T., Jr. (Eds.), *Chemical Perspectives of Microelectronic Materials*, Materials Research Soc., Pittsburgh, PA, 1989.

Harpp, K. S., Housecroft, C. E., Rheingold, A. L. and Shongwe, M. S., *J. Chem. Soc., Chem. Commun.*, 965 (1988).

Herberich, G. E., Buller, B., Hessner, B. and Oschmann, W., *J. Organomet. Chem.*, **195**, 253 (1980).

Herrmann, W. A., *Angew. Chem. Int. Ed. Engl.*, **25**, 56 (1986).

Hoffmann, R. and Lipscomb, W. N., *Inorg. Chem.*, **2**, 231 (1963).

Hong, F.-E., Coffy, T. J., McCarthy, D. A. and Shore, S. G., *Inorg. Chem.*, **28**, 3284 (1989a).

Hong, F.-E., Eigenbrot, C. W. and Fehlner, T. P., *J. Am. Chem. Soc.*, **111**, 949 (1989b).

Housecroft, C. E. and Rheingold, A. L., *J. Am. Chem. Soc.*, **108**, 6420 (1986).

Housecroft, C. E. and Rheingold, A. L., *Organometallics*, **6**, 1332 (1987).

Housecroft, C. E., *Polyhedron*, **6**, 1935 (1987).

Housecroft, C. E. and Fehlner, T. P., *Adv. Organomet. Chem.*, **21**, 57 (1982).

Housecroft, C. E., Buhl, M. L., Long, G. J. and Fehlner, T. P., *J. Am. Chem. Soc.*, **109**, 3323 (1987).

Huttner, G. and Knoll, K., *Angew. Chem.*, **99**, 765 (187).

Interrante, L. V., Carpenter, L. E., II, Whitmarsh, C., Lee, W., Garbauskas, M. and Slack, G. A., *Mat. Res. Soc. Symp. Proc.*, **73**, 359 (1986).

Jan, D.-Y. and Shore, S. G., *Organometallics*, **6**, 428 (1987).

Jefferies, P. M. and Girolami, G. S., *Chemistry of Materials*, **1**, 8 (1989).

Jensen, J. A., Gozum, J. E., Pollina, D. M. and Girolami, G. S., *J. Am. Chem. Soc.*, **110**, 1643 (1988).

Johnson, B. F. G., Eady, C. R. and Lewis, J., *J. Chem. Soc., Dalton Trans.*, 477 (1977).

Kaesz, H. D., Fellmann, W., Wilkes, G. R. and Dahl, L. F., *J. Am. Chem. Soc.*, **87**, 2753 (1965).

Kennedy, J. D., *Prog. Inorg. Chem.*, **34**, 211 (1986) and **32**, 519 (1984).

Khattar, R., Puga, J. and Fehlner, T. P., *J. Am. Chem. Soc.*, **111**, 1877 (1989).

Kidd, R. G., *NMR of Newly Accessible Nuclei* (P. Laszlo, Ed.), Vol. 2, pp. 50–77, Academic Press, New York, 1983.

Kuzma, Y. B., *Kristallokhimiia*, L'vov: Eischa shkola. Izd-va pri L'vov. un-te, 1983.

Lipscomb, W. N., *J. Less Common Met.*, **82**, 1 (1981).

Lundström, T., *Ark. Kemi*, **31**, 227 (1969).

Martinengo, S., Ciani, G., Sironi, A., Heaton, B. T. and Mason, J., *J. Am. Chem. Soc.*, **101**, 7095 (1979).

Matkovich, V. I. (Ed), *Boron and Refractory Borides*, Springer-Verlag, Berlin, 1977.

Meng, X., Rath, N. P. and Fehlner, T. P., *J. Am. Chem. Soc.*, **111**, 3422 (1989).

Meng, X., Fehlner, T. P. and Rheingold, A. L., *Organometallics*, **9**, 534 (1990).

Meng, X., Ph.D. Thesis, Univ. Notre Dame, 1990.

Mingos, D. M. P. and Johnston, R. L., *Structure and Bonding*, **68**, 30 (1987).

Noläng, B. I., Tergenius, L.-E. and Westman, I., *J. Less. Common Met.*, **82**, 303 (1981).

Paetzold, P., *Adv. Inorg. Chem.*, **31**, 123 (1987).

Pipal, J. R. and Grimes, R. N., *Inorg. Chem.*, **16**, 3255 (1977).

Pipal, J. R. and Grimes, R. N., *Inorg. Chem.*, **18**, 257 (1979a).

Pipal, J. R. and Grimes, R. N., *Inorg. Chem.*, **18**, 252 (1979b).

Post, B., *Boron, Metallo-Boron Compounds and Borane* (R. M. Adams, Ed.), pp. 301–371, Wiley-Interscience, New York, 1964.

Powell, D. F., Oxleyh, J. H. and Blocher, J. M., *Vapor Deposition*, Wiley, New York, 1966.

Rath, N. P. and Fehlner, T. P., *J. Am. Chem. Soc.*, **109**, 5273 (1987).

Rath, N. P. and Fehlner, T. P., *J. Am. Chem. Soc.*, **110**, 5345 (1988).

Rice, G. W. and Woodin, R. L., *J. Am. Ceram. Soc.*, **71**, C181 (1988).

Schmid, G., *Angew. Chem. Int. Ed. Engl.*, **17**, 392 (1978).

Schmid, G., *Structure and Bonding*, **62**, 51 (1985).

Schmid G., Vatzel V., Elzrodt G. and Pfel R., *J. Organomet. Chem.*, **86**, 257 (1975).

Seyferth, D. and Wiseman, G. H., *Ultrastructure Processing of Ceramics, Glasses, and Composites* (L. L. Hench and D. R. Ulrich, Eds), pp. 265–271, Wiley, New York, 1984.

Shore, S. G., Jan, D.-Y., Hsu, L.-Y. and Hsu, W.-L., *J. Am. Chem. Soc.*, **105**, 5923 (1983).

Spielvogel, B. F., *Advances in Boron and the Boranes* (J. F. Liebman, A. Greenberg and R. E. Williams, Eds.) pp. 329–342, VCH, New York, 1988.

Steigerwald, M. L., *Chemistry of Materials*, **1**, 52 (1989).

Vahrenkamp, H., *Adv. Organomet. Chem.*, **22**, 169 (1983).

Vites, J. C., Housecroft, C. E., Jacobsen, G. B. and Fehlner, T. P., *Organometallics*, **3**, 1591 (1984).

Vollhardt, K. P. C., Bercaw, J. E. and Bergman, R. G., *J. Organomet. Chem.*, **97**, 283 (1975).

Wade, K., *Adv. Inorg. Chem. Radiochem.*, **18**, 1 (1976).

Wayda, A. L., Schneemeyer, L. F. and Opila, R. L., *Appl. Phys. Lett.*, **53**, 361 (1988).

Werheit, H., DeGroot, K., Malhemper, W. and Lundström, T., *J. Less Common Met.*, **82**, 163 (1981).

Whitener, M. A., Bashkin, J. D., Hagen, K. S., Girerd, J.-J., Gamp, E., Edelstein, N. and Holm, R. H., *J. Am. Chem. Soc.*, **108**, 5607 (1986).

Whitmire, K. H., *J. Coord. Chem.*, **17**, 95 (1988).

Wijeyesehera, S. D. and Hoffmann, R., *Organometallics*, **3**, 949 (1984).

Wong, K. S., Scheidt, W. R. and Fehlner, T. P., *J. Am. Chem. Soc.*, **104**, 1111 (1982).

CHAPTER 13

Oxidative Addition of Carbon-Hydrogen Bonds to Soluble Complexes of Iridium and Osmium

T. C. FLOOD

13.1. INTRODUCTION

One hope harbored by organometallic chemists for many years was that their field would be the spawning ground for catalysts which could be "fine-tuned" to selectively "activate" and flexibly functionalize any specific C–H bond in solution at or below room temperature. The potential far-reaching benefits of such a family of catalysts is obvious. Over the last 10–15 years, significant

Electron Deficient Boron and Carbon Clusters, Edited by George A. Olah, Kenneth Wade, and Robert E. Williams.
ISBN 0-471-52795-5 © 1991 John Wiley & Sons, Inc.

progress has been made in observing activation of assorted C–H bonds, but it
has become evident that the goal of general homogeneous catalytic functionalization of specific C–H bonds, particularly aliphatic ones, is not going to be
easily accomplished. This is not to say that solutions will not be found, but none
are evident as yet. However, even if such general catalysis is not accomplished,
there still remain fascinating and scientifically fundamental and worthwhile
questions.

What does it take for a metal complex to disrupt a carbon-hydrogen bond?
The answer, as it is shaping up, is that there are several ways in which C–H
bonds may be activated. Early transition metals, lanthanides, and actinides in
closed-shell-configuration oxidation states (e.g., $Cp_2^*ThR_2$ [1], Cp_2^*LuR [2],
Cp_2^*YR [2], and Cp_2^*ScR [3], where $Cp^* = \eta^5\text{-}C_5Me_5$, and $(t\text{-}Bu_3SiNH)_3ZrR$
[4] undergo metathetical reactions with hydrocarbon C–H bonds. Apparently,
these reactions depend critically on the Lewis acidity of the metal center.

At the other end of the transition series, palladium and platinum are known
to catalyze H/D exchange of hydrocarbons in reactions where catalytic
intermediates are not well characterized [5–7].

It may be that maximum flexibility for subsequent functionalization of C–M
bonds could be achieved when the C–H bond would be actived via an
"oxidative addition" type of reaction, as shown in equation (13.1).

$$
L\!-\!\!\overset{\displaystyle L}{\underset{\displaystyle L}{M}}\!\!-\!L \;+\; R\text{-}H \;\rightleftharpoons\; L\!-\!\!\overset{\displaystyle \overset{H}{|}\;L}{\underset{\displaystyle L}{M}}\!\!-\!R \qquad (13.1)
$$

d^8, L = a variety of neutral d^6
and / or anionic ligands

A number of Cp or Cp*-containing molecules have been found which exhibit
oxidative addition behavior with hydrocarbons [8–12]. Some of the earliest and
best-defined examples of C–H oxidative addition reactions are among these.
Nevertheless, since Cp and Cp* ligands rather rigidly occupy three coordination
sites of the metal, they may be more restrictive than would be desirable for a
potentially catalytic species. One would really like to observe C–H reactions in
systems of interconverting square-planar-d^8 and octahedral-d^6 species, as in
equation (13.1), of the type that have proved most generally interesting for
homogeneous catalysis.

A number of significant studies related to equation (13.1) have appeared. In
fact, 25 years ago Chatt and Davidson [13] reported the intermolecular C–H
activation of a C–H bond in a DMPE (1,2-bis(dimethylphosphino)ethane)
ligand coordinated to ruthenium. This type of chemistry was further developed
for Ru and Fe by chemists at duPont [14], and most recently by Field [15].
Crabtree [16] and Felkin [17] provided the first clear examples of activation of
alkane C–H bonds by a path almost certainly to be oxidative addition, in these
cases by Ir and Re. Whitesides [18] has generated (DCPE)Pt (DCPE = 1,2-
bis(dicyclohexylphosphino)ethane) by thermolysis of (DCPE)Pt(H)(neopentyl)
and found that it activates a wide variety of C–H bonds, including those of

methane and other alkanes. Thus, there are a number of interesting studies of systems related to equation (13.1). Nevertheless, it is probably accurate to say that the scope, mechanisms, and structure-reactivity characteristics of such reactions are still far from thoroughly understood.

13.2. PLATINUM AND CARBON-CARBON ACTIVATION

Our investigations related to equation (13.1) began indirectly several years ago when we were studying C–C activations in the coordination sphere of platinum (equation (13.2)) [19–21].

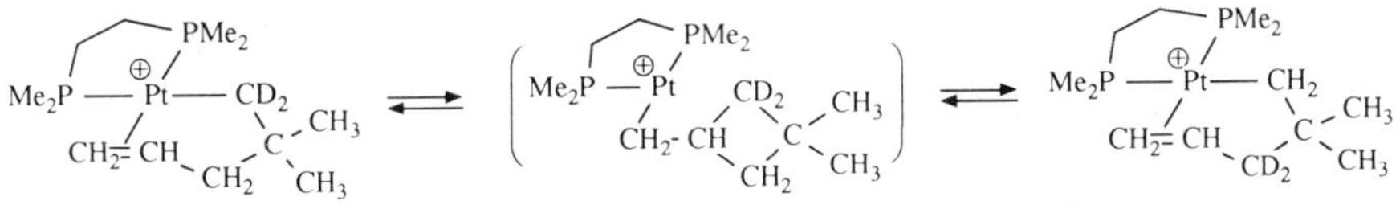

$$(13.2)$$

Two of the more probable mechanisms for this type of rearrangement are depicted in the context of a hypothetical deuterium labeling experiment in Scheme I, wherein a (2,2-dideuterio-1-methylcyclobutyl)carbinyl-ligated platinum complex would be heated to liberate methylpentadienes **1** and **2**. Path **a** is the simple "Cossee" mechanism [22] while path **b** is a modification of a mechanism first posited by Rooney and Green and co-workers [23] for Ziegler coordination polymerization. In principle, the two mechanisms could be distinguished by the **1** : **2** ratio, reflecting the presence or absence of a primary kinetic isotope effect (KIE). The problem that we faced was that there was at the

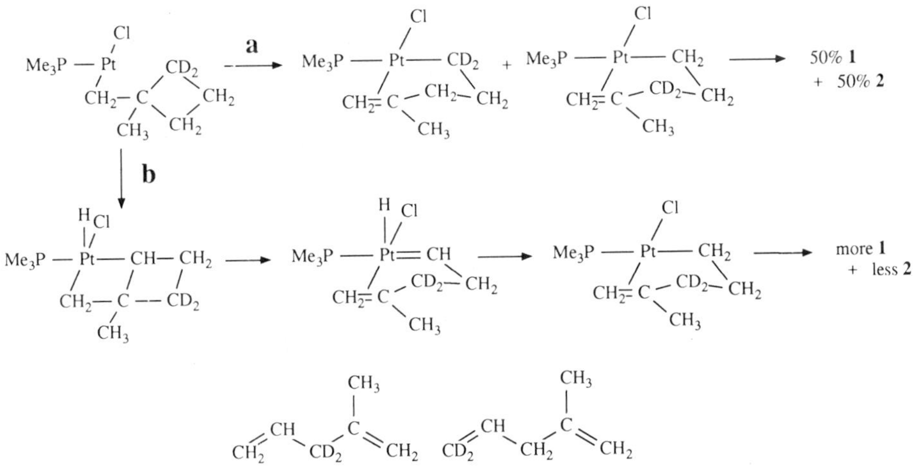

Scheme I.

time no direct information on the magnitude of such isotope effects for cyclometallation. To get some idea of the magnitude of the KIE of a γ-CH activation we decided to turn to a model reaction of an iridium complex.

13.3. IRIDIUM AND INTRAMOLECULAR CARBON-HYDROGEN ACTIVATION

A few years before, Tulip and Thorn had published the reaction of equation (13.3) [24].

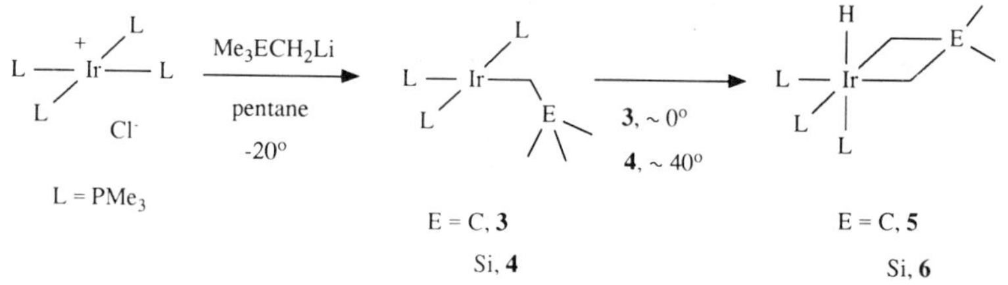

$$(13.3)$$

No mechanistic investigations had been undertaken. This system should offer the possibility of determining the KIE of a γ-C–H activation in a system rather analogous to the platinum species shown above, so we began an investigation of this reaction.

To design the isotope effect experiment, we considered the following precedents. Whitesides [25] had carefully studied the thermolysis of **7** to form the cyclometallated product **8** as shown in Scheme II. His results clearly established the need for initial dissociation of phosphine before C–H oxidative addition to Pt can occur. Presumably, there is a need for pentacoordination for the subsequent reductive elimination of neopentane. This suggests that, in the case of the Pt^{II}-Pt^{IV} redox couple, the reaction of equation (13.1) does not proceed simply as written, but requires ligand-dissociated intermediates.

There are indications that this requirement may be general. Milstein [26] has

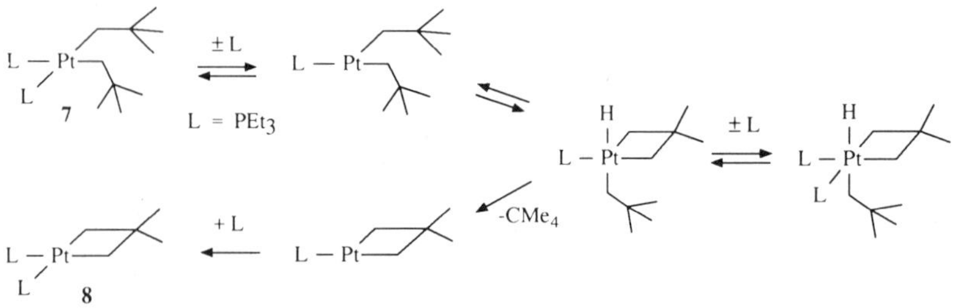

Scheme II.

found that the reductive elimination of acetone from **9** proceeds only after ligand dissociation (equation 13.4),

$$(13.4)$$

and Bresadola [27] has also observed the same phenomenon in the reductive elimination of C–H from Ir^{III} alkyl hydride complexes. Ligand dissociation is required for reductive elimination of two alkyl ligands from Pt^{IV} to form a C–C bond [28, 29]. The implication is, of course, that the oxidative addition of C–H or C–C bonds to tetra-coordinate-d^8 complexes must also proceed via tricoordinate intermediates. Thus, there is mounting evidence that the reaction of equation (13.1) will have a significant kinetic barrier when it occurs in a single step. In fact, we are unaware of any established example of an intermolecular $C(sp^3)$-H oxidative addition directly to a square-planar complex.

	R
3-d_6	$-CH_2C(CH_3)(CD_3)_2$
4-d_6	$-CH_2Si(CH_3)(CD_3)_2$
4-d_9	$-CH_2Si(CD_3)_3$

Thus, it appeared probable that the reaction of equation (13.3) would require ligand dissociation. While it is likely that the second step would be rate-limiting, it was not a certainty, and so we first prepared deuterated materials **3**-d_6 and **4**-d_6, and observed their cyclometallation. In these cases, the isotope effect would result in an intramolecular competition, and its measurement would be independent of the rate determining step. The products of these reactions are shown in equation (13.5).

$$(13.5)$$

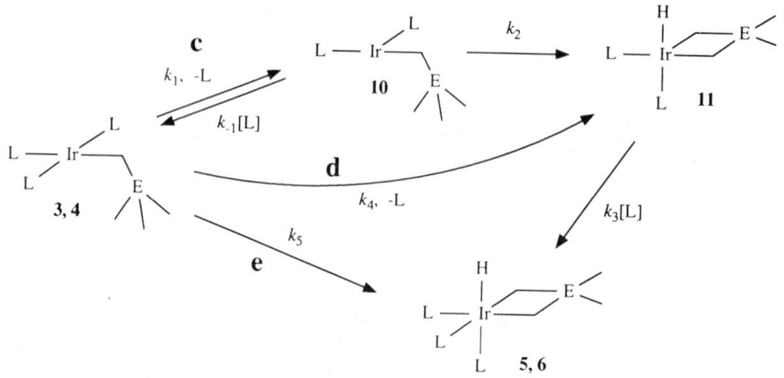

c

k_1, -L
k_{-1}[L]

L — Ir with L, and $\overset{10}{\underset{E}{}}$

k_2

$\overset{H}{\underset{L}{L — Ir}}$ E   **11**

L — Ir with L, L, E   **3, 4**

d

k_4, -L

k_3[L]

k_5

e

$\overset{H}{\underset{L}{L — Ir}}$ E   **5, 6**

Scheme III.

The lack of random scrambling of the label is a clear indication that the reaction is irreversible, and the label ratios correspond to an activation KIE about 3.

Thus, if the isotope effect for the platinum reaction of path **b** of Scheme I were similar to that for these iridium cases, the experiment of Scheme I should be meaningful. However, at this point we made some additional observations about the iridium cyclometallation which led our investigations in a different direction. The kinetics of the reaction of equation (13.5) were cleanly first order for both 3-d_6 and 4-d_6 and the rate was unaffected by the addition of up to 2.4 M free L (PMe$_3$). Three possible mechanisms for the cyclometallation, including the anticipated path, **c**, are shown in Scheme III. Mechanism **c**, since intermediates **10** and **11** would be present in very low concentration, would be expected to obey the usual, simple rate law of $-d[\mathbf{3}]/dt = k_{obsd}[\mathbf{3}]$ (and similarly for **4**) wherein $k_{obsd} = k_1 k_2 / (k_{-1}[L] + k_2)$. Thus, if k_2 were much larger than $k_{-1}[L]$, the k_1 step would be rate-determining, and there would be no inhibition by added phosphine, as observed. However, if k_1 were rate-limiting, then there should be no primary KIE on k_{obsd}. The rate of reaction of 4-d_9 compared to that of **4** revealed a kinetic isotope effect, KIE, $k_H/k_D = 3.2$ on k_{obsd}. Thus, mechanism **c** does not operate.

Because of precedents cited above [25–29] that suggest that a C–H oxidative addition should proceed via prior ligand dissociation, even in an intramolecular case [25], we sought to consider mechanisms alternative to the simple concerted path, path **e** of Scheme III. Although unconventional, the mechanism of path **d** is a possibility. Perhaps the close proximity of the C–H bond to the metal would result in its concurrent oxidative addition to the metal in the event of the dissociation of L. Since both paths **c** and **d** would be expected to show no inhibition by L and to exhibit the full isotope effect of C–H activation in k_{obsd}, other data were needed to distinguish between them. It was decided that the experiment outlined in Scheme IV should allow differentiation. The experiment depends on the ability of ^{31}P-NMR spectroscopy to distinguish between

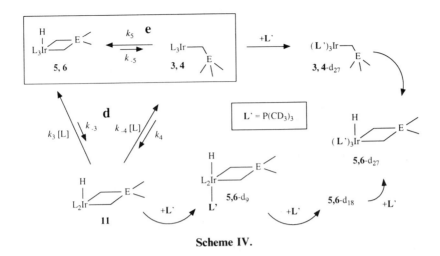

Scheme IV.

$P(CH_3)_3$ (L) and $P(CD_3)_3$ (L'). Fortunately, there is a large isotope effect on the ^{31}P chemical shifts. For example, uncoordinated L appears at $\delta - 62.3$ and L' at -65.2. The difference is not quite as large for coordinated L, but it is still pronounced, with $\Delta\delta$ generally close to 2 ppm for Ir-L vs. Ir-L'. Thus, incorporation of L' into any of the complexes is easily detected. The experiment also relies on two additional observations. First, it is important to note that when **3** or **4** is treated with a large excess of L' it immediately fully exchanges all three ligands with the labeled material. Second, the cycloaddition reverses at approx. 90°C, judging from the deuterium scrambling which occurs in **5-d$_6$** or **6-d$_6$** which occurs at that temperature.

The procedure involves heating **5** or **6** in the presence of a large excess of L' and determining the substitution pattern which results. Considering the **5-3** interconversion for example, by mechanism **c** when **5** would revert to **3**, the latter would immediately fully exchange with L' via an associative path. Then $(L')_3IrNp$, **3-d$_{27}$**, would cyclometallate again. Thus, as the exchange proceeds, only **5-d$_0$** and **5-d$_{27}$** would be present. If path **d** were operative, a single L would be lost from **5** to form **11**. Since the rate determining step in the reversal of **5** to **3** would be L-induced reductive elimination from **11** to form **3**, the fastest reaction of **11** would be to take up L' to form **5-d$_9$**. Exchange of ligands in **5** would therefore take place one L' at a time. Any significant quantity of isomers **5-d$_9$** and **5-d$_{18}$** would be clearly visible from ^{31}P-NMR spectra of the isotopomeric mixture, as confirmed from synthetic samples with mixed ligands.

The result of heating **5-d$_0$** with a large excess of L' at 95°C in cyclohexane at approx. two-thirds reaction was to lead to no detectable quantities of mixed isomers of **5** which is most consistent with a concerted path (path **e**) for the **3-5** interconversion. Thus, our results clearly establish that in the cases of **3** and **4** no unsaturation is required beyond the usual 16-electron configuration for intramolecular C–H oxidative addition [30].

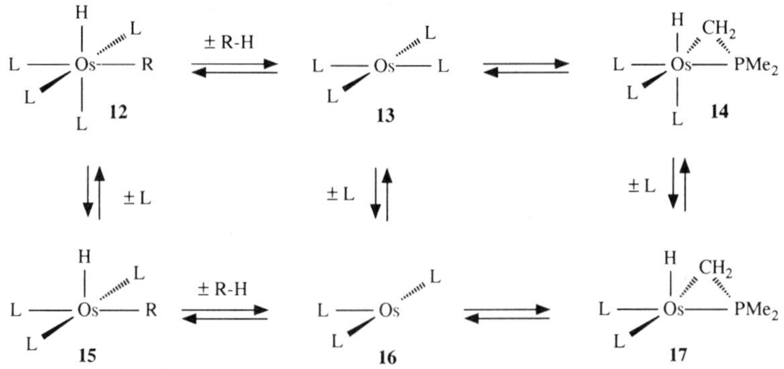

Scheme V.

13.4. ACID-CATALYZED HYDROGENOLYSIS OF OSMIUM ALKYL BONDS

The observation that $(Et_3P)_2Pt(neopentyl)_2$ requires loss of L to exhibit intramolecular C–H oxidative addition while $L_3Ir(CH_2EMe_3)$ does not, raised some intriguing questions some of which are summarized in Scheme V. Would a d^8, tetracoordinate osmium(0) complex such as **13** also cyclometallate, in this case to form **14** from its own PMe_3 ligand, or would it be reactive enough to undergo intermolecular reactions with hydrocarbons to yield **12**? Would a 14-electron intermediate such as **16** be required in order to form either **12** or **14**? Since complexes such as **13** and **16** could not be synthesized directly, we decided that we might begin with an osmium(II) alkyl hydride such as **12** or **14** and see if intermediates like **13** or **15–17** could be generated from it, and what would be the resulting chemistry be?

We began with **14** which had been reported from Helmut Werner's group [31]. Thermolysis of this molecule under a variety of conditions revealed it to be remarkably stable, more about which will be said below. Photolysis was very messy and unproductive. We did observe, though, that **14** undergoes hydrogenolysis at 80°C (equation (13.16)), albeit slowly, in a reaction that is inhibited by L as it builds up or is inhibited by added L.

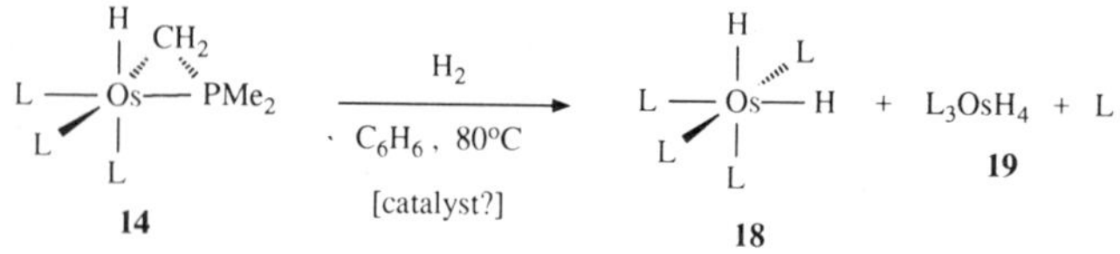

$$(13.6)$$

The conventional mechanistic assumption for formation of dihydride **18** and tetrahydride **19** would be as shown in Scheme VI.

Preliminary observations of this system revealed some unexpected features which led us to examine the mechanism of the transformation more closely. We

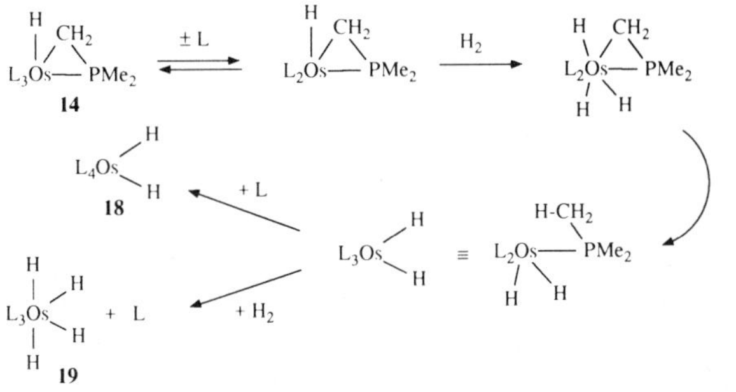

Scheme VI.

established that at 80°C no reaction occurs by the conventional path. Instead, the reaction is entirely acid catalyzed. We uncovered two distinct paths. The reaction of equation (13.6) is catalyzed by species derived from interaction of water with the osmium alkyl [32]. We discuss this path shortly. The second path [33] involves catalysis by protonic acids and leads only to formation of dihydride **18**; no **19** is formed. We were led to the hypothesis of acid catalysis by the observation that when special care is taken to keep the reaction mixture dry and basic, no hydrogenolysis whatsoever is observed over a period of weeks. In addition, heating of **14** with $P(CD_3)_3$, L', for weeks results in no exchange of L with L', ruling out a dissociative path. The mechanism of Scheme VII was proposed as a guiding hypothesis for an acid catalyzed reaction. Although we now know that this mechanism is not in fact the one responsible for the reaction of equation (13.6), all but one of the features of the mechanism can be substantiated independently and the path does operate. The complex $[L_4OsH_3]OTf$, designated **18-H$^+$(OTf$^-$)** where OTf$^-$ = trifluoromethane-sulfonate or triflate, serves as an efficient catalyst for the hydrogenolysis of **14** as shown in equation (13.7). Added L significantly retards the reaction, and under these conditions the rate of disappearance of **14** is independent of [**14**].

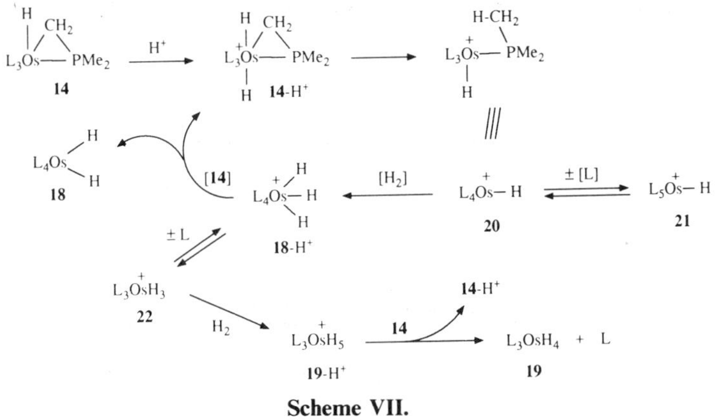

Scheme VII.

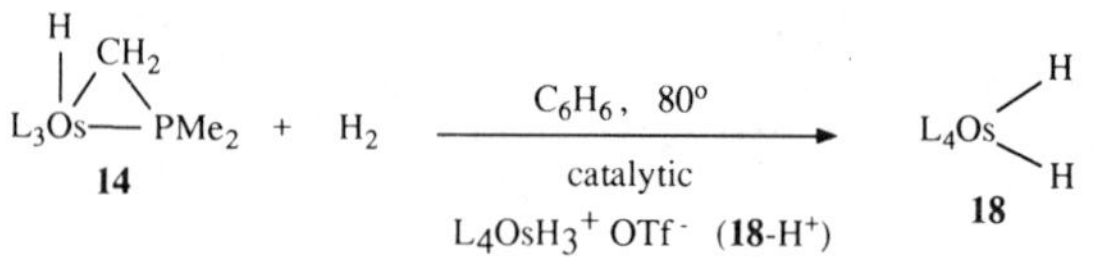

$$
\underset{\mathbf{14}}{\overset{\overset{\displaystyle H}{\underset{\displaystyle |}{}}\overset{\displaystyle CH_2}{\diagup\diagdown}}{L_3Os\!-\!PMe_2}} + H_2 \xrightarrow[\substack{\text{catalytic} \\ L_4OsH_3^+\,OTf^-\ (\mathbf{18\text{-}H^+})}]{C_6H_6,\ 80°} \underset{\mathbf{18}}{L_4Os\!\diagup\!\diagdown\!\overset{H}{\underset{H}{}}}
$$

$$(13.7)$$

Mechanism (VII) requires that **14** be readily protonated by acids to form **14-H$^+$** which then should reductively eliminate the phosphine methyl group to form the reactive intermediate **20**. Indeed, **14** is readily protonated even by weak acids, yielding products resulting from complexation by an appropriate ligand. Two examples are shown in equations (13.8) and (13.9),

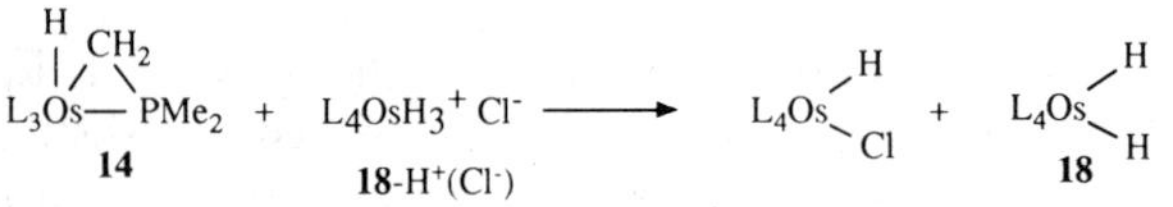

$$
\underset{\mathbf{14}}{L_3Os\!-\!PMe_2} + \underset{\mathbf{18\text{-}H^+(Cl^-)}}{L_4OsH_3^+\,Cl^-} \longrightarrow L_4Os\!\diagdown\!\overset{H}{\underset{Cl}{}} + \underset{\mathbf{18}}{L_4Os\!\diagdown\!\overset{H}{\underset{H}{}}}
$$

$$(13.8)$$

$$
\underset{\mathbf{14}}{L_3Os\!-\!PMe_2} + H\!-\!\overset{+}{P(CH_3)_3}\ ^-OTf \longrightarrow \underset{\mathbf{21}}{L_5Os\!-\!H^+\,{}^-OTf}
$$

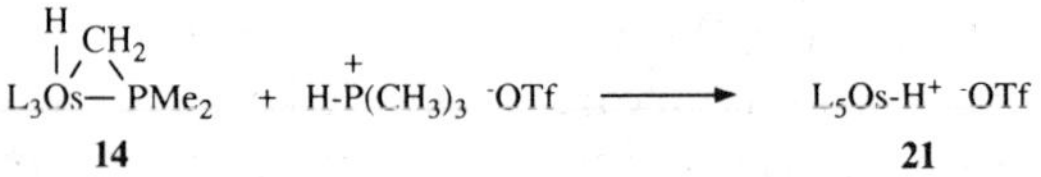

$$(13.9)$$

and the generalized proposed mechanism is shown in Scheme VIII. Both of these reactions are rapid at room temperature.

An impression of the basicity of **14** and **18** can be gained from the observations that the ^{31}P-NMR spectrum of **14** or **18** in THF at ambient temperature is unaltered by the presence of H_2O, but in THF, **18** is quantitatively protonated by the triflic acid complex of 1,8-bis(dimethylamino)naphthalene ("proton sponge"), the latter having a pK_a of 12.4 in water [34].

The ability of **18-H$^+$(Cl$^-$)** to protonate **14** is particularly significant since **18-H$^+$** is the species postulated to protonate **14** in the catalytic cycle of mechanism (13.7). Although the reaction of equation (13.8) is complete in the time required

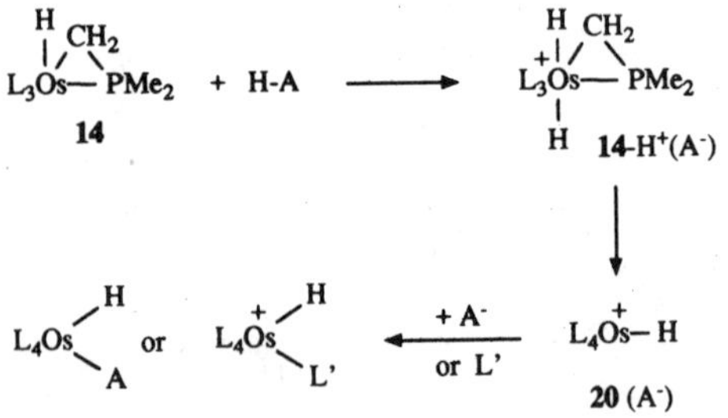

Scheme VIII.

to obtain the NMR spectrum, it may not be fast on the NMR time scale. A mixture of **18** and **18**-H$^+$ in THF shows no broadening of resonances in its phosphorus spectrum, even up to 90°C, and 1 or 2 kcal/mol of thermodynamic driving force of additional basicity of **14** over **18** is unlikely to substantially reduce what appears to be a significant kinetic barrier to proton transfer between the metals.

Referring to the mechanism of Scheme VII (mechanism VII), the **20**-to-**21** conversion is very likely to be reversible. Otherwise, upon addition of free L, all of the available catalytic acid would be removed from the system in the form of **21**, which is very probably too low in acidity to function as a catalyst in this scheme. In fact, the hydrogenolysis is inhibited by excess L, but not completely stopped. The **20-21** interconversion can be demonstrated indirectly, in conjunction with the **18**-H$^+$-**20** interconversion, by observation of the reaction of equation (13.10).

$$\overset{+}{L_5OsH}(OTf^-) + H_2 \underset{THF}{\overset{80°C}{\rightleftharpoons}} \overset{+}{L_4OsH_3}(OTf^-) + L$$
$$\quad\;\; \mathbf{21} \qquad\qquad\qquad\qquad\qquad\qquad \mathbf{18\text{-}H^+}$$

$$(13.10)$$

The equilibrium between **21** and **18**-H$^+$ is established in a few hours at 80°C, and is reversibly shifted to the right at higher temperature.

It is possible that the chemistry described so far might be associated with the particular properties of **14**, perhaps owing to its small ring. We therefore sought to determine whether the catalysis is general, with at least one other more representative substrate, cis-L$_4$Os(H)Me, **23**. The preparations of **23** and several other cis-L$_4$Os(H)R (**12**) are described below. The chemistry of methyl hydride complex **23** closely parallels that of **14**. For example, **23** undergoes the same reaction with **18**-H$^+$(Cl$^-$) as shown for **14** in equation (13.8). Also, a 1.3:1 mixture of **23** and **18**-H$^+$(OTf$^-$) in THF at 23°C rapidly forms a mixture in which the **18**-H$^+$ has converted an equivalent of **23** to hydridotriflate **24** in high yield (equation 13.11).

$$L_4Os\overset{H}{\underset{CH_3}{<}} + L_4OsH_3^+\,OTf^- \xrightarrow{-CH_4} L_4Os\overset{H}{\underset{OTf}{<}} + L_4Os\overset{H}{\underset{H}{<}}$$
$$\quad\; \mathbf{23} \qquad\quad \mathbf{18\text{-}H^+(OTf^-)} \qquad\qquad\qquad \mathbf{24} \qquad\quad \mathbf{18}$$

$$(13.11)$$

Overall, most of the steps of Scheme VII are precedented in the reactions of equations 13.7–13.11, for both **14** and **23**. All that remained was to provide a more quantitative demonstration of the catalysis. When hydrogenolysis of **23** is carried out as in equation (13.12) in the presence of added L, to maintain behavior zero-order in **23** at reasonable rates, and with different amounts of **18**-H$^+$(OTf$^-$) as catalyst, a series of zero-order plots as shown in Figure 13.1 are

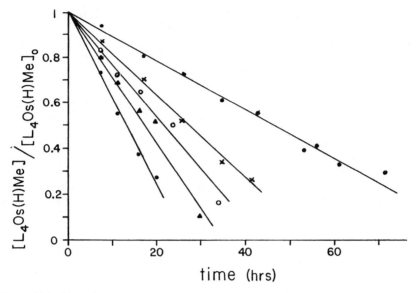

Figure 13.1. Plot of concentrations of **23** vs. time for the hydrogenolysis shown in equation (13.12) at different concentrations of $[L_4OsH_3](OTf)$, **18-H$^+$(OTf$^-$)**.

obtained. THF was used as solvent rather than benzene because of the greater solubility of **18-H$^+$(OTf$^-$)**.

$$(13.12)$$

A plot of the zero-order rates of disappearance of **23**, k_{obsd}, for the individual reactions of Figure 13.1 as a function of the concentration of **18-H$^+$(OTf$^-$)** can be seen from Figure 13.2 to be linear, clearly establishing the acid catalysis.

Two mechanisms are commonly invoked for the hydrogenolysis of metal alkyl bonds. One involves initial oxidative addition of H_2 to an unsatured metal, followed by alkane reductive elimination (as in Scheme VI) [35]. The second proceeds via a direct four-centered transition state between the M–C and M–H bonds, and seems to require substantial electrophilicity of the metal [36–40]. We have shown that by catalysis with a protic acid a coordinatively saturated, 18-electron alkyl complex such as **14** or **23** can undergo hydrogenolysis without prior ligand dissociation or the need for an existing acidic orbital.

Of course, this catalytic path requires an acid of sufficient strength to at least transiently protonate the metal alkyl. Protonation of the metal center results in a higher oxidation state, a higher coordination number, and the presence of

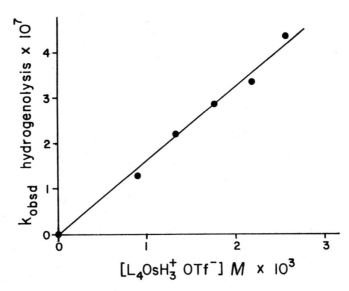

Figure 13.2. Plot of rates of disappearance of **23** vs. concentration of $[L_4OsH_3](OTf)$, **18**-$H^+(OTf^-)$, for the hydrogenolyses of **23** shown in equation (13.12) and Figure 13.1.

positive charge at the osmium center. These factors lead to facile reductive elimination. In general, the resulting cation $[(ligands)MH_2]^+$ will be acidic enough to propagate the catalysis by proton transfer to (ligands)MR since the latter will always be slightly more basic than (ligands)MH. This does not mean that the catalysis will be extremely rapid, since proton transfers between metals need not be rapid [41], and in any event, the proton transfer need not be the rate determining step. In fact, this step cannot be rate limiting in the present case since the reaction is zero-order in **14** or **23**.

With regard to the original observation of hydrogenolysis of **14** shown in equation (13.6), it is to be noted that in addition to the generation of **18**, up to 40% of the tetrahydride **19** and an equivalent quantity of L were formed. On the other hand, the formation of **19** was not observed under any conditions in the hydrogenolysis of **14** or **23** when catalyzed by **18**-H^+. This suggests that the dissociation of L from **18**-H^+ to form $L_3OsH_3^+$, **22**, the step postulated in Scheme VII to account for the formation of **19**, does not occur. Note that all of the other steps in Scheme VII involve osmium with at least four phosphine ligands, and there is no obvious way in which **19** could form in this scheme aside from the intermediacy of **22**. Thus, other possible mechanisms needed to be considered.

The absence of tetrahydride **19** in hydrogenolyses catalyzed by **18**-H^+ led us to consider the possibility that the hydrogenolysis of equation (13.6) might be catalyzed by traces of water, and that this catalysis might be different from that discussed above. So, we checked the reaction of **14** with water (equation 13.13).

$$\underset{\mathbf{14}}{\text{L}_3\text{Os}-\text{P}\overset{\text{H}}{\diagdown}} \xrightarrow[\text{THF}]{\overset{\text{H}_2\text{O}}{\text{warm}}} \underset{\mathbf{25}}{\text{L}-\overset{\text{H}}{\underset{\text{L}}{\text{Os}}}-\text{OH}} + \underset{\mathbf{26} \text{ trace}}{\text{L}-\overset{\text{H}}{\underset{\text{L}}{\text{Os}}}-\text{O}-\overset{\text{H}}{\underset{\text{L}}{\text{Os}}}-\text{L}}$$

(13.13)

The reaction was rapid in THF at 80°C and afforded the *cis*-hydroxy hydride **25**, with traces of a material identified as the anhydride **26**. Other osmium alkyls, such as methyl hydride **23**, react similarly. Recently we have found that **14** reacts with a variety of phenols and anilines to yield the O–H and N–H activation products analogous to **25**. In connection with the chemistry of **25** it is useful to note the rates of exchange of several $\text{L}_4\text{Os(H)X}$ with $\text{P(CD}_3)_3$ (L′) (Table 13.1). The most reactive of these species is the neopentyl hydride **27** which undergoes L dissociation with a $t_{1/2}$ of 0.5 h at 80°C. Then as the size of X diminishes, so does the rate of dissociation, to the point that methyl hydride **23** exhibits a $t_{1/2}$ of approx. 4 days at 110°C. In sharp contrast to **23**, hydroxy hydride **25**, with its OH group about the same size as CH_3, dissociates L with $t_{1/2}$ 1.6 h at 80°C. Thus, the hydroxide ligand exerts a substantial labilizing effect on the phosphine ligands, reminiscent of effects seen in the now classic studies of base catalyzed hydrolysis of $[\text{XCo(NH}_3)_5]^{2+}$ [42].

As we suspected, **25** undergoes hydrogenolysis as shown in equation (13.14) to yield not only **18**, but also tetrahydride **19** and an equivalent quantity of L.

TABLE 13.1. Half-lives and sites of exchange of $\text{P(CD}_3)_3$ (L′) into various *cis*-$\text{L}_4\text{Os(H)X}$ complexes.

$$\text{L}_4\text{Os}\overset{\text{H}}{\underset{\text{X}}{\diagup}} \xrightarrow[k_6]{\text{L′} = \text{P(CD}_3)_3} \text{L′L}_3\text{Os}\overset{\text{H}}{\underset{\text{X}}{\diagup}}$$

X	Temp °C	$t_{1/2}$ (k_6)	Sites exchanged
27 -CH$_2$CMe$_3$	80	32 min	a
28 -CH$_2$SiMe$_3$	80	73 hrs	a
29 -CH$_2$CH$_2$CH$_2$CH$_2$CH$_3$	80	68 days	a
	110	22 hrs	a
23 -CH$_3$	110	4 days	all
25 -OH	80	1.6 hrs	a, c
30 -O-C$_6$H$_5$	80	8 days	a, c

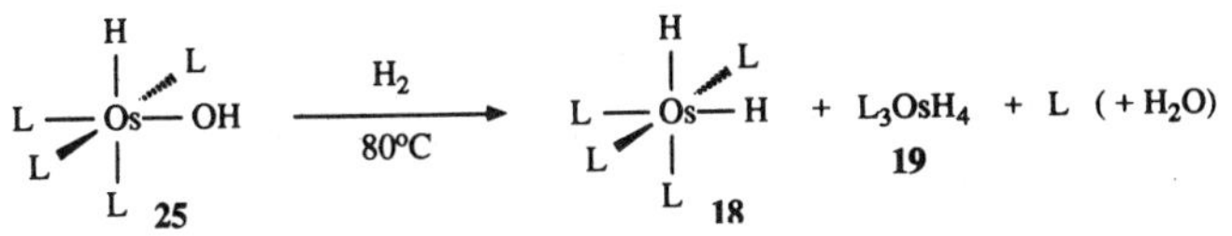

$$L \overset{\displaystyle H}{\underset{\displaystyle L}{\overset{|}{\underset{|}{Os}}}}{\overset{\nearrow L}{\diagdown}} OH \quad \xrightarrow[80°C]{H_2} \quad L \overset{\displaystyle H}{\underset{\displaystyle L}{\overset{|}{\underset{|}{Os}}}}{\overset{\nearrow L}{\diagdown}} H \; + \; L_3OsH_4 \; + \; L \; (+H_2O)$$

$$\mathbf{25} \qquad\qquad\qquad\qquad \mathbf{18} \qquad\qquad \mathbf{19}$$

$$(13.14)$$

This reaction is inhibited by added L. Although the catalysis is still under investigation, we believe that the cycle is as shown in Scheme IX. Thus, the acidity of water provides a path for loss of the metal alkyl group, and the resulting hydroxy hydride **25** has a facile path for hydrogenolysis because of the labilizing effect of the hydroxide ligand. We are seeking other examples of this dual role of O–H and potentially of N–H bonds.

The chemical precedents described herein clearly outline two distinct catalytic paths for the hydrogenolysis of osmium alkyls such as **14** and **23**. Similar paths involving Lewis and Brønsted acids can be imagined for a number of homogeneous and heterogeneous catalytic reactions of hydrocarbons. It may be a general phenomenon that catalytic processes of basic metal complexes in homogeneous solution and of metal catalysts on solid supports are themselves acid catalyzed.

13.5. REACTION CHEMISTRY IN THE OSMIUM SYSTEM

The study of catalyzed hydrogenations described above led to the means of preparing cis-$L_4Os(H)R$ complexes (**12** in Scheme V) that we had been seeking. Hydrido triflate **24** can be alkylated with a variety of lithium reagents in good yield as shown in equation (13.15).

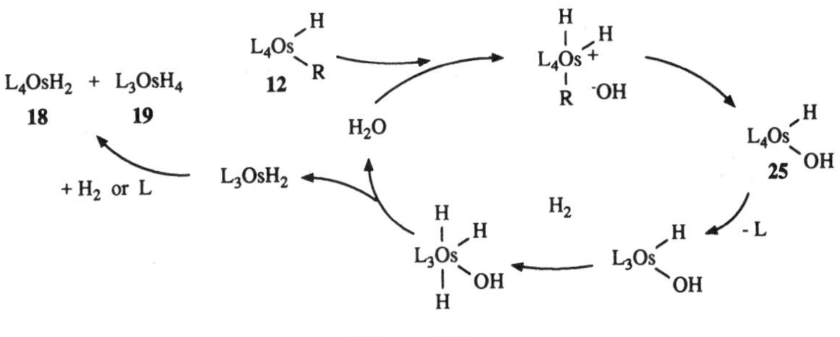

Scheme IX.

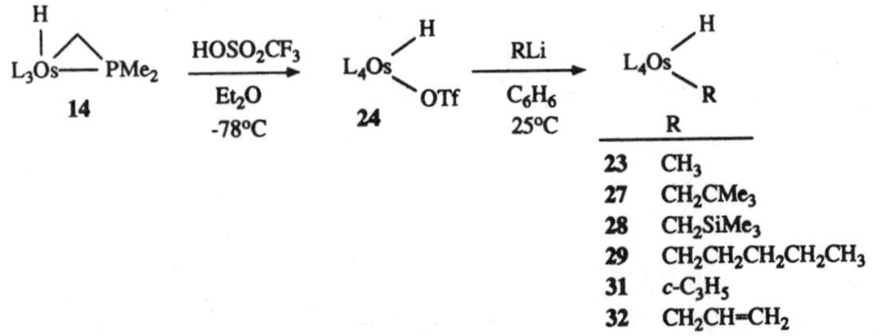

$$(13.15)$$

The first osmium alkyl which was prepared, neopentyl hydride **27**, happened to be the most reactive, as mentioned above in the context of ligand dissociation rates. It has been found to react with a wide variety of hydrocarbons, many of which exhibit C–H activation chemistry often in high yield. Scheme X shows a summary of many of these reactions, all of which proceed at 80°C with half lives of from 2 h to approx. 4 days. We have directed substantial effort toward understanding the mechanisms of many of these transformations.

13.6. OSMIUM AND THE ACTIVATION OF BENZENE

Benzene is activated quantitatively on heating **27** in benzene as solvent at 80°C [43]. The reaction is cleanly first order in **27**, is strongly inhibited by added phosphine, and shows kinetic dependence on the concentration of benzene.

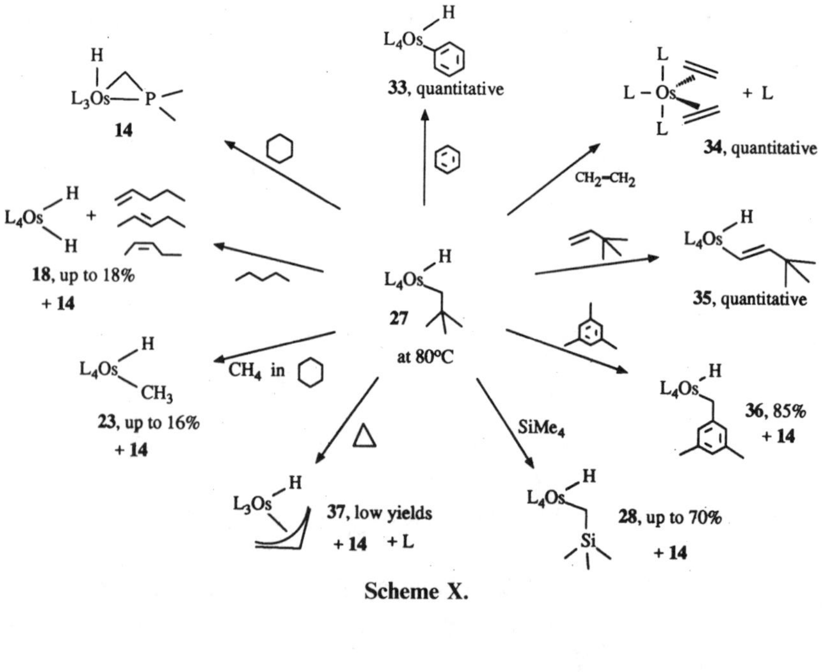

Scheme X.

When C_6D_6 is activated, there is crossover of approx. 13% of deuterium into the liberated neopentane and approx. 13% H remains in the hydride position of **33**. These data point to a mechanism as shown in Scheme XI, where the activation step involves uptake of benzene by unsaturate **38** to yield the seven-coordinate, d^4, 18-electron, Os^{IV} intermediate **39**. Comparison of reaction rates in C_6H_6 vs. C_6D_6 yields a KIE on k_{obsd} of 2.2, while it can be shown that the activation step itself possesses $k_{B(H)}/k_{B(D)}$ of 3.6. This indicates that $k_{-6}[L]$ and $k_B[C_6H_6]$ are about the same size, and corroborates the Os^{IV} mechanism.

Heating of **27** in other aromatic solvents also results in very clean conversion to $L_4Os(H)(aryl)$ with relative rates as shown below. The extremes in rates of reaction of these substrates differ by only about a factor of 10. Toluene is activated in the meta and para positions with a statistical ratio, but no ortho or alpha reaction products were detected.

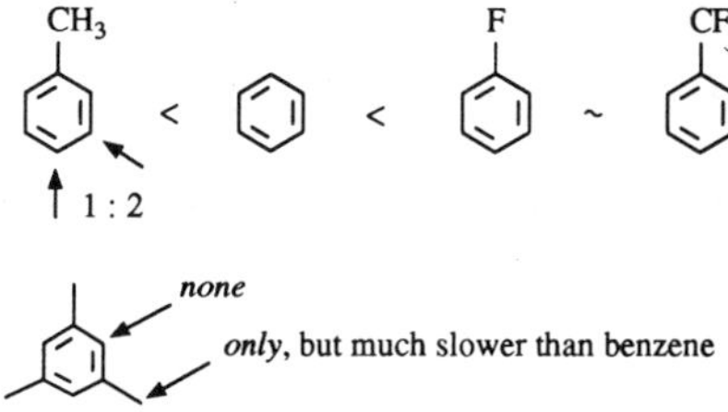

Mesitylene, on the other hand, reveals only alpha activation, and at a much slower rate than for the other arenes. We believe that this reaction proceeds by a different mechanism, as described below.

We had occasion to use $L_4Os(H)[CH_2C(CH_3)(CD_3)_2]$, **27**-$d_6$, in a reaction with benzene, and discovered that the deuterium in the liberated neopentane-d_6 was largely positionally randomized. This led us to thermolize both **27**-d_6 and the analogous silicon-containing complex, **28**-d_6, in alkane solvent. ^1H- and ^2H-NMR revealed the migration of deuterium randomly around the aliphatic ligand and into the Os–H position. The migration was strongly inhibited by a

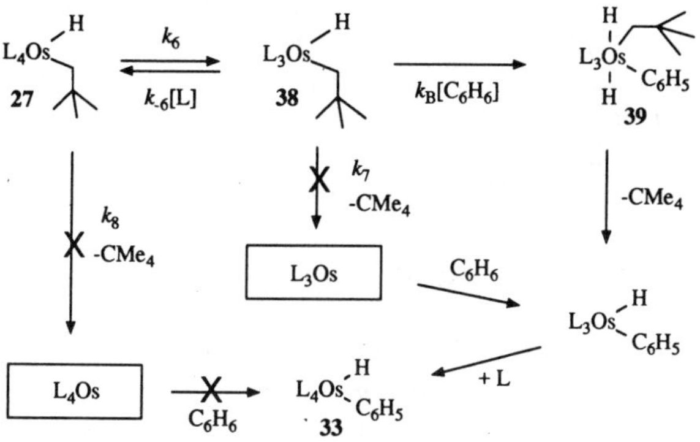

Scheme XI.

large excess of free L. These data are most consistent with the internal C–H activation (cyclometallation) of the unsaturated intermediate **38** to form the transient intermediate **40**, as shown in equation (13.16).

$$\text{L}_4\text{Os} \overset{\text{H}}{\diagup} \underset{\textbf{27}}{\bigwedge} \quad \xrightarrow[\text{cycloalkane}]{80°C} \quad \underset{\textbf{14}}{\text{L}_3\text{Os}} \overset{\text{H}}{-} \text{PMe}_2$$

$$(13.16)$$

Thus, there is no inherent, significant, enthalpic barrier to the oxidative addition of saturated C–H bonds to Os^{II} intermediate **38**.

13.7. PHOSPHINE CYCLOMETALLATION

Thermolysis of **27** at 80°C in cyclopentane or cyclohexane leads to formation of only **14** (equation (13.17)) [44].

$$\underset{\textbf{27, 28}}{\overset{\text{H}}{\underset{\text{L}}{\text{L}-\text{Os}}}} \underset{k_{-6}\,[\text{L}]}{\overset{k_6}{\rightleftharpoons}} \underset{\textbf{38}}{\overset{\text{H}}{\text{L}-\text{Os}}} \rightleftharpoons \underset{\textbf{40}}{\overset{\text{H}}{\text{L}-\text{Os}}}$$

$$(13.17)$$

No deuterium is incorporated into **14** when C$_6$D$_{12}$ is used as solvent. The pyrolysis of **27** is retarded by the addition of free L. It is important to point out at the outset that while pyrolysis of **27** in arenes does not generate any concentration of free L detectable by ^{31}P-NMR, pyrolysis in alkanes invariably does generate some L, generally in the range of from 7×10^{-4} to 4×10^{-3} M, and most usually approx. 2×10^{-3} M. Complex **27** is typically employed at concentrations of approx. $3-9 \times 10^{-2}$ M. The free ligand is generated immediately in the reaction and then its concentration does not change. Since rates are sensitive to [L], it is important to note this concentration when discussing relative rates of reactions. The source of free L is not known, but other than its inhibitory effect, its generation does not cause any problems that we can detect.

As [L] is increased by the addition of L, the rate does not go to zero, but approaches a finite limiting value above approx. 0.1 M L. This relationship is shown in the plot of k_{obsd} vs. 1/[L] in line I of Figure 13.3, for which the intercept is the limiting rate, 1.10 (0.08) $\times 10^{-6}$ s^{-1}. Thus, it appears that there are two mechanisms operating, one inhibited by L and the other independent of L.

13.7.1. The Phosphine-Independent Path

Several of the most probable mechanisms for formation of **14** are shown in Scheme XII. Based on data and arguments presented in this section, we have concluded that path **i** is responsible for the L-independent cyclometallation reaction.

Path **i** would clearly be independent of L. Any of paths **f**, **g**, or **h** could be L-independent in the event that k_6 were rate determining. The k_6 path can be ruled out as rate-determining, however, since we have measured k_6, as mentioned above, from the rate of exchange of **27** with added free L' ($P(CD_3)_3$). The first-order rate constant for this exchange is $k_6 = 7.25 \times 10^{-4}\,s^{-1}$ at 80°C, about 300 times faster than the formation of **14** with [L] approx. $2 \times 10^{-3}\,M$. Thus, $k_{-6}[L]$ must be larger than k_7, k_9, or $k_{10}[L]$ and k_6 cannot be rate limiting. Since k_6 is not rate limiting, paths **f** and **g** would both be inhibited by phosphine but path **h** would not. The lack of L-dependence for path **h** comes from the fact that k_{-6} and k_{10} would both be associated with a first-order dependence on [L]. Thus, the two most reasonable candidates for the mechanism for the L-independent formation of **14** would be paths **h** and **i**. We can differentiate between these two paths based on two lines of evidence, isotope effects and isotopic crossover.

Figure 13.3 includes a plot, line II, of k_{obsd} vs. $1/[L']$ for thermolysis of $[(CD_3)_3P]_4Os(H)Np$, **27**-d_{36}, at 80°C. The intercept for this plot is 8.1 $(0.7) \times 10^{-7}\,s^{-1}$. This and the intercept for the **27**-d_0 plot, line I, give a KIE of 1.36 (0.15) for the L-independent path.

There is evidence that the isotope effect for the C–H oxidative addition step itself is much larger than that seen in k_{obsd} for the L-independent formation of **14**. When one heats **27** at 80°C in alkane solvent in the presence of a large excess of a 3:1 mixture of L' and L, the four Ls coordinated to **27** are completely

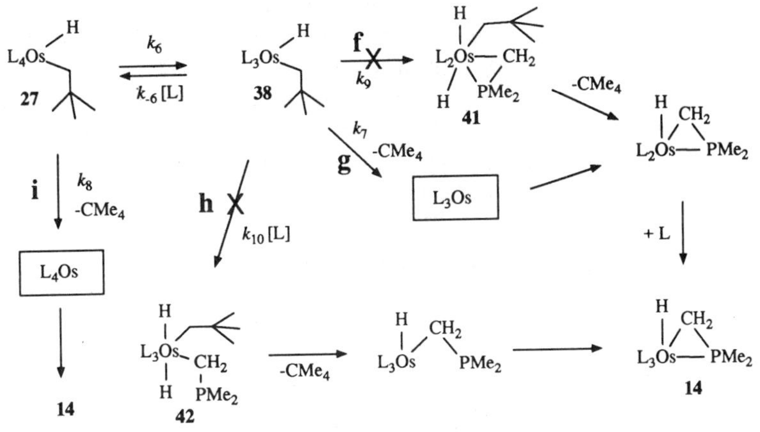

Scheme XII.

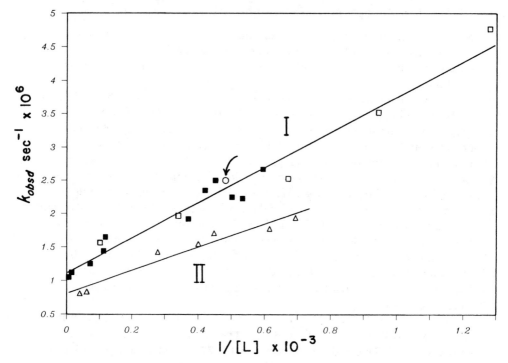

Figure 13.3. Plot of k_{obsd} for the disappearance of **27** vs. 1/[L] for the thermolysis of **27** at 80°C in the following solvents. For line I: alkane, ■; tetramethylsilane, □; and tetramethylsilane-d_{12}, O (designated by an arrow). For line II: the points △ are for reaction of **27**-d_{36} in alkane solvent. Scatter of the data arises from difficulty in measuring low concentrations of free ligand accurately (i.e., uncertainty in the abscissa).

randomized with the added L'/L mixture very early in the reaction. Then, cyclometallation occurs with an internal competition regardless of the rate-determining step. Scheme XIII illustrates path **i**, but the same conclusion would obtain for path **h**. Product **14** exchanges neither L nor Os–H under the reaction conditions, since **14**-d_0 can be heated indefinitely at 80°C with high concentrations of L' without any effect. In fact, exchange of **14**-d_0 with L' occurs with a significant rate only above 150° (see below). The thermolysis of **27** with L/L' was

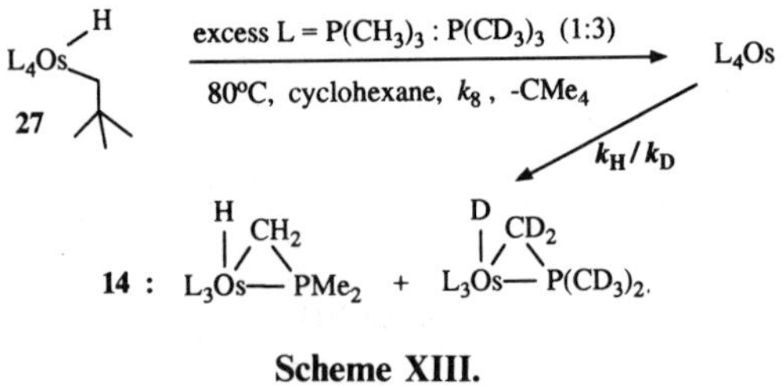

Scheme XIII.

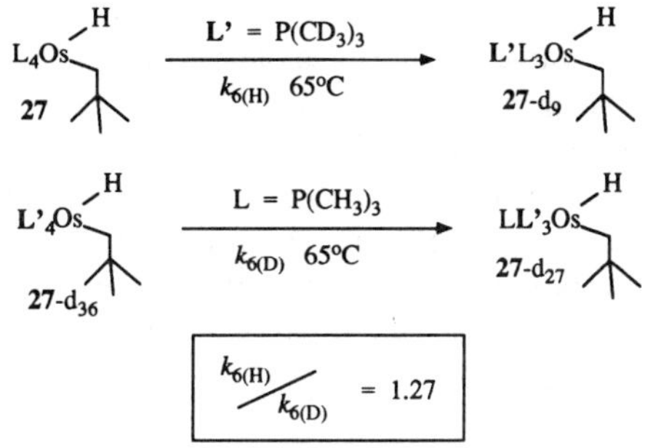

Scheme XIV.

conducted two times and yielded $k_H/k_D = 4.7$ (0.2) for the cyclometallation step itself.

Since, as mentioned above, k_{obsd} for cyclometallation is much slower than the rate of L dissociation, k_6, the full KIE of k_{10} (4.7) would necessarily be apparent in k_{obsd} for path **h**. The fact that this is so clearly not the case rules out path **h**.

Parenthetically, the small isotope effect on k_{obsd} for path **i** (k_8) probably has its origins in steric effects (see below). Reductive elimination of neopentane from **27** must have a strong steric driving force since methyl hydride **23**, which is significantly less crowded than **27**, does not form **14** in an alkane solvent, with or without added L, except above 150°C. Thus, the slightly smaller size of deuterium compared to protium [45] could result in reduced crowding which would increase the barrier to reductive elimination of neopentane. If this idea is correct, then since we have shown that initial dissociation of L from **27** is driven by steric crowding, there might be a steric isotope effect on dissociation of L as well. The rate of substitution by free, unlabeled L on **27**-d$_{36}$ (Scheme XIV) was therefore determined at 65°C, yielding $k_6 = 8.25$ (0.27) $\times 10^{-5}$ s^{-1}. The rate of substitution of L' (PMe$_3$-d$_9$) on **27**-d$_0$ was also determined by the identical procedure at 65°C, and was 1.05 (0.02) $\times 10^{-4}$ s^{-1} (both numbers are the average of two determinations). Thus, there is a KIE on the rate of dissociation of L from **27** (k_6) of 1.27 (0.05) at 65°C. So it seems quite plausible that this size isotope effect could operate on the k_8 step as well.

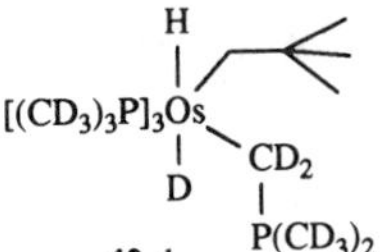

An additional indication of the absence of path **h** comes from the fact that an Os^{IV} intermediate such as **42** would probably show isotopic crossover under appropriate conditions. Use of **27**-d$_{36}$ as substrate should result in incorporation of some deuterium into the liberated neopentane via **42**-d$_{36}$. The mass spectrum of the neopentane product revealed no deuterium content above natural abundance. Of course, the absence of crossover is negative evidence, but in the case of benzene-d_6 activation via Os^{IV} intermediate **39** approx. 13% crossover was reproducibly observable [43].So it is reasonable to assume the likelihood of observing it were such a mechanism to operate here.

The most reasonable alternative mechanism for the [L]-independent re-action which accommodates the data is path **i**. Activation parameters were determined for this reaction over the range of 80–110°C, and yielded values of $E_a = 32.9$ (1.5) kcal/mol, ΔH (80°C) = 32.2 (1.5) kcal/mol, and ΔS (80°C) = 5.0 (4.4) e.u.

13.7.2. The Phosphine-Inhibited Path

At concentrations of L below approx. 2×10^{-3} M, most of the **14** from the thermolysis of **27** at 80°C forms via a path which is sensitive to [L]. The most probable mechanisms for this reaction are paths **f** and **g** in Scheme XII, either of which would show inhibition by L. The data presented below lead us to conclude that it is mechanism **g** which operates. These paths can be distin-guished by several lines of evidence. Use of kinetic isotope effects to establish the rate-determining step is considered here first.

The slope of each line in Figure 13.3 is $k_6 k_n / k_{-6}$, where k_n will be either k_7 or k_9 (the kinetic equations are discussed below). Thus, the slope of line I (for **27**) divided by that of line II (for **27**-d_{36}) gives the cumulative isotope effect on these rate constants, and is 1.53 (0.18). Dividing this number by the isotope effect on k_6 (1.27, discussed above) yields the KIE on k_n / k_{-6} which is 1.20 (0.15). Since the isotope effect on k_{-6} should be small, that of k_n must not be very different from 1.

The KIE for the cyclometallation step itself was measured in the following way (Scheme XV). Exchange of the ligands of **27** with enough L′ to give a 4:1 exchanged L′/L ratio was carried out at 80° for 48 h, sufficient time for the complete randomization of the mixture of ligands throughout **27** rather than just in the mutually *trans* positions. Exchanged **27** (80% L′) was isolated and submitted to thermolysis at 80°C in cyclohexane with no added L or L′. The

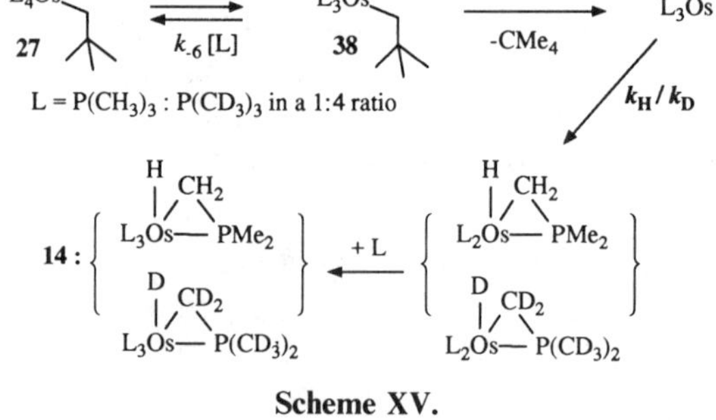

Scheme XV.

KIE can be calculated directly from the intensities of the two ^{31}P resonances of the unique cyclometallated ligand centered at δ -70.2 (for L) and -72.0 (for L'). This treatment yielded $k_H/k_D = 3.1$ (0.3). This value is a composite of the isotope effect on path **i** and that of the [L]-dependent path, **f** or **g**. After correction for the contribution of path **i**, it was calculated that k_H/k_D for the phosphine-inhibited path is 2.2 (0.4). An independent determination on a second sample gave the same answer; 2.1 (0.4). The errors are somewhat large because of all the variables in the calculation.

The fact that the KIE for k_n/k_{-6} (assuming k_{-6} is approx 1) is near 1 and that the isotope effect measured for the cyclometallation step itself is 2.2 is consistent with path **g** (k_7) being responsible for the formation of **14**, but inconsistent with the operation of path **f** (k_9).

13.8. THE INTERMOLECULAR ACTIVATION OF SATURATED C–H BONDS

13.8.1. Thermolysis of 27 in Tetramethylsilane

Thermolysis of **27** in neat SiMe$_4$ at 80°C results in a very slow and very clean reaction, first-order in **27**, leading to formation of **28** but with competitive formation of **14** (equation (13.18).

$$(13.18)$$

This reaction is inhibited by added phosphine and the rate depends on the amount of free phosphine added or spontaneously generated, as shown in Figure 13.1. At concentrations of L above approx. 0.1 M, the formation of **28** is completely inhibited, but **14** continues to form at the rate of the L-independent path.

When Si(CD$_3$)$_4$ is used as solvent, the rate of disappearance of **27** (k_{obsd}) is indistinguishable from that in nondeuterated Si(CH$_3$)$_4$. There is, however, an effect of changing from SiMe$_4$ to SiMe$_4$-d$_{12}$ solvent on the **14/28** ratio (e.g., Table 13.2, reactions 4 and 5). There is also an effect of phosphine concentration on this ratio (Table 13.2, cf. columns **b** and **c**). Scheme XVI shows some of the possible mechanisms for formation of **28**. This scheme should be superimposed on Scheme XII since both **28** and **14** are formed simultaneously. In the remainder of this section, the mechanisms of these two schemes are considered in turn as possible sources of **28** and **14**. It is shown that only paths **g** and **i** are consistent with the experimental data.

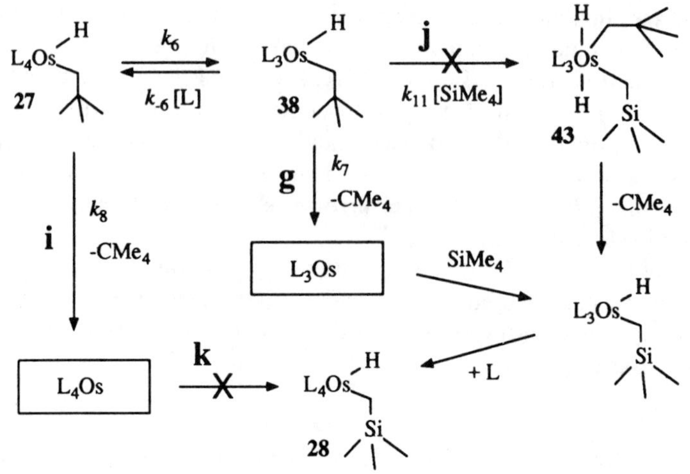

Scheme XVI.

TABLE 13.2. Thermolysis of $L_4Os(H)Np$ (27) at 80°C in $SiMe_4$ and $SiMe_4$-d_{12} solvent showing the dependence of the $L_3Os(H)(\eta^2\text{-}CH_2PMe_2)$ to $L_4Os(H)CH_2SiMe_3$ (14/28) ratio on [L], on isotopic substitution of solvent, and on the reaction path.

	a	b	c	d	e
rxn	$k_{obsd} \times 10^6$	[L] $\times 10^4$	**14/28**	%rxn via k_7	14/28 via k_7 only
1	28.0	1.0 [a]	0.5	96	0.43
2	4.77	7.8	1.2	77	0.71
3	3.52	11	1.1	68	0.45
4	2.53	15	1.5	56	0.37
5	2.50 [b]	21	3.4	56	1.43
6	1.97	29	2.2	44	0.42
7	1.57	100	4.0	29	0.45

[a] Calculated from the best line shown in Figure 13.3 since it could not be determined by ^{31}P-NMR.
[b] In $Si(CD_3)_4$ solvent.

Consider first the possibility that the reaction of equation (13.18) were to proceed by paths **f**, **i** and **j**. Applying the steady state assumption to the intermediates involved, the corresponding rate expression would be that given in equation (13.19)

$$k_{obsd} = k_8 + \frac{k_6 (k_9 + k_{11} [SiMe_4])}{k_{-6} [L] + k_9 + k_{11} [SiMe_4]}$$

(13.19)

where k_8 is the limiting rate constant in the presence of added L for reaction of **27** in alkanes (path **i**, 1.1×10^{-6} s^{-1}). Now, if we consider the points in the plot of line I of Figure 13.3 for only those reactions carried out in alkane solvent, from the slope it can be shown (see below) that k_9, and, since it would be about the same size as k_9, $k_{11} [SiMe_4]$ must both be negligible compared to $k_{-6} [L]$ for concentrations of L that we observe. Thus, equation (13.19) reduces to equation (13.20). The latter is given in such a form as to emphasize the fact the rates of reactions of **27** with its own ligands and with solvents should be additive for the mechanism of paths **f** and **j**.

$$k_{obsd} = k_8 + \frac{k_6 k_9}{k_{-6} [L]} + \frac{k_6 k_{11} [SiMe_4]}{k_{-6} [L]}$$

(13.20)

For those reactions in SiMe$_4$, if only the rates of formation of **14** were plotted, then these points should fall on the line in Figure 13.3. Conversely, if the total rate of disappearance of **27**, k_{obsd}, were to be plotted for the SiMe$_4$ reactions, these points should fall on a line with a substantially greater slope than that of the reactions carried out in alkane solvents. In addition, use of SiMe$_4$-d$_{12}$ should affect not only the **14/28** ratio, it should affect the overall rate.

In fact, plotting of the total rate for the SiMe$_4$ reactions gives points falling on line I. As mentioned above, use of SiMe$_4$-d$_{12}$ results in a rate indistinguishable from that in SiMe$_4$. This means that both **28** and **14** come from a common intermediate in reactions after the rate determining step, and is inconsistent with mechanisms proceeding via paths **f** and **j**.

Assuming now that only the mechanisms of paths **g** and **i** are operating, the rate expression would be as shown in equation (13.21).

$$k_{obsd} = k_8 + \frac{k_6 k_7}{k_{-6} [L] + k_7}$$

(13.21)

If k_7 is negligible compared to $k_{-6} [L]$ for most [L], then equation (13.21) reduces to equation (13.22).

$$k_{obsd} = k_8 + \frac{k_6 k_7}{k_{-6} [L]}$$

(13.22)

As mentioned above, the slope (divided by k_6) and intercept of Figure 13.3 give $k_8 = 1.10 \times 10^{-6}\,\text{s}^{-1}$, and $k_7/k_{-6} = 3.67\ (0.25) \times 10^{-6}\,\text{M}$. For all [L] above approx. $10^{-4}\,\text{M}$, k_7 clearly is negligible with respect to k_{-6}[L].

This expression fully accounts for the data since the total rate of disappearance of **27**, k_{obsd}, will be independent of the solvent with which $L_3\text{Os}$ reacts. There will also be no isotope effect on k_{obsd} upon use of $\text{SiMe}_4\text{-d}_{12}$ as solvent.

There is an additional check to differentiate path **g** from paths **f** and **j**. An Os^{IV} intermediate such as **41** or **43** should show isotopic crossover under appropriate conditions. Use of $(\text{L}')_4\text{OsH(Np)}$, **27**-d$_{36}$, as substrate should result in incorporation of some deuterium into the liberated neopentane via **41**-d$_{27}$. Use of $\text{SiMe}_4\text{-d}_{12}$ solvent in reaction with unlabeled **27** should also form some neopentane-d$_1$ via **43**-d$_{12}$.

41-d$_{27}$ **43**-d$_{12}$

We have checked for crossover in the case where $\text{Si(CD}_3)_4$ was used as solvent. The mass spectrum of the neopentane product revealed no neopentane-d$_1$ above natural abundance. As mentioned above, the absence of crossover is negative evidence, but crossover which was observed in the case of benzene-d$_6$ activation via Os^{IV} intermediate **39** [43], analogous to **43**, suggests the likelihood of observing it were such a mechanism to operate here.

Knowing the total rate and k_8, one can calculate from k_{obsd} the percentage of each path which contributes to a given reaction. Table 13.2, column **d** shows the percent reaction which proceeds via the k_7 path in SiMe_4 solvent. Furthermore, since we observe that the k_8 path forms only **14** and not **28** (i.e., path **k** in Scheme XVI does not operate) then the **14/28** ratio can be calculated for the k_7 path only. This ratio is shown in Table 13.2, column **e**. There is some scatter, but it is clear that this ratio is essentially constant at approx. 0.42 and independent of L over this concentration range of L; that is, SiMe_4 as neat solvent undergoes C–H activation by $L_3\text{Os}$ about 2.4 times as fast as the latter attacks its own ligand.

The data in Table 13.2, column **e** suggest two things. First, the constancy of the **14/28** ratio as a function of [L] strongly suggests that $L_3\text{Os}$ reacts more rapidly with C–H bonds than it does to take up L to form $L_4\text{Os}$ over this range of concentrations of L. Even though $L_4\text{Os}$ is undoubtedly crowded, the barrier to coordination of $L_3\text{Os}$ by L is probably low, so the barrier to C–H activation must also be quite low.

Second, the **14/28** ratio of 1.43 for the reaction run in $\text{SiMe}_4\text{-d}_{12}$ (column **e**, entry 5) when compared to that of the other reactions implies that there is an isotope effect for the C–H activation of SiMe_4 of approx. 3.4.

The magnitude of the isotope effect can be confirmed more directly. Use of a 15/85 molar mixture (determined by mass spectrometry) of SiMe_4 and SiMe_4-d$_{12}$ as solvent for the pyrolysis of **27** formed a mixture of **28**-d$_0$/**28**-d$_{12}$. The observed ratio corresponded to a k_H/k_D for the C–H activation step itself of 3.6. Since that on k_{obsd} when SiMe$_4$-d$_{12}$ is used as solvent is 1, clearly the step of k_{11} is not rate determining. The most reasonable alternative is that the reaction proceeds via L$_3$Os, the k_7 path.

For comparison to complex **27**, the activation parameters were determined for the dissociation of L from **28** by thermolysis of **28** in the presence of excess L′ over the temperature range of 66–99°C, and yielded values of $E_a = 34.6$ (0.2) kcal/mol, ΔH (80°C) = 33.9 (0.2) kcal/mol, and ΔS (80°C) = 14.5 (0.6) e.u.

13.8.2. Thermolysis of 27 in Mesitylene

Thermolysis of **27** in neat mesitylene at 80°C resulted in slow and clean conversion to the product of benzylic C–H activation (**36**) in competition with **14** (equation (13.23)) in an 85/15 ratio.

$$(13.23)$$

The reaction was again first order in **27** ($k_{obsd} = 3.0$ (0.15) $\times 10^{-6}$ s^{-1}, free L generated $= 1.1 \times 10^{-3}$ M), and was strongly inhibited by added L. A key point is the site of attack by osmium on mesitylene. ^{1}H- and ^{31}P-NMR spectra showed good evidence of only one isomeric product, that resulting from benzylic attack, was formed. While we have not examined this reaction in detail, its rate at a given [L] is most consistent with C–H activation via L$_3$Os.

13.8.3. Reaction Path i

We have independent evidence that L$_4$Os is the intermediate on path **i**, as postulated, and that it does not activate C–H bonds intermolecularly, that is, for example, that path **k** (Scheme XVI) does not operate. We have recently prepared L$_5$Os and examined some of its chemistry [46]. One interesting feature of its reactivity that is particularly relevant here is that its pyrolysis in benzene or in neohexene or in SiMe$_4$ solvent results in formation of **14** only; no phenyl hydride (**33**), neohexenyl hydride (**35**), or **28** is formed. The reaction is first order in L$_5$Os, and is inhibited by added L. We conclude that the reaction of equation (13.24) takes place, and that L$_4$Os is not capable of intermolecular C–H activation. In addition, it is clear that L$_4$Os, while probably rather crowded, is not crowded enough to undergo dissociation of L to form L$_3$Os during its short lifetime. Were the latter to occur, then in SiMe$_4$ solvent some **28** would form.

$$(13.24)$$

13.8.4. Activation of Methane by Osmium

Successful activation of C–H bonds in $SiMe_4$ and mesitylene led us to believe that the system was very close to alkane activation by L_3Os. Since the system was so sensitive to steric bulk, perhaps methane could approach the metal well enough to undergo oxidative addition. Thermolysis of **27** at 80°C in cycloalkane solvent under a pressure of methane estimated to be between 42 and 68 atm (at ambient temperature) yielded **14** and **23**, with the latter in yields of up to 16% (equation (13.25)) [47]. Formation of **23** was completely suppressed by the addition of free phosphine, but formation of **14** was not, as discussed above.

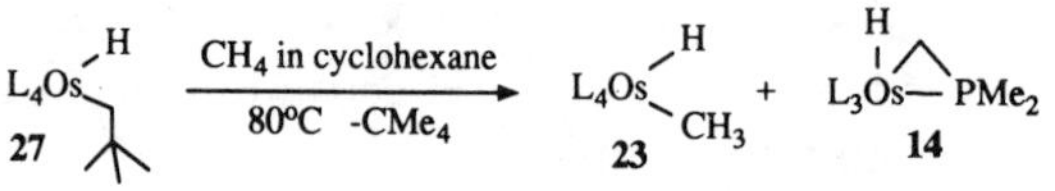
L4Os
H
27
CH4 in cyclohexane
80°C -CMe4
L4Os
H
CH3
23
+
H
L3Os— PMe2
14

(13.25)

The origin of the methyl group was established by carrying out the thermolysis of **27** under $^{13}CH_4$ which yielded [*methyl*-^{13}C]**23**. The ^{31}P-NMR resonances of the latter were identical with those of authentic material prepared by the alkylation reaction of equation (13.15) using [^{13}C]methyllithium. In addition, the ^{13}C-NMR spectrum of authentic [*methyl*-^{13}C]**23** was identical to the especially intense methyl group resonance of **23** from the activation reaction.

In view of the fact that the unsaturated intermediate $L_3Os(H)Np$, **38** does activate benzene, alkenes, and, most importantly, its own neopentyl group, it seemed possible that it might also accommodate the particularly small methane molecule. Thus, methane might undergo oxidative addition by the same mechanism as benzene (path **n** in Scheme XVII) rather than that by which the larger $SiMe_4$ and mesitylene molecules are activated (path **g**–**m**). The typical yield of approx. 10–15% **23** precluded any dependable determination of kinetic

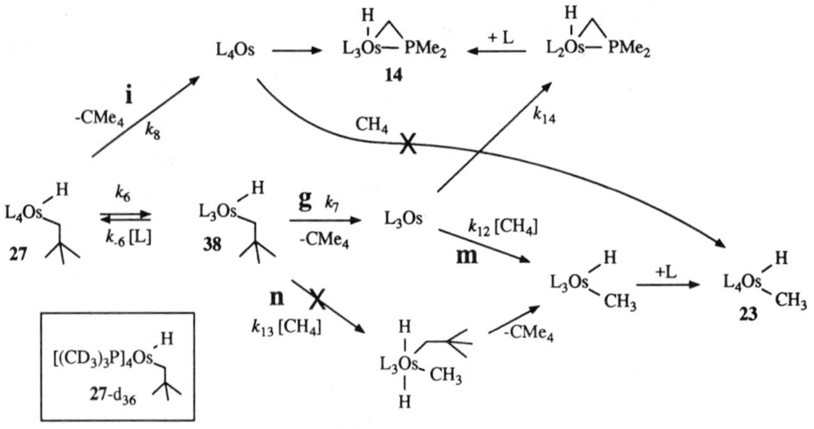
L4Os
H
L3Os— PMe2
14
+ L
H
L2Os— PMe2
i
-CMe4
k8
CH4
k14
L4Os
H
27
k6
k-6 [L]
L3Os
H
38
g k7
-CMe4
L3Os
k12 [CH4]
m
L3Os
H
CH3
+L
L4Os
H
CH3
23
n
k13 [CH4]
H
L3Os
CH3
H
-CMe4
[(CD3)3P]4Os
H
27-d36

Scheme XVII.

dependence on methane concentration or any isotopic crossover test for the mechanism, so a different approach was used. The basis of this approach is outlined in Scheme XVII. In thermolysis of $[(CD_3)_3P]_4Os(H)Np$, 27-d_{36}, there will be a primary kinetic isotope effect only on k_{14}, the cyclometallation of L_3Os. Of course, there will also be a primary effect on the cyclometallation of L_4Os, but we know that this intermediate does not activate methane. If methane activation proceeds via path g–m, then there will be an increase in the ratio of 23/14 that results from L_3Os when 27-d_{36} is used. If 23 forms via path n, that is, via addition to 38, then the 23/14 ratio will be unchanged by use of 27-d_{36} since that ratio would be determined by the $k_{13}[CH_4]/k_7$ ratio and not k_{14}. Naturally, corrections must be made for the portion of 14 which forms via L_4Os (path i with k_8) and for smaller secondary isotope effects on k_8, k_6, k_{-6}, and k_7. Figure 13.3, line II shows a plot of k_{obsd} vs. $1/[L']$ for thermolysis of 27-d_{36} from which the appropriate corrections can be made for small isotope effects on the present isotopic perturbation experiment. The average of four pyrolyses of 27-d_{36} yielded a clear change in the 23/14 ratio from which k_H/k_D on k_{14} could be calculated to be 2.4. This is quite good agreement with the independent measurement of this isotope effect of 2.2 described above. Thus, it seems clear that in spite of its small size methane, like the other saturated substrates, is activated by L_3Os and not 38.

13.8.5. Evidence for Activation of Primary C–H Bonds in Linear Alkanes [48]

When conducted in cycloalkane solvents, thermolysis of 27 forms no L_4OsH_2 (18) side product. However, in linear hydrocarbons, including pentane, hexane and octane, 18 grows in with a first-order rate up to a concentration of approx. 15% of the product mixture (equation (13.26)).

$$\text{(13.26)}$$

Quantitative analysis of alkene generated detects all three pentene isomers in about the same total amount as the 18 formed. At concentrations of free L high enough to fully inhibit path g, no 18 is observed. These data suggested (Scheme XVIII) that L_3Os may indeed attack the primary C–H bonds of linear hydrocarbons to form $L_4Os(H)(1^\circ\text{-R})$ products such as 29, but that the latter, or the $L_3Os(H)(1^\circ\text{-R})$ intermediates leading to them (44), decompose at 80°C to 18 at a rate which is faster than the disappearance of 27. Corroboration of primary C–H activation comes from the fact that heating of 27 in n-C_5D_{12} solvent at 80°C results in formation of 14 and L_4OsD_2 (18-d_2). The same thermolysis carried out in the presence of 0.1 M L leads to no observable 18-d_2, only 14.

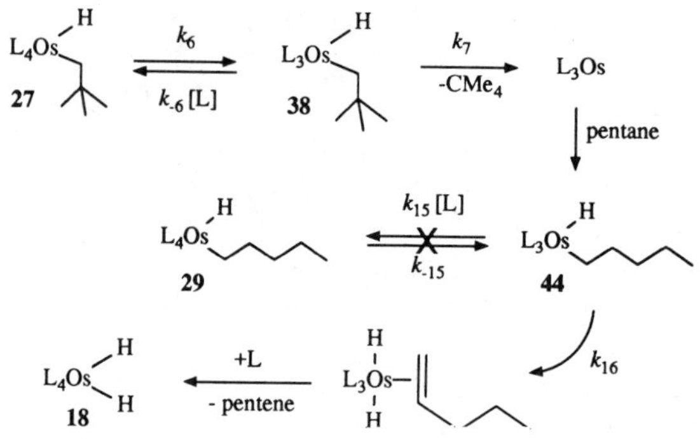

Scheme XVIII.

Further information regarding Scheme XVIII is obtained from thermolysis of independently prepared pentyl hydride **29**. At 80°C, **29** decomposes to **18** with a half-life extrapolated to be about two months (equation (13.27)).

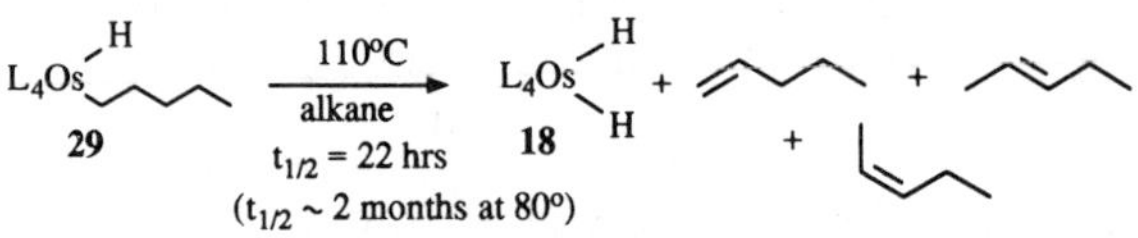

$$(13.27)$$

At 110°C, the half-life is about a day. If **29** were to form during the heating of **27** in pentane solvent, clearly it would persist. This apparent dilemma is cleared up by carrying out the thermolysis of **29** in neat phosphine (equation (13.28)), in this case L′ (P(CD$_3$)$_3$).

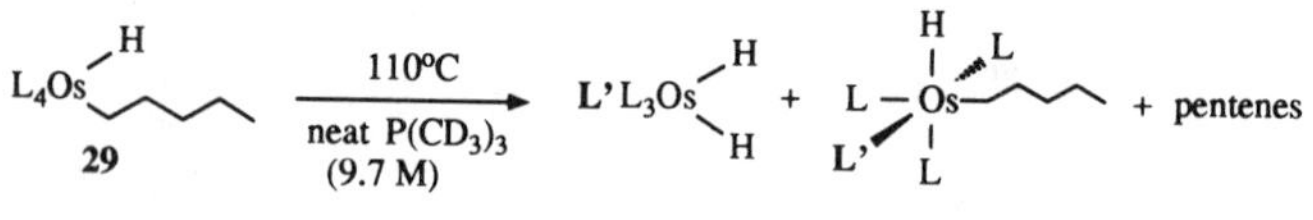

$$(13.28)$$

In neat L′ at 110°C, **29** incorporates L′ with a half-life for the first incorporation of about 22 h, as shown above in Table 13.1. One may notice that this is about the same rate as the thermal decomposition of **29** shown in equation (13.27). Thus, even in 9.7 M L′, dihydride **18**-d$_9$ forms at about the same rate as **29** takes up L′. That is, even in 9.7 M L′, k_{16} is competitive with k_{15}[L]. So, it is not surprising that at 10^{-3} M concentrations of L where pentane activations by thermolysis of **27** are conducted, uptake of L by **44** to form **29** is not competitive with formation of **18**.

13.8.6. High Temperature Reactions of Methyl Hydride 23

We have previously shown that the fastest reaction of unsaturated intermediate **38** is reversible intramolecular activation of its neopentyl group to form osmacyclobutane **40**. Because of the rapid scrambling of deuterium label that results from this process in a species such as $L_4Os(H)[CH_2C(CD_3)_2(CH_3)_2]$, it is not possible to detect whether any path exists for scrambling of alpha-hydrogens of an alkyl group with the metal hydride. Since **23** has only alpha-hydrogens, it would be the logical place to look for this process. Accordingly, $L_4Os(H)(CD_3)$, **27**-d_3, was prepared and submitted to thermolysis. After one week at 125°C in hexane solvent, no migration of the deuterium to any other location in the molecule or the solvent was noted by ^{2}H-NMR. It is known that L dissociates from **23** with $k_6 = 1.9 \times 10^{-6} \, s^{-1}$ at 105°C [43]. Thus, at 125°C the lack of reactivity is not due to a lack of ligand dissociation. When the same sample was heated at 160°C for 9 h, approx. 30% conversion to **14** had taken place, and the label was partially scrambled into the hydride of **23** or **14** or both, as determined by ^{2}H-NMR. (The hydride resonances for the two compounds have the same chemical shift.) At approx 60% conversion to **14**, the label was essentially random between the OsMe and OsH sites (equation (13.29)).

$$L_4Os\begin{smallmatrix}H\\CD_3\end{smallmatrix} \xrightarrow[\substack{\text{alkane}\\60\% \text{ reaction}}]{160°C} L_3\overset{\overset{H}{|}}{Os}-PMe_2 + L_4Os\begin{smallmatrix}H\\CD_3\end{smallmatrix} + L_4Os\begin{smallmatrix}D\\CD_2H\end{smallmatrix}$$

23-d_3 **14**   1 : 3

(13.29)

Completion of the reaction revealed no deuterium in the hydride or phosphine regions of **14**, and no enhancement above natural abundance in solvent. Thus, the scrambling had occurred between the OsMe and OsH sites of **23** only.

There are two possible mechanisms for this alpha-scrambling which are shown as paths **r** and **s** in Scheme XIX. We cannot strictly distinguish between the two possibilities of reversible alpha elimination (path **s**) and reversible reductive elimination to a methane complex (path **r**), but the fact that label scrambling only occurs when **14** is forming suggests the latter. The fact that there is no detectable crossover, that is, no deuterium appears in **14**, suggests the formation of **14** via reductive elimination of methane followed by oxidative addition of L(Os°) rather than oxidative addition of L followed by reductive elimination of methane (Os^{IV}).

13.8.7. High Temperature Reactions of Werner's Complex, 14

Initially, we had assumed that cyclometallated material **14** was generally thermodynamically preferred over all of the intermolecular C–H activation products. This was eventually shown to be false, at least for benzene, when **14** was heated at 178°C in benzene (equation (13.30)).

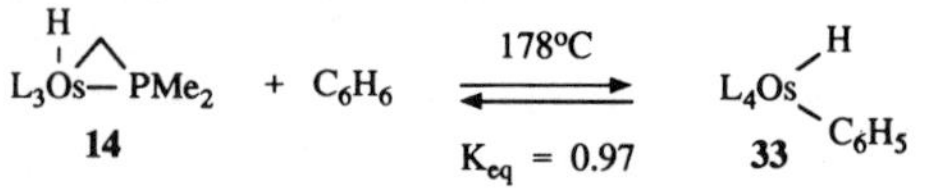

$$L_3Os-PMe_2 \;(H) \quad \mathbf{14} \quad + \; C_6H_6 \quad \underset{K_{eq}=0.97}{\overset{178°C}{\rightleftharpoons}} \quad L_4Os \begin{smallmatrix} H \\ C_6H_5 \end{smallmatrix} \quad \mathbf{33}$$

$$(13.30)$$

At this temperature, **14** underwent a first-order conversion to phenyl hydride complex **33**. Heating pure **14** in cyclohexane-d$_{12}$ with 1.0 M benzene resulted in about a 1:1 ratio of **14/33** ($K = 0.98$), with no incorporation of deuterium into either material detectable by ^{2}H-NMR.

In contrast to the high temperature reaction of **14** with benzene, when **14** was heated in neat SiMe$_4$ at 178°C for 7 days, no change was seen. In light of the benzene result, this is likely to reflect a thermodynamic preference for **14** and not a kinetic problem for the formation of **28**.

In order to determine whether reductive elimination of the C–H bond from **14** to form L$_4$Os or dissociation of L from **14** is its initial thermal process, **14** was heated in cycloalkane solvent at 155°C in the presence of large excess of L′. Incorporation of L′ into **14** occurred at all positions at the same rate, including the cyclometallated site, with a half-life of approx. 7 days. These exchange data are most consistent with the operation of path **t** in Scheme XX. We know that at about 50°C L$_5$Os exchanges with added L′ more rapidly than it converts to **14** [46]. Thus, when **14** would form L$_4$Os at 155°C, L$_4$Os would probably completely exchange with the excess L′ via reversible formation of L$_5$Os, and cyclometallation of the fully exchanged (L′)$_4$Os would yield only **14-**d$_{36}$. This would give incorporation of L′ into **14** at all four sites at the same rate.

In contrast, if path **u** in Scheme XX were to operate, intermediate **45** very probably would take up L′ to return to **14** much faster than it would undergo reductive elimination to form L$_3$Os; that is, $k_{-17}[L]$ probably would be much larger than k_{18}. This would mean that **14** would incorporate L′ much more rapidly into the three terminal positions than in the cyclometallated position. In fact, to obtain L′ in the cyclometallated position at all, the L$_3$Os would need to exchange L/L′ via L$_4$Os and almost certainly via L$_5$Os, and then undergo dissociation of L twice to return to L$_3$Os, and then from there return to **45**, and then to **14**. We know that at 80°C L$_4$Os cyclometallates directly to **14** rather than to undergo dissociation of L to form L$_3$Os, since heating of L$_5$Os (which

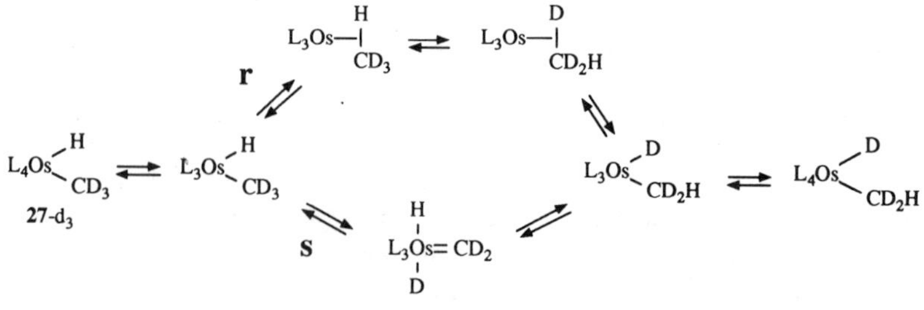

Scheme XIX.

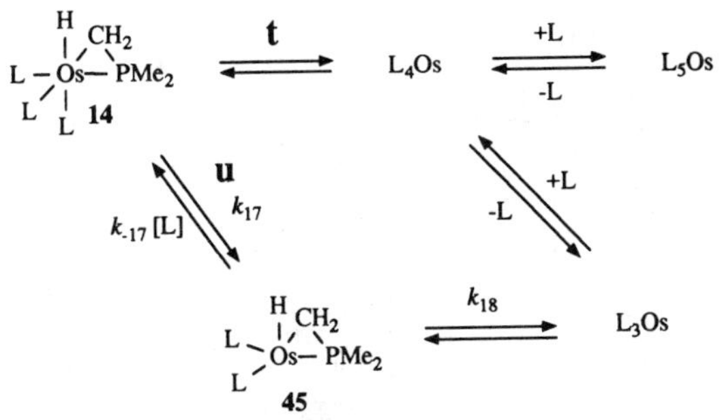

$$
\begin{array}{ccccc}
\underset{\displaystyle \underset{14}{\overset{\displaystyle L}{\big|}}}{\overset{\displaystyle \overset{H}{\big|}\;CH_2}{L-Os-PMe_2}}
& \underset{\longleftarrow}{\overset{t}{\longrightarrow}}
& L_4Os
& \underset{-L}{\overset{+L}{\rightleftharpoons}}
& L_5Os \\
\end{array}
$$

$k_{-17}\,[L] \diagdown \quad \Big\Vert\; k_{17}$ $\qquad\qquad\qquad$ $\overset{+L}{\diagup}\;\; -L\;\diagdown$

$$
\begin{array}{ccc}
\underset{\displaystyle \underset{45}{\overset{\displaystyle L}{\diagup}}}{\overset{\displaystyle \overset{H}{\big|}\;CH_2}{L\!\!\diagdown\!Os-PMe_2}}
& \overset{k_{18}}{\underset{\longleftarrow}{\longrightarrow}}
& L_3Os \\
\end{array}
$$

Scheme XX.

produces low concentrations of L_4Os) in neat $SiMe_4$ forms only **14** and no **28**; and thermolysis of **27** in the same solvent with added L (which proceeds via L_4Os) gives no **28**. (Recall, **28** forms only from L_3Os in $SiMe_4$.) It seems unlikely that the preference at 80°C for L_4Os to cyclometallate to form **14** rather than to dissociate L to form L_3Os would be completely reversed at 155°C, as would be required by L′ incorporation into the cyclometallated site of **14** by path **u** in Scheme XX.

13.9. TYING IT ALL TOGETHER

As mentioned at the beginning, significant advances in the understanding of C–H "activations" (oxidative addition reactions of C–H bonds or metathetical C–H exchanges) by soluble complexes of the transition metals have been made in recent years [1–18], especially in Cp- or Cp*-containing systems (Cp = cyclo-C_5H_5; Cp* = cyclo-$C_5(CH_3)_5$). There has not been as much well-defined mechanistic work on non-(Cp or Cp*)-containing complexes. The L_4Os^{II} system described herein has exhibited a rich variety of chemistry including the activation of assorted types of C–H bonds. It is important as an example of a soluble, mononuclear, non-cyclopentadienyl-containing system of metal complexes whose chemistry is clean and unusually diverse. A summary of many of the paths which we have detected in our mechanistic studies so far is given schematically in the qualitative free energy diagram (80°C) of Figure 13.4.

Issues of central interest for mechanisms of oxidative addition of C–H bonds include: (1) the coordination number and geometry of the reactive intermediate, (2) the orientations, populations, symmetry properties, and energies of the reactive orbitals at the metal, and (3) the competition between intermolecular C–H activation and intramolecular ligand reaction. The first two of these issues are, of course, closely related.

It is generally believed that the essential features of the oxidative addition reaction of a C–H bond to a metal are as shown in Figure 13.5 [49]. The metal is required to possess a vacant, acidic (low-lying) molecular orbital (MO) of σ

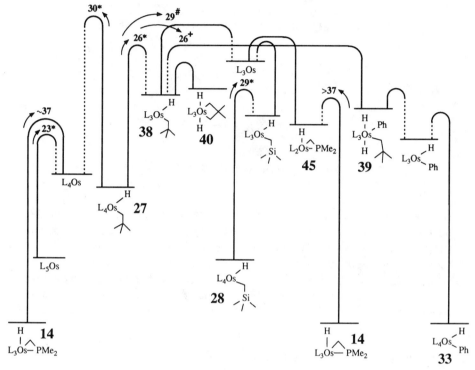

Figure 13.4. Qualitative free energy diagram for some of the reactions of **27** and related species at 80°C. Bimolecular steps are represented by dashed lines approaching the transition state. Reactant concentrations are not standard, but are those found or estimated under a given set of reaction conditions. Notes for the figure: [*]Activation parameters have been determined; [#][L] = 7×10^{-4} M; [+][L] = 7.4×10^{-5} M, [C_6H_6] = 11.2 M.

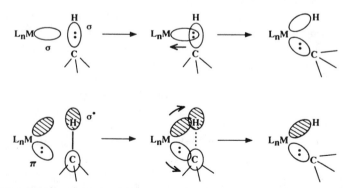

Figure 13.5. Proposed reaction coordinate for the oxidative addition of C–H bonds to transition metal centers. The top series of structures traces σ-symmetry changes and the bottom series π-symmetry changes.

symmetry, and a basic (filled), π-symmetry MO. The initial interaction of the hydrocarbon with the metal center is presumably that of electron density donation from a C–H σ bond to the acidic metal σ MO. As this acid-base coordinate bonding develops and the hydrocarbon comes closer to the metal, the metal π-symmetry basic orbital begins to donate electron density back into the empty C–H antibonding orbital. These two interactions lead to the formation of the two quasi-"symmetric and antisymmetric" sigma bonding combinations of the C–M–H bonds.

In our investigations with this osmium system, we have accumulated evidence for at least three distinct reactive intermediates: **38**, L_4Os, and L_3Os. We have good evidence [43] that **38** is stereochemically rigid, and so probably is square pyramidal, as would be expected for a five-coordinate, d^6, third-row metal [50]. The orbital at the vacant octahedral site is the LUMO and there are two filled π-symmetry MOs directed at this site which are essentially the d_{xz} and d_{yz} metal orbitals (Figure 13.6). Consistent with this picture, intermediate **38** (and the analogue with CH_2SiMe_3 in place of Np) has been shown to cyclometallate its own alkyl group to form **40** (and its silicon analogue) and to attack benzene and other arenes easily. The intramolecular C–H activation of the neopentyl group in **38** clearly shows that it has no inherent enthalpic problem in the activation of sp^3 C–H bonds. We presume that the absence of any intermolecular reaction of this intermediate with saturated substrates is associated with the substantial bulk of the four PMe_3 ligands and the neopentyl group. The alkane simply cannot approach close enough to interact at a competitive rate.

It is interesting to note that while L_4Os and L_3Os are both reactive enough to attack the C–H bonds of their own phosphine ligands, we have no evidence for the analogous process in **38**; that is, there is no indication of the formation of **41**. This may suggest that **38** is inherently less reactive than either L_4Os or L_3Os.

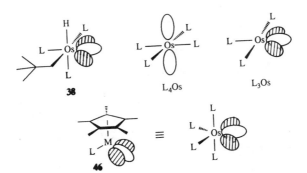

Figure 13.6. Reactive orbitals of some intermediates relevant to this work. σ symmetry lobes are unshaded, while both lobes of π symmetry orbitals are shaded. Species **38** and L_3Os each have a second π-symmetry orbital, orthogonal to the one shown, which has been omitted for clarity. Also, L_4Os shows only the empty p_z; the filled d_{z^2}, colinear with the p_z, and the filled π-symmetry d_{xz} and d_{yz} orbitals are not shown.

Alternatively, it is possible that $L_3Os(H)Np$ is predominantly in the cyclometal-lated form, **40**, and that its lifetime as **38** is too short to allow formation of **41**. This last possibility is less tenable in view of the observation that **38** reacts so readily with benzene and alkenes.

The fact that **38** does intermolecularly attack the $C(sp^2)$–H bond in benzene but not $C(sp^3)$–H bonds is most easily explained by invoking the intermediacy of an arene π complex with **38**, although we have no evidence which bears directly on this issue. Jones et al. have data strongly implicating the inter-mediacy of η^2-arene complexes in reactions of $Cp^*Rh(PMe_3)$ with arenes [11].

Intermediate L_4Os would presumably prefer to be square planar. In a square-planar complex, the two degenerate filled orbitals, essentially d_{xz} and d_{yz}, are the reactive π-symmetry orbitals. The essentially p_z which is co-linear with the filled d_{z^2} orbital, is the reactive acidic σ orbital. In this case, the acid-base interaction of the empty metal p_z with the C–H σ bond is offset by the repulsive interaction of that σ bond with the filled metal d_{z^2} orbital [49]. Thus, the hydrocarbon cannot get close enough to the metal for back-donation from the metal π-donor orbital to the C–H antibonding orbital to become important, and the oxidative addition does not occur.

Parenthetically, it is generally appreciated that one of the useful features of the Cp or Cp* ligand is that, necessarily occupying three facial coordination sites, it constrains d^8 CpML (16-electron) intermediates to a non-square-planar geometry, as in **46**. Calculations show [51] that the preferred geometry of **46** is the bent one shown in Figure 13.6. In this geometry, there is a filled orbital of π-symmetry with respect to the vacant metal coordination site, and, of course, the acidic σ orbital is the vacant site itself [51]. Thus, the orbital configuration is ideal for C–H activation in this geometry.

In any event, while L_4Os, as a four-coordinate, d^8 complex, would probably prefer to be square planar for electronic reasons, it is undoubtedly distorted from this geometry due to the congestion of the four bulky ligands. Substantial puckering is evident in the X-ray crystal structure of $[(Me_3P)_4Rh]^+Cl^-$ which is distorted toward tetrahedral geometry (the two *trans*-P-Rh-P angles are 148° and 151°) [52], and $[(MePh_2P)_4Ir]^+BF_4^-$ is similarly distorted (both *trans*-P-Ir-P angles are 151°) [53]. In spite of its probable distortion, it appears that L_4Os will not undergo intermolecular reactions with alkanes, arenes, or $SiMe_4$, so it may still have the essential orbital features of a square-planar complex which were described above. Alternatively, it may be that the four PMe_3 ligands are bulky enough to block access to the metal, even if the complex is rather distorted. Intramolecular C–H oxidative addition of L_4Os is discussed below. Our data convince us that all of the intermolecular activations of sp^3 CH bonds proceed via three coordinate L_3Os.

L_3Os is probably either "T"-shaped or "Y"-shaped [54]. For simplicity we will assume the former. In this geometry, the vacant site of what would otherwise be a square-planar array is the position of the LUMO, and there are two filled π-symmetry orbitals, essentially d_{xy} and d_{xz}. Among the species L_4Os, L_3Os, and **38**, it is L_3Os which is least hindered, and on that basis it may be expected to be the most reactive. In addition, since both the σ acidity and π basicity of the metal are important, it is possible that, as an Os° complex, L_3Os may have significantly greater π-basicity and so greater reactivity than **38**.

The importance of understanding the competition between intramolecular ligand activation and intermolecular C–H activation has been generally appreciated [12, 16a, 55]. In Crabtree's Ir-based alkane activations, the PPh_3 ligands, normally so prone to cyclometallation, are innocuous apparently because of the reversibility of their interaction with the metal center [16b]. In the $Cp^*Ir(PMe_3)$ system of Bergman [9], the $Cp^*Ir(CO)$ system of Graham [10], and the $Cp^*Rh(PMe_3)$ system of Jones [11, 55] cyclometallation of the Cp^* or PMe_3 ligand is not observed. Apparently, the additional energy required to bend the ligand over to allow C–H activation offsets the favorable entropy for intramolecular activation and allows the intermolecular C–H activation to completely dominate the reactivity of these intermediates. This is possible in these cases because of the very low activation energies for the intermolecular reactions. In the case of the unsaturated intermediate $Cp^*Re(PMe_3)_2$, however, the greater steric congestion about the metal may slow intermolecular reactions with hydrocarbons, and also may bend the PMe_3 methyl groups a bit closer to the metal. In this case, cyclometallation of the phosphine ligand takes place competitively with the intermolecular reaction [12].

Our data clearly establish that both L_4Os and L_3Os have a pronounced tendency to cyclometallate. In the case of L_4Os, this is to the exclusion of any intermolecular reaction with $SiMe_4$, alkanes, or even benzene or neohexene [46], the last two of which undergo facile C–H activation with **38**. As mentioned above, L_4Os is probably significantly distorted from planarity. This may in part account for the molecule's facile cyclometallation. Also, the forced propinquity of the CH bond to the metal would result in the penetration of that bond beyond the zone of net repulsive interaction (between the filled metal d_{z^2} orbital and the C–H bonding electrons) to the point that π-backbonding to the CH antibonding orbital could become important, and the oxidative addition could proceed. These arguments would also be consistent with the faster cyclometallation of the neopentyl Ir complex **3** compared to the silylmethyl complex **4**. Since the Si–C bond is approx. 0.3 Å longer than the C–C bond, the neopentyl group of **3** will be held closer to the crowded metal center than will the CH_2SiMe_3 group of **4**, and **3** will be more congested and probably more distorted. Thus, the greater reactivity of **3** over **4** may be the result of the greater distortion of **3** and/or the greater penetration of the C–H bond into the orbitals of the iridium. In the case of **3** and **4**, there is an additional possibility of an electronic effect of Si on the C–H bond being activated or on the activation transition state. We cannot determine the relative importance of these factors as yet.

It is interesting to compare Whitesides' [25] result of Scheme II with that of our iridium and osmium investigations. Neopentyl platinum complex **7** does not undergo any intramolecular C–H oxidative addition below the temperature required for dissociation of phosphine; L_3IrR complexes **3** and **4** are nicely stable and isolable, but facilely cyclometallate γ-CH bonds without loss of L;

L$_4$Os cannot be isolated and extremely rapidly cyclometallates what is effectively a β-CH bond of its ligand. It is likely that the steric congestion is similar for **7, 3, 4**, and L$_4$Os, and the orbital/ligand proximity discussed above for the cyclometallation of L$_4$Os should apply to a greater or lesser extent to all of these. The most obvious change across this series is the basicity of the metal, and this may be the overriding factor determining rates of C–H oxidative addition to square-planar metal complexes. Whether this same factor will be dominant in tricoordinate, 14-electron complexes remains to be seen.

Although we know little quantitatively about the relative energies of the species L$_4$OsHR, a qualitative free energy picture is given in Figure 13.4. The reactivity of **27** contrasts sharply with the thermal stability of most of the other L$_4$OsHR species that we have prepared. It is certain that **27** lies well above **14** and **33**, and that **14** and **33** are at about the same energy (at least at 178°C with 1 M C$_6$H$_6$). Steric compression between the neopentyl group and the four phosphines is responsible for **27** being at higher energy. For example, L$_4$Os(H)Me, **23**, exhibits a half-life for dissociation of L (measured by exchange with L′) [43] of 4 days at 110°C, and a half-life for conversion to **14** of 1 week at 150°C. In comparison, the same two processes for **27** exhibit half-lives of 32 min and 3 days, respectively, both at 80°C. This higher reactivity of **27** allows observation of relatively clean and mild chemistry of what would otherwise be a highly inert octahedral Os^{II} system.

A particularly interesting comparison would be between the activation free energies for the reductive elimination of neopentane from saturated **27** vs. from unsaturated **38**. Of course, we cannot make this comparison quantitatively, because we do not know the rate of uptake of L by **38**, but it is clear that the barrier is significantly lower for elimination from the five-coordinate intermediate. This gives important additional support to the generality of a trend seen in the reductive elimination chemistry of complexes of Rh, Ir, and Pt [25–29], and is the first example where both processes have been clearly observed simultaneously.

We have observed secondary kinetic isotope effects in two reactions of [(CD$_3$)$_3$P]$_4$Os(H)Np, **27**-d$_{36}$: the dissociation of L, and the reductive elimination of neopentane. The k_H/k_D appears to be about 1.3–1.4 in both cases. Although electronic effects cannot be ruled out, we believe that these isotope effects are most likely to have their origins in steric effects. It is known from microwave spectroscopy that the C–D bond is approx. 0.003 Å shorter than the C–H bond [45], and both of the reactions involved here are certainly driven by steric crowding. The isotope effect per-hydrogen would thus be $\sqrt[36]{1.4}$ or 1.009. Steric isotope effects larger than this have been reported [56].

We have measured several deuterium KIEs of C–H activation reactions in these systems; for oxidative addition of the C–H bond of benzene to the five-coordinate Os^{II} complex **38** (3.3), for activation of SiMe$_4$ and its own phosphine ligand by L$_3$Os (3.6 and 2.2, respectively), activation of its own ligand by L$_4$Os (4.7), and cyclometallation of **4** (4.8). These numbers have been very useful in the delineation of mechanistic alternatives, but we believe that speculation as to the significance of their relative magnitudes is premature.

To date, the goal of using the M–C bound organic fragment obtained via $C(sp^3)$–H disruption for some useful catalytic reaction remains largely elusive for soluble, late transition metal systems. Nevertheless, important insight is being gained into the requirements for C–H reactivity with metal complexes. We are currently synthesizing molecules similar to **3** and **27** using other organic ligands and other phosphorus ligands with the goal of acquiring information regarding the relative importance of metal acidity and basicity and other factors in the mechanisms of X–C and X–H (X = C, O, and N) oxidative-addition and reductive-elimination reactions.

ACKNOWLEDGMENTS

I am deeply indebted to the able efforts of a talented group of graduate students and post doctoral research associates who have worked on the above projects: listed chronologically; John A. Statler, Steven P. Bitler, Susan P. Ermer, Peter J. Desrosiers, Dr. Ronald S. Shinomoto, T. Gregory, P. Harper, and Mark A. Deming. This work was supported by the National Science Foundation (CHE-8406900, CHE-8705228). Loans of heavy metal salts by Johnson Matthey Co. are gratefully acknowledged.

REFERENCES

1. Fendrick, C. M. and Marks, T. J., *J. Am. Chem. Soc.*, **106**, 221 (1984).

2. Watson, P. L., *J. Am. Chem. Soc.*, **105**, 6491 (1983).

3. Thompson, M. E., Baxter, S. M., Bulls, A. R., Burger, B. J., Nolan, M. C., Santarsiero, B. D., Schaefer, W. P. and Bercaw, J. E., *J. Am. Chem. Soc.*, **109**, 203 (1987).

4. Cummins, C. C., Baxter, S. M. and Wolczanski, P. T., *J. Am. Chem. Soc.*, **110**, 8731 (1988).

5. Shilov, A. E. and Shteinman, A. A., *Coord. Chem. Rev.*, **24**, 97 (1977).

6. Sen, A., *Acc. Chem. Res.*, **21**, 421 (1988).

7. Olah, G. A., *Acc. Chem. Res.*, **20**, 422 (1987).

8. Green, M. L. H., *Pure Appl. Chem.*, **50**, 27 (1978).

9. Janowicz, A. H. and Bergman, R. G., *J. Am. Chem. Soc.*, **105**, 3929 (1983).

10. Hoyano, J. K., McMaster, A. D. and Graham, W. A. G., *J. Am. Chem. Soc.*, **105**, 7190 (1983).

11. Jones, W. D. and Feher, F. J., *J. Am. Chem. Soc.*, **108**, 4814 (1986).

12. Wenzel, T. T. and Bergman, R. G., *J. Am. Chem. Soc.*, **108**, 4856 (1986).

13. Chatt, J. and Davidson, J. M., *J. Chem. Soc.*, 843 (1965) (cf. Cotton, F. A., Hunter, D. L. and Frenz, B. A., *Inorg. Chim. Acta*, **15**, 155 (1975)).

14. Tolman, C. A., Ittel, S. D., English, A. D. and Jesson, J. P., *J. Am. Chem. Soc.*, **100**, 4080 (1978).

15. Baker, M. V. and Field, L. D., *J. Am. Chem. Soc.*, **109**, 2825 (1987).

16. (a) Crabtree, R. H., *Chem. Rev.*, **85**, 245 (1985); (b) Crabtree, R. H., Demou, P. C., Eden, D., Mihelcic, J. M., Parnell, C. A., Quirk, J. M. and Morris, G. E., *J. Am. Chem. Soc.*, **104**, 6994 (1982).

17. Cameron, C. J., Felkin, H., Fillebeen-Khan, T., Forrow, N. J. and Guittet, E., *J. Chem. Soc., Chem. Commun.*, 801 (1986).

18. (a) Hackett, M., Ibers, J. A. and Whitesides, G. M., *J. Am. Chem. Soc.*, **110**, 1436 (1988); (b) Hackett, M. and Whitesides, G. M., *J. Am. Chem. Soc.*, **110**, 1449 (1988).

19. Flood, T. C. and Statler, J. A., *Organometallics*, **3**, 1795 (1984).

20. Flood, T. C. and Bitler, S. P., *J. Am. Chem. Soc.*, **106**, 6076 (1984).

21. Bitler, S. P., Ermer, S. P., Statler, J. A. and Flood, T. C., manuscript in preparation.

22. Arlman, E. J. and Cossee, P., *J. Catal.*, **3**, 99 (1964).

23. Ivin, K. J., Rooney, J. J., Stewart, C. D., Green, M. L. H. and Mahtab, R., *J. Chem. Soc., Chem. Commun.*, 604 (1978).

24. Tulip, T. H. and Thorn, D. L., *J. Am. Chem. Soc.*, **103**, 2448 (1981).

25. (a) DiCosimo, R., Moore, S. S., Sowinski, A. F. and Whitesides, G. M., *J. Am. Chem. Soc.*, **104**, 124 (1982); (b) Foley, P., DiCosimo, R. and Whitesides, G. M., *J. Am. Chem. Soc.*, **102**, 6713 (1980).

26. Milstein, D., *J. Am. Chem. Soc.*, **104**, 5227 (1982).

27. (a) Basato, M., Longato, B., Morandini, F. and Bresadola, S., *Inorg. Chem.*, **23**, 3972 (1984); (b) Basato, M., Morandini, F., Longato, B. and Bresadola, S., *Inorg. Chem.*, **23**, 649 (1984).

28. Clark, H. C. and Manzer, L. E., *Inorg. Chem.*, **12**, 362 (1973).

29. Brown, M. P. and Puddephatt, R. J., *J. Chem. Soc., Dalton Trans.*, 2457 (1974).

30. Harper, T. G. P., Desrosiers, P. J. and Flood, T. C., *Organometallics*, **9** (1990).

31. (a) Werner, H. and Gotzig, J., *Organometallics*, **2**, 547 (1983); (b) Gotzig, J., Werner, R. and Werner, H., *J. Organomet. Chem.*, **285**, 99 (1985).

32. Deming, M. A., Desrosiers, P. J., Shinomoto, R. S. and Flood, T. C., Unpublished results and work in progress.

33. Desrosiers, P. J., Shinomoto, R. S., Deming, M. A. and Flood, T. C., *Organometallics*, **8**, 2861 (1989).

34. Alder, R. W., Bowman, P. S., Steele, W. R. S. and Winterman, D. R., *J. Chem. Soc., Chem. Commun.*, 723 (1968).

35. Reamey, R. H. and Whitesides, G. M., *J. Am. Chem. Soc.*, **106**, 81 (1984).

36. (a) Bercaw, J. E. and Brintzinger, H. H., *J. Am. Chem. Soc.*, **91**, 7301 (1969); (b) Brintzinger, H. H., *J. Organomet. Chem.*, **171**, 337 (1979).

37. Lin, Z. and Marks, T. J., *J. Am. Chem. Soc.*, **109**, 7979 (1987).

38. Gell, K. I., Posin, B., Schwartz, J. and Williams, G. M., *J. Am. Chem. Soc.*, **104**, 1846 (1982).

39. Evans, W. J., Meadows, J. H., Wayda, A. L., Hunter, W. E. and Atwood, J. L., *J. Am. Chem. Soc.*, **104**, 2008 (1982).

40. Mayer, J. M. and Bercaw, J. E., *J. Am. Chem. Soc.*, **104**, 2157 (1982).

41. Edidin, R. T., Sullivan, J. M. and Norton, J. R., *J. Am. Chem. Soc.*, **109**, 3945 (1987).

42. Basolo, F. and Pearson, R. G., *Mechanisms of Inorganic Reactions*, Chap. 3, Wiley, New York, 1968.

43. (a) Desrosiers, P. J., Shinomoto, R. S. and Flood, T. C., *J. Am. Chem. Soc.*, **108**, 1346 (1986); (b) Desrosiers, P. J., Shinomoto, R. S. and Flood, T. C., *J. Am. Chem. Soc.*, **108**, 7964 (1986).

44. Shinomoto, R. S., Desrosiers, P. J., Harper, T. G. P. and Flood, T. C., *J. Am. Chem. Soc.*, **112**, 704 (1990).

45. (a) Laurie, V. W. and Herschbak, D. R., *J. Chem. Phys.*, **37**, 1687 (1962); (b) Gordy, W. and Cook, R. L., *Microwave Molecular Spectra*, pp. 709–714, Wiley, New York, 1984.

46. Ermer, S. P., Shinomoto, R. S., Deming, M. A. and Flood, T. C., *Organometallics*, **8**, 1377 (1989).

47. Harper, T. G. P., Shinomoto, R. S., Deming, M. A. and Flood, T. C., *J. Am. Chem. Soc.*, **110**, 7915 (1988).

48. Harper, T. G. P., Deming, M. A., Shinomoto, R. S. and Flood, T. C., manuscript in preparation.

49. Saillard, J.-Y. and Hoffmann, R., *J. Am. Chem. Soc.*, **106**, 2006 (1984).

50. Elian, M. and Hoffmann, R., *Inorg. Chem.*, **14**, 1058 (1975).

51. Hofmann, P. and Padmanabhan, M., *Organometallics*, **2**, 1273 (1983).

52. Jones, R. A., Real, F. M., Wilkinson, G., Galas, A. M. R., Hursthouse, M. B., *J. Chem. Soc., Dalton Trans.*, 511 (1980).

53. Clark, G. R., Skelton, B. W. and Waters, T. N., *J. Organomet. Chem.*, **85**, 375 (1975).

54. Komiya, S., Albright, T. A., Hoffmann, R. and Kochi, J. K., *J. Am. Chem. Soc.*, **98**, 7255 (1976).

55. Jones, W. D. and Feher, F. J., *J. Am. Chem. Soc.*, **107**, 620 (1985).

56. (a) Hirsch, J. A., *Concepts in Theoretical Organic Chemistry*, pp. 157–159, Allyn and Bacon, Boston, 1974; (b) Mislow, K., Graeve, R., Gordon, A. J. and Wahl, G. H., *J. Am. Chem. Soc.*, **86**, 1733 (1964); (c) Sherrod, S. A. and Boekelheide, V., *J. Am. Chem. Soc.*, **94**, 5513 (1972).

CHAPTER 14

Hypercarbon Cluster Chemistry of Hydrocarbons

G. A. OLAH

14.1. INTRODUCTION

Electron deficient clusters by definition contain a group of three or more atoms bound by multicenter bonds with too few electrons for the bonding to involve only two electron two center (i.e., Lewis type) bonds.

As carbon is a small first row element unable to extend its valence shell, hypervalency cannot exist in carbon compounds, but hypercoordination can. Indeed a variety of carbon clusters exist in which carbon is simultaneously bound to five, six, and even eight atoms.

Hypercoordinate carbon compounds (to be brief we suggested the use of the name hypercarbon compounds [1]) contain one or more carbon atom bound not only by 2e-2c, but also 2e-3c (or $\geqslant$ 3c) bonds.

In this chapter the hypercarbon cluster chemistry of hydrocarbons will be discussed, tracing it from structural-mechanistic studies to synthetic reactions.

Organic chemistry is the chemistry of carbon compounds. Organic chemists, however, in their textbooks and discussions still nearly exclusively include only limiting four-valent compounds. It is the purpose of this chapter to show not only that hypercarbon compounds (binding simultaneously five or more ligands) are an integral, although still neglected part of organic (i.e., carbon) chemistry, but their significance is rapidly advancing from structural-mechanistic studies, to practical synthetic reactions and industrial processes.

When discussing hypercarbon clusters and their chemistry, first it is necessary

Electron Deficient Boron and Carbon Clusters, Edited by George A. Olah,
Kenneth Wade, and Robert E. Williams.
ISBN 0-471-52795-5 © 1991 John Wiley & Sons, Inc.

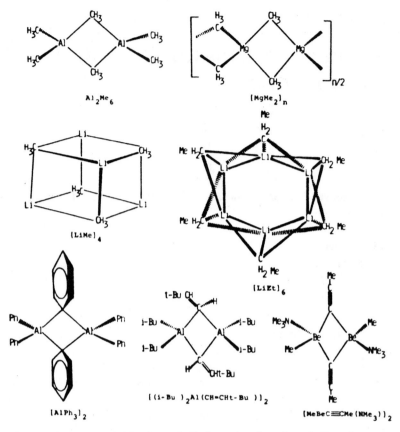

Figure 14.1. Bridged metal alkyls, aryls, alkenyls and alkynyls.

to point out that a large variety of elemental organo compounds were for a long time known to contain carbon atoms simultaneously coordinated to five, six, or even eight ligands. These compounds have been well reviewed and thus need not be discussed here in any detail. The interested reader is referred to the recent monograph on *Hypercarbon Chemistry* [1]. In bridged metal alkyls (aryls, alkenyls, and alkynyls) illustrative examples of which are shown in Figure 14.1, the bridging carbon atom is coordinated to five (or six) atoms at least two of which are metal atoms. In carboranes (Figure 14.2) again similar five or six coordinate carbon atoms are part of the cluster structure. In metal carbides, the hypercarbon bonds with five, six, or even eight metal atoms (Figure 14.3).

Not only bridged and cage hypercoordinate organometallic compounds exist, but also open systems. Recently a remarkably stable penta gold-coordinate analog of CH_5^+ (vide infra) was prepared by Schmidbauer et al. and its X-ray structure was obtained [2].

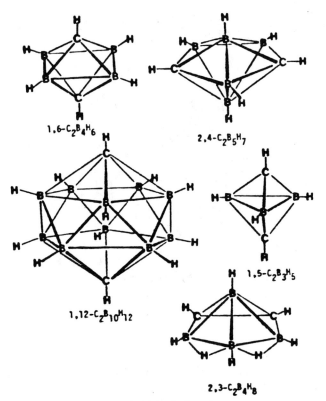

Figure 14.2. Carboranes.

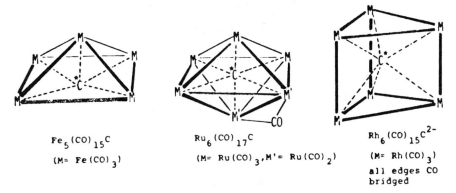

Figure 14.3. Metal carbides. ∗ denotes hypercarbons.

$$CH_2[B(OCH_3)_2]_2 \xrightarrow[\text{HMPA, 24 h, 20°C}]{[(C_6H_5)_3PAuCl]/CsF} [\{(C_6H_5)_3PAu\}_5C]^{\oplus}BF_4^{\ominus}$$

In contrast to the C_s symmetry of the parent CH_5^+ ion (vide infra) strong Au–Au interaction makes the C_{3v} form preferred.

Schmidbauer and his co-workers [3] also succeeded in preparing the hexacoordinate $[(Ph_3P)Au]_6C^{2+}$ complex, a stable organometallic derivative of CH_6^{2+} (vide infra). Again strong metal-metal interaction stabilizes the remarkable CR_6^{2+} analog.

Whereas hypercarbon containing metal-carbon clusters, metal carbides, carboranes, metallocarboranes, and carbon bridged organometallics are well recognized and extensively studied, purely organic, that is, only carbon and hydrogen containing higher coordinated systems were for a long time controversial and their existence was questioned.

Before starting the discussion of hypercarbon cluster compounds and their role in hydrocarbon chemistry it is worth pointing out that the parent of electron-deficient hypercoordinate cluster compounds is H_3^+. It was as early as 1911 that Thompson [4] first suggested its existence and over the years extensive theoretical [5] as well as more recently direct spectroscopic [6] evidence was obtained proving its 2e-3c bonded nature.

Studying solutions of D_2 in superacids, such as HF-SbF$_5$ (as well as H_2 in DF-SbF$_5$) Olah et al. observed facile H-D exchange [7]

involving protonation (deuteration) to isotopomeric H_3^+ ions. When Xe was added to the system the rate of the H-D exchange reaction abruptly decreased, indicating that proton (deuteron) transfer to Xe, having a higher proton affinity than hydrogen, takes place preferentially.

Extensive studies on borane and carborane cluster systems (discussed previously in this book), as well as the recognition of the isoelectronic and isosteric nature of trivalent boron and carbon, foreshadowed the fact that hypercoordinate boron and carbon systems should also show a similar relationship. It was the study of carbocations (i.e., electron deficient cationic carbon compounds) which established the existence and chemistry of hypercoordinate hydrocarbon systems.

14.2. THE CONCEPT OF CARBOCATIONS AND STUDY OF PERSISTENT (LONG-LIVED) HYPERCOORDINATE CARBONIUM IONS

Extensive studies helped to develop the general concept of carbocations [8] (the generic name suggested for all cations of carbon compounds in accordance with the anions being named carbanions). Based on experimental studies it became clear that two limiting classes of carbocations exist: trivalent ("classical") carbenium ions, of which CH_3^+ is the parent and penta (or higher) coordinate ("nonclassical") carbonium ions, with CH_5^+ the parent.

Lewis' concept that a chemical bond consists of a pair of electrons shared between two atoms became the foundation of structural organic chemistry, and chemists still tend to name compounds as anomalous when their structures cannot be depicted in terms of such bonds alone. Carbocations with too few electrons to allow a pair for each "bond" came to be referred to as "nonclassical", a label still used even though it is now recognized that, like many other compounds, they adopt the delocalized structures appropriate for the number of electrons they contain.

It should be emphasized that within these limiting categories there is a whole range of intermediate nature involving partial delocalization or bridging.

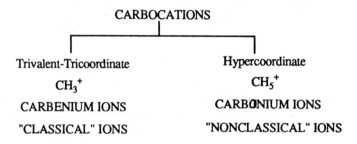

Expansion of the carbon octet via 3d-orbital participation does not seem possible [9]; there can be only eight valence electrons in the outer shell of carbon. Thus, the covalency of carbon cannot exceed 4. Penta- (or higher) coordination implies five (or more) ligands within reasonable bonding distance from the central carbon atom [10]. The transition states for S_N2 reactions represent such cases, but involve 10 electrons around the carbon center. Charge-charge repulsions in the S_N2 transition state forces the entering and leaving substituents as far apart as possible leading to a trigonal bypyramidal arrangement with a long 4e-3c bond allowing little possibility of a stable intermediate [11]. In contrast, S_E2 substitution reactions involve only eight electrons around the carbon atom with a 2e-3c interaction. These reactions have in the past mainly involved organometallic compounds, for example, organomercurials [12] and it was only with the advent of superacid induced reactions of saturated hydrocarbons that true S_E2 reactions proceeding through five coordinate carbocations were realized.

The direct observation of stable penta- (or higher) coordinate carbonium ions ("nonclassical ions") with eight electrons around the carbocationic center became possible during studies of long-lived ions in superacid solvent systems.

Neighboring group interactions with the vacant p-orbital of a trivalent carbenium ion center can contribute to stabilization of the ions via charge delocalization. Such phenomena can involve atoms with unshared electron pairs (n-participation), C–H and C–C hyperconjugation (σ-participation), bent σ-bonds (as in cyclopropylcarbenium ions) and π-electron systems (π-conjugative

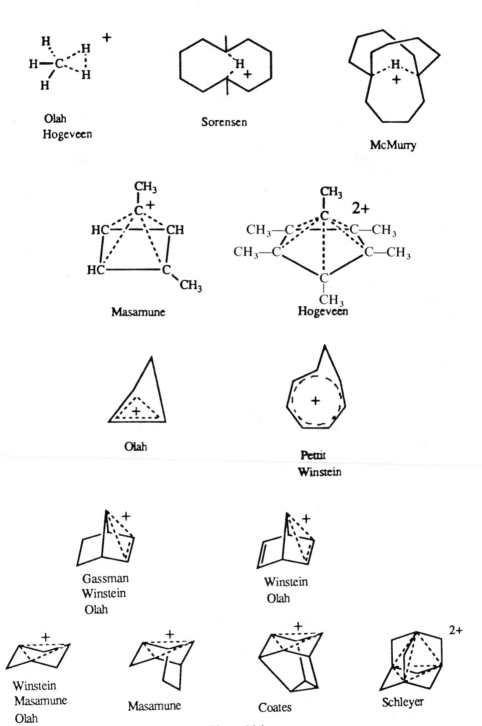

Figure 14.4.

stabilization). Thus, trivalent carbenium ions can show varying degrees of delocalization without becoming higher coordinate carbonium ions. The defined limiting classes do not exclude varying degrees of delocalization, but in fact imply a spectrum of carbocation structures [8].

In contrast to the well-defined trivalent ("classical") carbenium ions, "nonclassical" carbenium ions have long been poorly defined. A lively controversy has centered on the classical-nonclassical ion problem [13]. The extensive use of "dotted lines" in depicting carbonium ion structures has been (rightly) criticized by Brown who carried, however, the criticism to question the existence of any σ-delocalized (nonclassical) ion. For these ions, it was suggested at the time, if they would exist, a new bonding concept yet unknown to chemists must be involved. This is obviously not the case as "nonclassical" ions are simply hypercoordinate carbocations involving besides 2e-2c also 2e-3c (or multicenter) bonding.

The discovery of a significant number of hypercoordinate carbocations ("nonclassical" ions) first on the basis of solvolytic studies and subsequently as observable, stable ions in superacidic media [15], shows that carbon hypercoordination is also a general phenomenon in electron deficient purely hydrocarbon systems (see Figure 14.4).

The most controversial of the hypercoordinate carbocations was the 2-norbornyl cation, around which the classical-nonclassical ion controversy centered [16]. Solvolytic rate studies per se could not unequivocally resolve the question. As the ion, however, is readily prepared and stable under superacidic conditions (even crystalline salts could be isolated and studied by solid state NMR, ESCA and X-ray crystallography), the structure of the ion by now is firmly established and proved Winstein's suggestion of a symmetrical nonclassical ion.

The symmetrically bridged, nonclassical nature of the long-lived 2-norbornyl cation was established by such structural studies as:

^{1}H- and ^{13}C-NMR spectroscopy   Olah et al. [17]
Isotopic perturbation of resonance NMR studies   Saunders [17]
Theoretical IGLO NMR shift calculations   Schindler [18]
Solid state NMR   Yanonni and Myhre [19]
Raman spectroscopy   Olah et al. [17]
X-ray photoelectron spectroscopy   Olah et al. [17],
       Johnson and Clark [20]
X-ray crystallography   Laube [21]

14.3. ELECTROPHILIC REACTIONS OF ALKANES

The formation of the σ-delocalized norbornyl cation via ionization of 2-norbornyl precursors in low nucleophilicity, superacidic media, can be considered as an analog of an intramolecular Friedel–Crafts alkylation, but of a saturated system. Indeed deprotonation gives nortricyclane.

This realization led to the study of possible intermolecular electrophilic reactions of saturated hydrocarbons. Not only protolytic reactions, but a broad range of other electrophilic reactions (alkylation, formylation, nitration, halogenation, oxygenation, etc.) were found to be feasible under superacidic reaction conditions.

Protolytic (deuterolytic) hydrogen-deuterium exchange is the simplest electrophilic substitution of alkanes. Protonation (and protolysis) of alkanes is readily achieved with superacids. The protonation of methane itself takes place [22].

The methonium ion CH_5^+ ion is readily formed in the gas phase in mass spectral studies at high source pressure [23]. Chemical ionization mass spectrometry frequently uses CH_5^+ as the convenient protonating agent.

Spectroscopic observation of CH_5^+ in solution is difficult, as the concentration of the ion even in superacidic media at any time is extremely low. In superacidic media at low temperature ESCA studies were carried out [24]. Matrices of superacids such as HSO_3H-SbF_5 or HF-SbF_5 saturated with methane were studied at $-180°C$. The observed carbon 1s binding energy, differing by less than 1 eV from that of methane, is attributed to CH_5^+. Neutral methane itself has practically no solubility in superacids at the low temperature of the experiments and at the applied high vacuum (10^{-9} Torr), it would be purged out from the system. The observed binding energy of CH_5^+ is in accord with theoretical calculations indicating that charge density is heavily on the hydrogen atom and the five coordinate carbon carries little charge. The CH_5^+ system was extensively investigated by theoretical (quantum mechanical) studies which showed the C_s symmetry form to be the most stable [25]

When methane is reacted with deuterated superacids (or heavy methane with protic acids) hydrogen-deuterium exchange is observed to occur with ease. The next higher homolog alkonium ion, the ethonium ion $C_2H_7^+$ was obtained under similar conditions by protonating ethane [22c]. Protonation can take place either in the C–C or C–H bonds, but the interconversion of the resulting homolog ions is a facile, low energy process. This is consistent with observed H-D exchange of labeled systems as well as formation of methane as by-product of protolytic C–C bond cleavage.

Lee et al. [26] recently reported IR spectroscopic study of the gaseous $C_2H_7^+$ ion and observed both isomeric ions.

Carbon can be involved not only in a single two-electron three-center bond formation, but was shown also capable in some carbodications to simultaneously participate in two 2e-3c bonds. Diprotonated methane (CH_6^{2+}) [27] and ethane ($C_2H_8^{2+}$) [28], as well as the dimer of the methyl cation ($C_2H_6^{2+}$) [29] are illustrative.

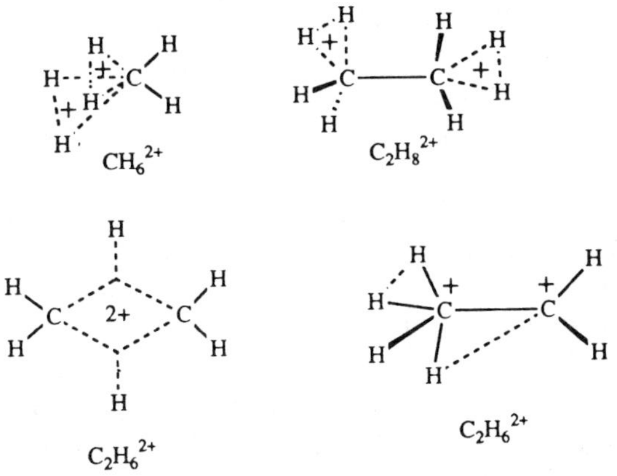

CH_6^{2+} $C_2H_8^{2+}$

$C_2H_6^{2+}$ $C_2H_6^{2+}$

The parent molecular ions of alkanes, such as $CH_4^{+\cdot}$, observed in mass spectrometry, according to theoretical calculations also prefer hypercarbon structures [30].

$CH_4^{+\cdot}$

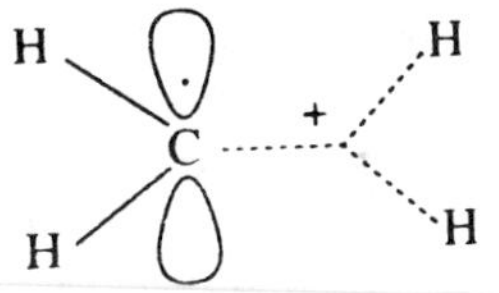

Even the CH_4^{2+} ion was recently discussed in similar terms of a planar C_{2v} structure [31].

Acid catalyzed isomerization reactions of alkanes, as well as alkylation and condensation reactions are initiated by protolytic ionization. Available evidence indicates nonlinear, but not necessarily triangular transition states [32a, 33].

$$R_3C{-}H + H^+ \longrightarrow [R_3C{\cdots}H{\cdots}H]^+ \longrightarrow R_3C^+ + H_2$$

Linear

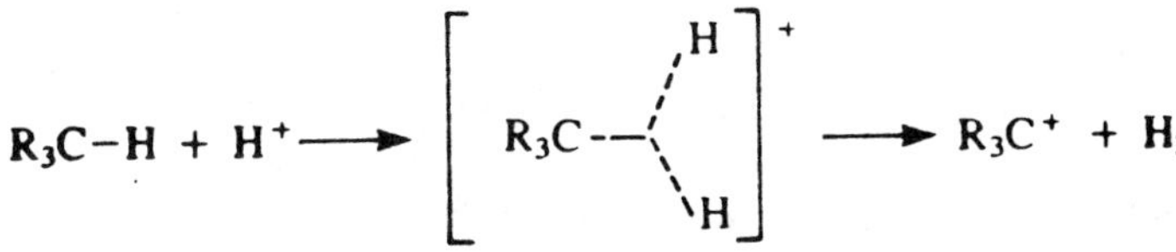

Nonlinear

The reverse reaction of the protolytic ionization of hydrocarbons to carbocations, that is, the reaction of carbocations with molecular hydrogen, also involves five-coordinate carbonium ions [32].

$$R_3C^+ \quad + \quad \begin{matrix} H \\ | \\ H \end{matrix} \quad \rightleftharpoons \quad \left[R_3C \cdots\cdots \diagdown\!\!\!\diagup \begin{matrix} H \\ \\ H \end{matrix} \right]^+ \quad \rightleftharpoons \quad R_3CH \ + \ H^+$$

The reaction of butane and cyclohexane are illustrative of superacid catalyzed protolytic isomerization [33].

In the isomerization of butane a decisive role is played by protonated cyclopropanes in the rearrangement of the intermediate *sec*-butyl cation into the *tert*-butyl cation (Scheme I). The edge- and corner-protonated forms with three center bonds of type (a) and (b) should be differentiated from methyl bridged ion (c).

The skeletal rearrangements, however, must always take place with participation of the incipient primary isobutyl cation. Protonated cyclopropanes alone cannot account for it, even though they are known to be involved in the hydrogen and carbon exchange (scrambling) processes.

Intramolecular alkylation processes involving C–C or C–H bonds play a decisive role in ring contraction (or expansion) isomerizations, as for example, in the cyclohexane-methylcyclopentane isomerization. Again, either C–H delocalized protonated cyclopropanes or C–C delocalized carbon bridged carbonium ion transition states are involved. The cyclohexane-methylcyclopentane isomerization involves initial formation of the cyclohexyl (methylcyclopentyl) cation (i.e., protolysis of a C–H bond). It should be mentioned that in the acid-catalyzed isomerization of cyclohexane up to 10% hexanes are also formed. This is indicative of C–C bond protolysis (Scheme II).

The superacid catalyzed cracking of hydrocarbons (a potentially significant practical application) can involve not only formation of trivalent carbocationic sites leading to subsequent β-cleavage, but also direct C–C bond protolysis [1] (Scheme III).

Whereas superacid catalyzed hydrocarbon transformations were first explored in liquid media (HF/BF$_3$, HF/SbF$_5$, HF/TaF$_5$. FSO$_3$H/SbF$_5$, etc.), subsequently solid acid catalyst systems, such as those based on Nafion-H, longer chain perfluorinated alkanesulfonic acids, fluorinated graphite intercalates, etc. were also developed and utilized for heterogenous gas and liquid phase reactions [34]. The superacidic nature of zeolite catalysts was also explored in case of ZSM-5 [35].

Not only protolytic reactions, but a whole range of varied electrophilic reactions of alkanes can be carried out under superacidic conditions [1, 33]. They follow the general patterns shown in Scheme IV.

Alkylation of isoalkanes is particularly instructive. The industrially significant alkylation of isobutane with isobutylene to isooctane is de facto alkylation

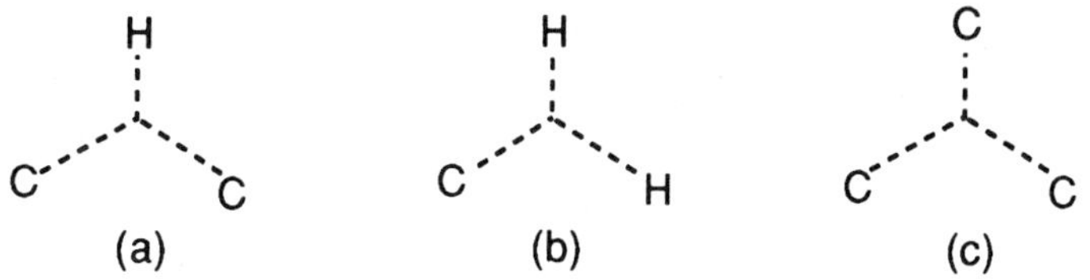

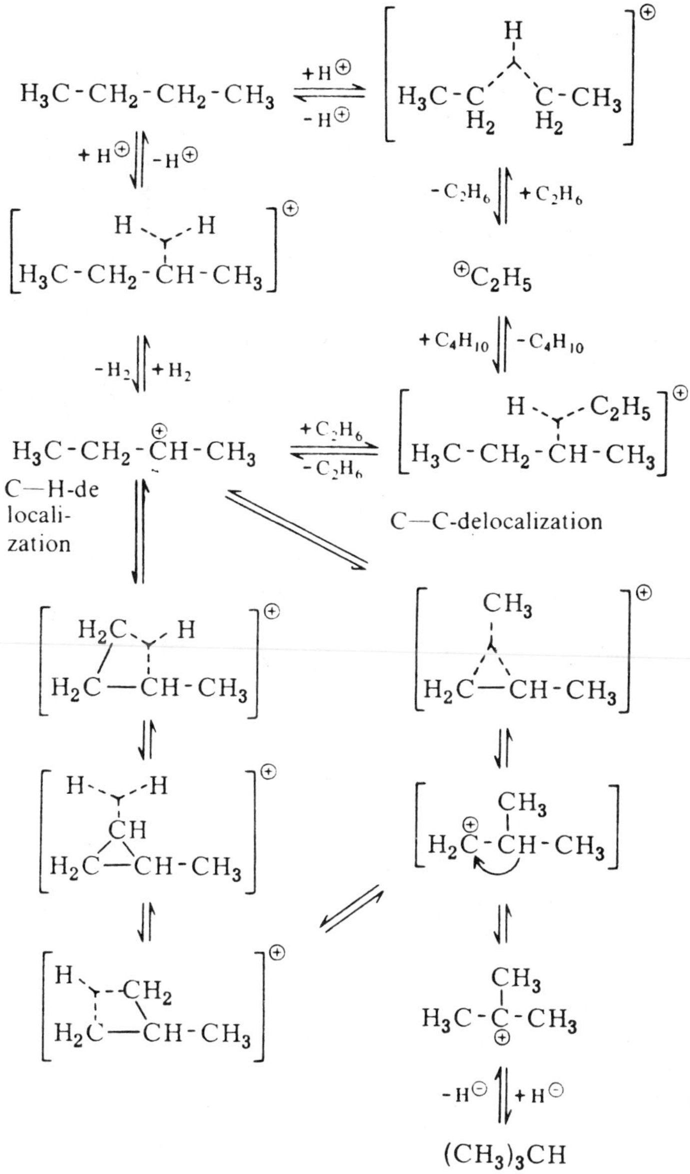

(a)

(b)

(c)

Scheme I

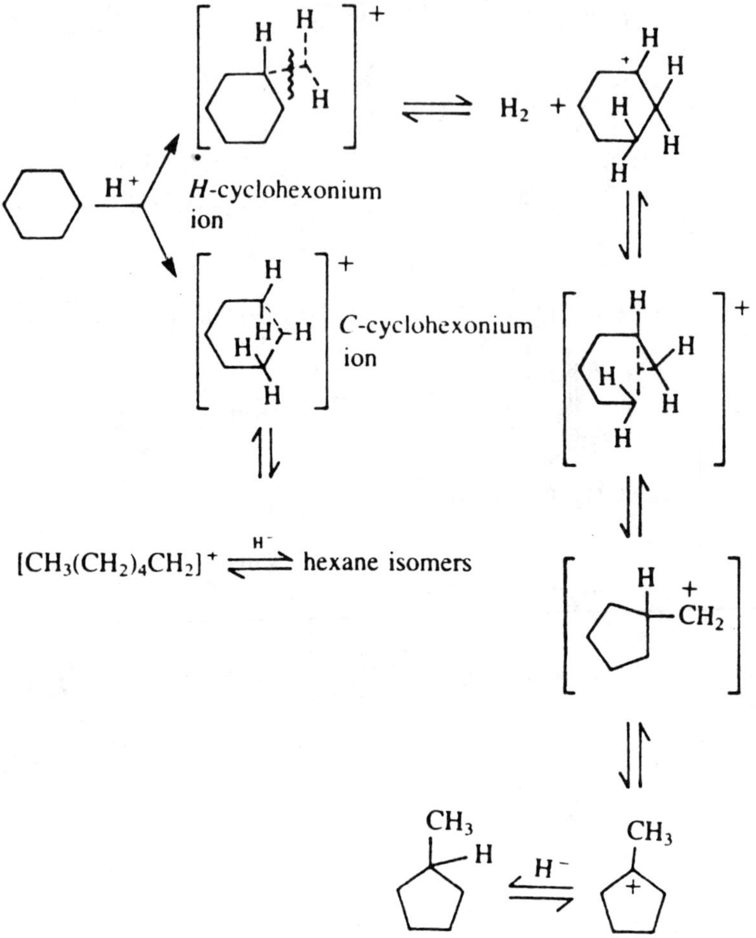

$$[CH_3(CH_2)_4CH_2]^+ \underset{}{\overset{H^-}{\rightleftharpoons}} \text{hexane isomers}$$

Scheme II.

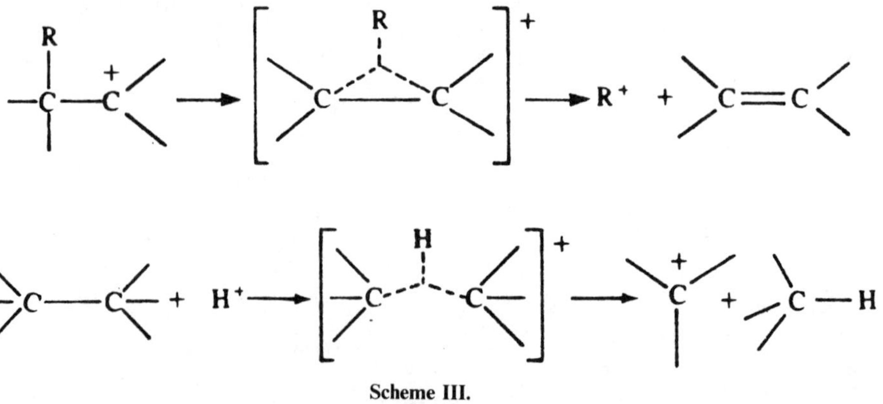

Scheme III.

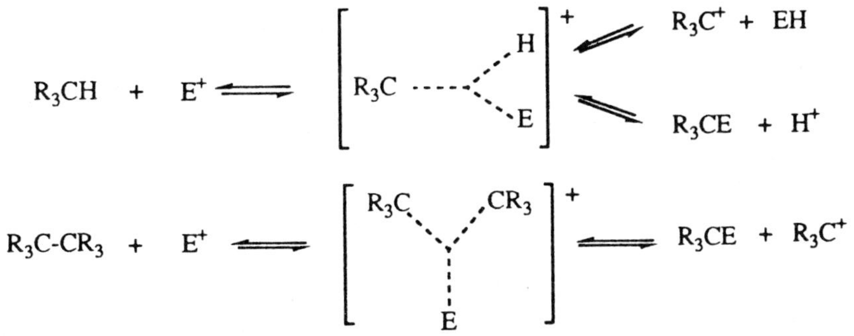

$$R_3CH + E^+ \rightleftharpoons \left[R_3C \cdots \begin{matrix} H \\ E \end{matrix} \right]^+ \begin{matrix} \rightleftharpoons R_3C^+ + EH \\ \rightleftharpoons R_3CE + H^+ \end{matrix}$$

$$R_3C\text{-}CR_3 + E^+ \rightleftharpoons \left[\begin{matrix} R_3C \quad CR_3 \\ E \end{matrix} \right]^+ \rightleftharpoons R_3CE + R_3C^+$$

$$E^+ = D^+, H^+, R^+, NO_2^+, HAL^+, HCO^+, \text{etc.}$$

Scheme IV.

of the reactive isobutylene and not of the saturated hydrocarbon. Isobutane only acts as a source of the *tert*-butyl cation, formed via intermolecular Bartlett–Nenitzescu–Schmerling hydride transfer (Scheme V).

When *tert*-butyl cation is reacted with isobutane under superacidic conditions the major fact reaction is still hydride transfer, but a detectable amount of 2,2,3,3-tetramethylbutane, the σ-alkylation product is also obtained [1, 36] (Scheme VI). With sterically less crowded systems σ-alkylation can become predominant.

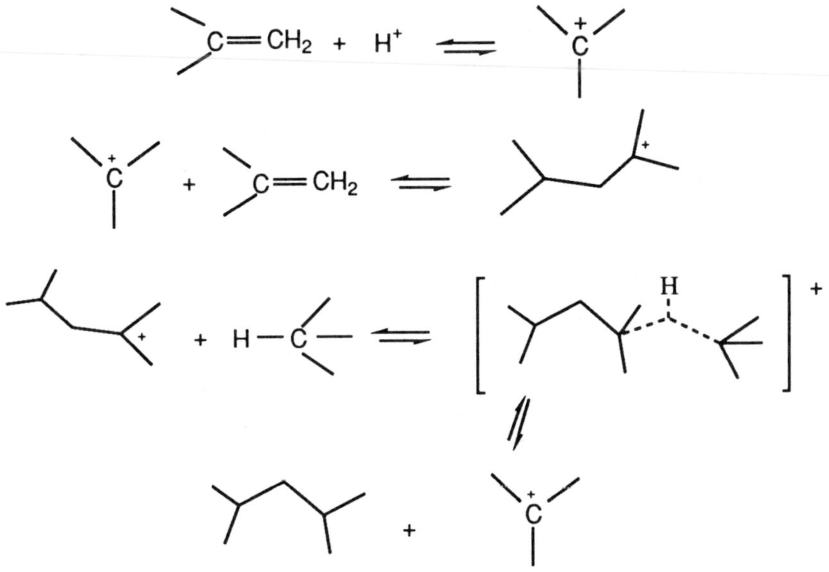

Scheme V.

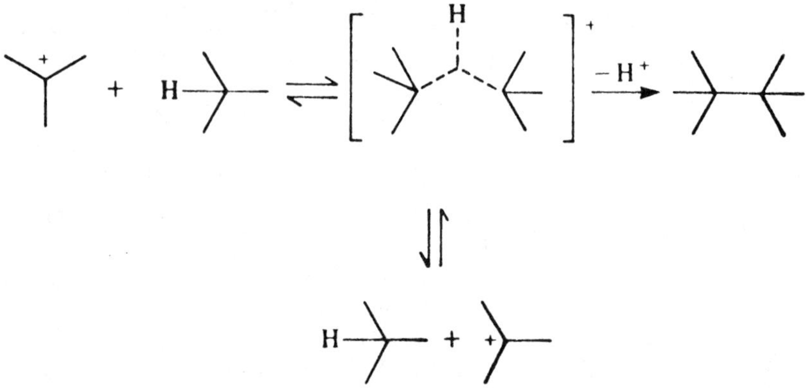

Scheme VI.

Butylation of adamantane with isobutylene similarly yields, besides isomeric butyladamantanes resulting from adamantylation of the olefin, 1-*tert*-butyladamantane, the σ-alkylation product [37] (Scheme VII). As neither adamantylation of isobutylene or any isomerization process can yield this product, despite the sterically unfavorable tertiary-tertiary interaction C–H bond σ-alkylation is indicated, although adamantylation of the olefin by the 1-adamantyl cation is still predominant (as in the case with other olefins).

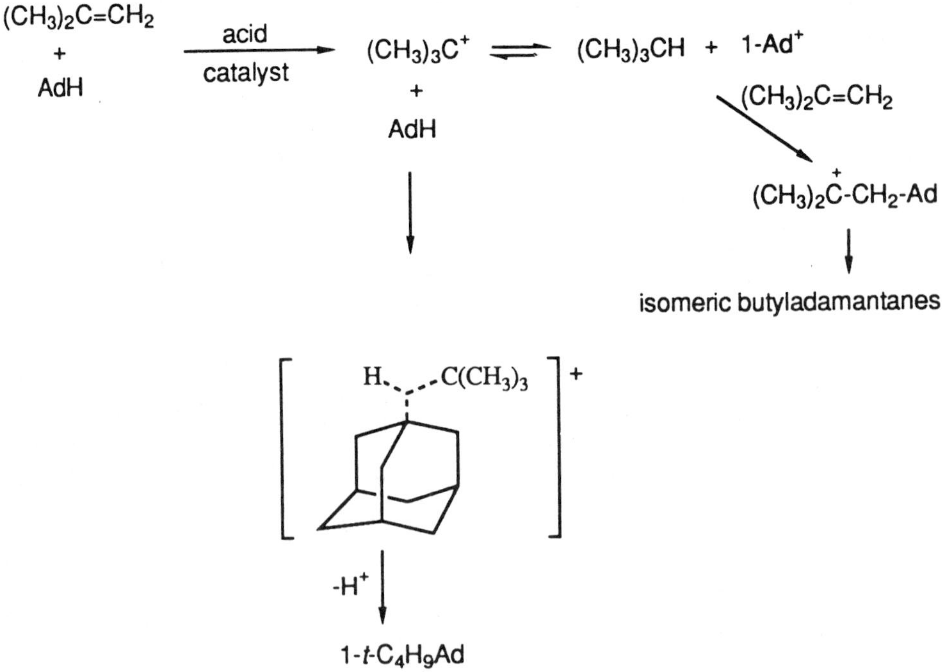

Scheme VII.

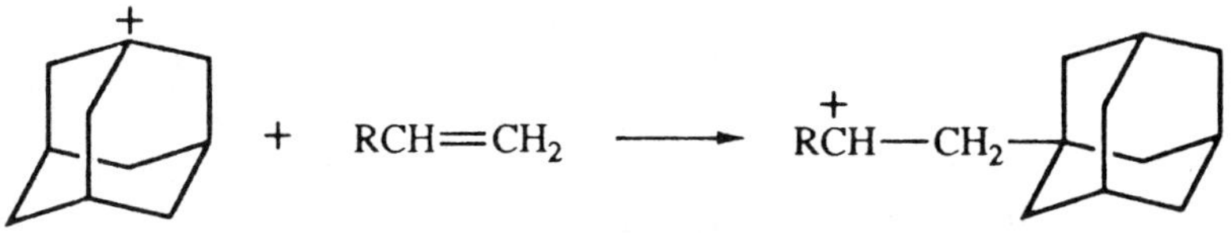

Oxidative protolytic condensation of alkanes is also possible under superacidic conditions. Whereas the condensation of two molecules of methane to ethane (and hydrogen) is exothermic by 16 kcal/mol, the reaction becomes feasible in the presence of strong oxidizing agents (either components of the superacid itself, i.e., SbF_5, or with oxidants such as O_2, S_x, Se, halogens) [38].

When olefins, such as ethylene, are reacted over superacids with excess methane, σ-alkylation takes place as shown by experiments using $^{13}CH_4$ [39].

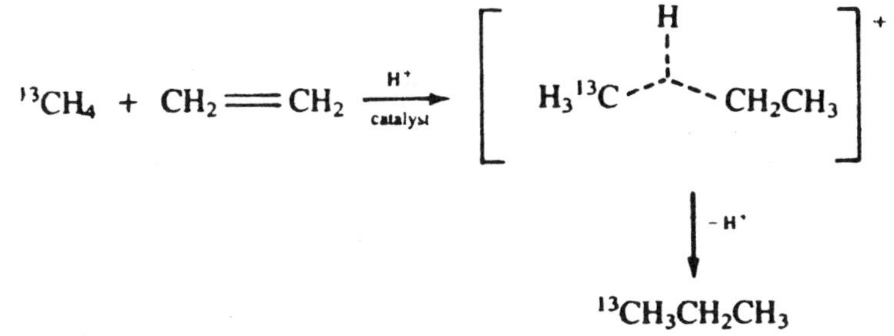

A fundamental difference exists between conventional acid catalyzed and superacidic hydrocarbon chemistry. In the former, olefins always play the key role, whereas in the latter hydrocarbon transformation can take place without the involvement of olefins with the intermediacy of five coordinate carbocations.

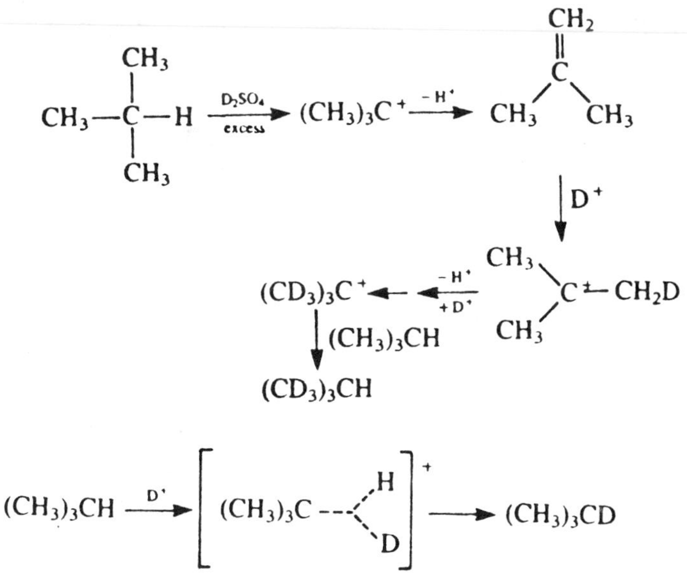

Scheme VIII.

Illustrative of this difference is the hydrogen-deuterium exchange of isobutane in D_2SO_4, as studied by Otvos et al. [40] contrasted with that in superacidic DF-SbF$_5$ and DSO$_3$F-SbF$_5$, as studied by Hoffman et al. and Olah et al., respectively [41]. In the former case all nine methyl hydrogen exchange, whereas in the latter these are not affected and only CH exchange is observed accompanying protolytic ionization (Scheme VIII).

Siskin's study of the HF/TaF$_5$ catalyzed ethylation of excess ethane with ethylene in a flow system is equally revealing [42]. Only butane was obtained with complete absence of isobutane, indicative of absence of alkylation of the alkene, which would lead to inevitable formation of isobutane.

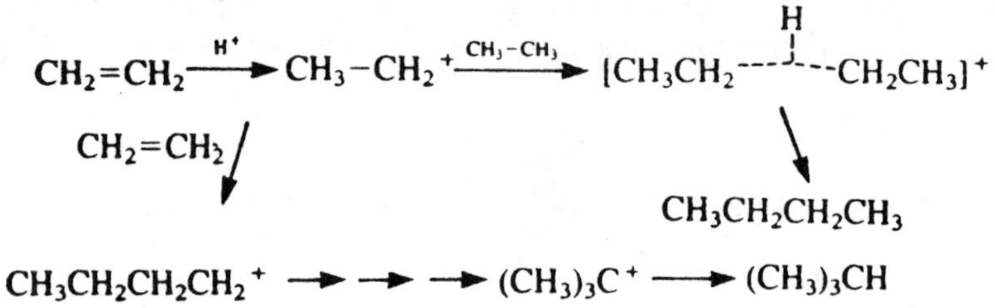

Nitration of alkanes with nitronium salts can also be achieved [43] involving insertion into C–H (or –C–C) σ-bonds. Representative is the nitration of adamantane.

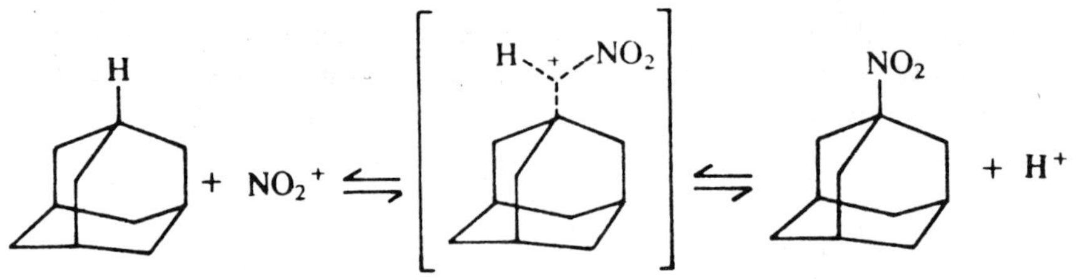

Superacid catalyzed carbonylation of adamantane [44] was also carried out giving besides some 1-adamantanol 1-adamantanecarboxylic acid (the Kock–Haaf reaction product) and $\sim 20\%$ of 1-adamantylcarboxaldehyde, the electrophilic formylation product (Scheme IX).

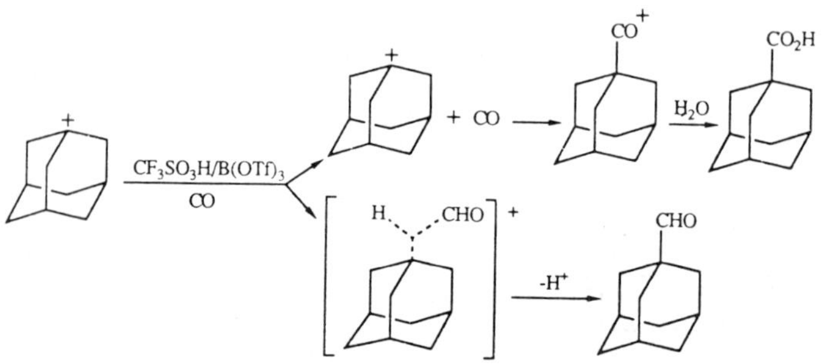

Scheme IX.

The superacid catalyzed electrophilic oxygenation of saturated hydrocarbons with hydrogen peroxide (i.e., $H_3O_2^+$) and ozone (i.e., HO_3^+) allow the efficient preparation of oxygenated derivatives [45] (Scheme X).

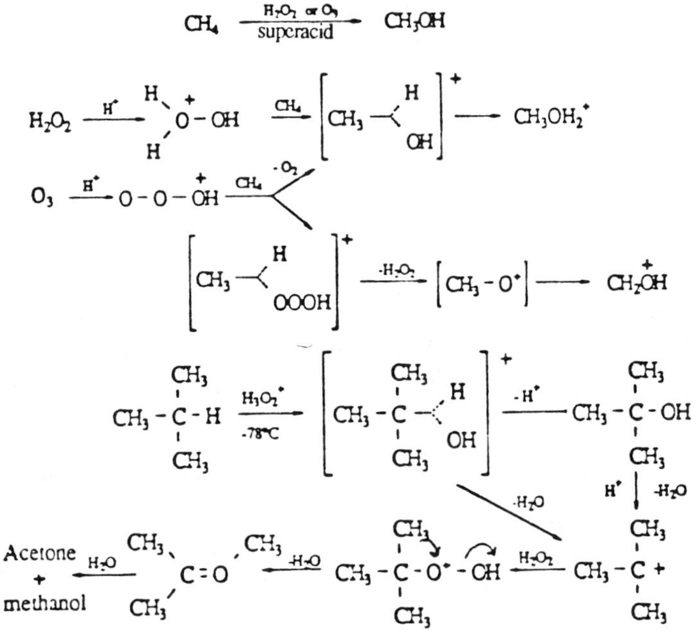

$$CH_4 \xrightarrow[\text{superacid}]{H_2O_2 \text{ or } O_3} CH_3OH$$

Scheme X.

As the protonation of ozone removes its dipolar nature, the electrophilic chemistry of HO_3^+, a very efficient electrophile, has no relevance to conventional ozone chemistry. Illustrated in Scheme XI is the superacid catalyzed reaction of isobutane with ozone giving formaldehyde and methyl alcohol, the aliphatic equivalent of the industrially significant Hock-reaction of cumene.

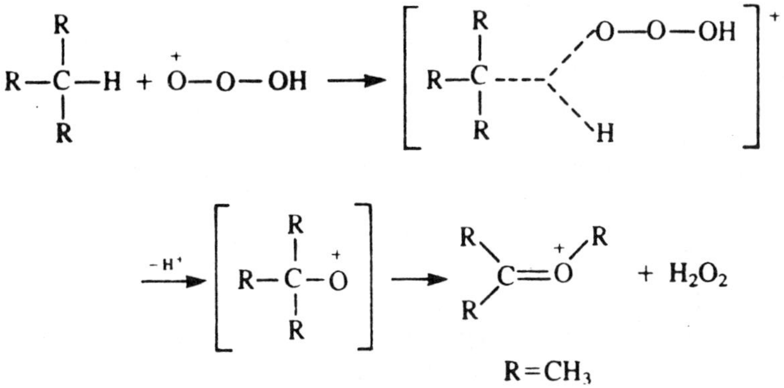

$$R = CH_3$$

Scheme XI.

Superacid catalyzed electrophilic sulfuration of alkanes with elementary sulfur (S_8) was also achieved giving predominantly dialkyl sulfides [46].

$$CH_3CH_2CH_3 \xrightarrow[CF_3SO_3H]{S_8} [(CH_3)_2CH]_2S$$

Electrophilic chlorination (bromination) of methane gives methyl halides in high selectivity [47].

$$CH_4 \xrightarrow[X=Cl,\ Br]{X_2,\ cat} \left[H_3C\text{---}\!\!\overset{H}{\underset{X}{\langle}} \right]^+ \xrightarrow{-H^+} CH_3X \underset{}{\overset{CH_3X}{\rightleftharpoons}} CH_3\overset{+}{X}CH_3$$

Combining the halogenation of methane with catalytic hydrolysis to methyl alcohol allows the preparation of the latter without going through syn-gas [37a, 47]. Byproduct HX, particularly in the case of HBr, is readily oxidized and thus recycled making the reaction catalytic in halogen. Overall a highly selective oxidative conversion of methane to methyl alcohol is thus achieved.

$$CH_4 + Br_2 + H_2O \longrightarrow CH_3OH + 2\,HBr$$

$$2\,HBr + 1/2\,O_2 \longrightarrow Br_2 + H_2O$$

$$\overline{CH_4 + 1/2\,O_2 \longrightarrow CH_3OH}$$

Methyl alcohol and methyl halides can be readily condensed to ethylene (propylene) and through them to higher hydrocarbons. The reaction takes place not only over shape selective zeolite catalysts such as ZSM-5 [48], but also with supported bifunctional acid-base catalysts, such as WO_3/Al_2O_3 [49].

The mechanism of this condensation involves an oxonium ylide pathway.

It should be pointed out the electrophilic insertion reactions into C–H (and C–C) bonds are not unique to carbocationic processes under low nucleophilicity superacidic conditions. The fast developing area of C–H (and C–C) bond activation by organometallic complexes, such as iridium complexes and other transition metal systems (rhodium, osmium, rhenium, etc.) is based on fundamentally similar electrophilic insertions [50]. These reactions, however, can not so far be made catalytic, although further work may change this. A wide variety of reactions of hydrocarbons with coordinatively unsaturated metal compounds and fragments involving hypercarbon intermediates (transition states) is also recognized ranging from hydrometallations to Ziegler–Natta polymerization [3], but these are outside the scope of the present discussion.

Hypercarbon chemistry has come a long way from the once controversial structural problems of some carbocations, such as the norbornyl cation, to the realization of a wide variety of significant, even practical reactions of alkanes involving electrophilic transformation of C–H and C–C single bonds. Much further chemistry remains to be explored.

ACKNOWLEDGMENTS

I am grateful to my colleagues mentioned in the references for their contributions which made our work possible. Financial support over the years was provided by the National Science Foundation, the National Institutes of Health, and the Loker Hydrocarbon Research Institute.

REFERENCES

1. G. A. Olah, G. K. S. Prakash, R. E. Williams, L. D. Field and K. Wade, *Hypercarbon Chemistry*, Wiley-Interscience, New York, 1987.
2. F. Scherbaum, B. Huber, C. Krüger and H. Schmidbaur, *Angew. Chem.*, **101**, 1600 (1989).
3. F. Scherbaum, A. Grokmann, G. Müller and H. Schmidbaur, *Angew. Chem.*, **101**, 464 (1989).
4. J. J. Thompson, *Philos. Mag.*, **21**, 225 (1911).
5. E. Pollak and C. Schlier, *Acc. Chem. Res.*, **22**, 223 (1989) and references therein.
6. (a) T. Oka, *Phys. Rev. Lett.*, **43**, 531 (1980); (b) J. T. Shy, J. W. Farley, W. E. Lamb, Jr. and W. H. Wing, *Phys. Rev. Lett.*, **45**, 535 (1980); (c) A. Carvington and I. R. McNab, *Acc. Chem. Res.*, **22**, 218 (1980).
7. G. A. Olah, J. Shen and R. H. Schlosberg, *J. Am. Chem. Soc.*, **95**, 4957 (1973).
8. G. A. Olah, *J. Am. Chem. Soc.*, **94**, 808 (1972).
9. Despite occasional claims such as R. J. Gillespie, *J. Chem. Soc.*, 1002 (1952); *J. Chem. Phys.*, **21**, 1893 (1953) or discussion in *Science*, October 28, 403 (1983), but rebutted by M. J. S. DeWar, *J. Chem. Soc.*, 2885 (1953); J. J. Jaffe, *J. Chem. Phys.*, **21**, 1893 (1953).
10. E. L. Muetterties and R. A. Schum, *Q. Rev.*, **20**, 245 (1966).
11. J. C. Martin, *Science*, **221**, 509 (1985), for interesting studies of a claimed pentavalent carbon intermediate.
12. For a review and critical discussion, see F. R. Jensen and B. Rickborn, *Electrophilic Substitution of Organomercurials*, McGraw-Hill, New York, 1968.
13. P. D. Bartlett, *Non-Classical Ions*, W. A. Benjamin, New York, 1965.
14. H. C. Brown, with commentary by P. v. R. Schleyer, *The Non-Classical Ion Problem*, Plenum Press, New York, 1971.
15. G. A. Olah, G. K. S. Prakash and J. Sommer, *Superacids*, Wiley-Interscience, New York, 1985, and references therein.
16. (a) S. Winstein, *Q. Rev.*, **23**, 141 (1969) and references therein; (b) H. C. Brown, *Chem. Eng. News*, 86 (1967) and references therein; (c) H. C. Brown, *Acc. Chem. Res.*, **16**, 432 (1983); (d) C. A. Grob, *Acc. Chem. Res.*, **16**, 426 (1983); (e) C. Walling, *Acc. Chem. Res.*, **16**, 448 (1983); (f) G. A. Olah, G. K. S. Prakash and M. Saunders, *Acc. Chem. Res.*, **16**, 440 (1983).
17. G. A. Olah, G. K. S. Prakash and M. Saunders, *Acc. Chem. Res.*, **16**, 440 (1983).
18. M. Schindler, *J. Am. Chem. Soc.*, **109**, 1020 (1987).
19. G. S. Yannoni, V. Macho and P. C. Myhre, *J. Am. Chem. Soc.*, **104**, 907, 7380 (1982).
20. S. A. Johnson and D. T. Clark, *J. Am. Chem. Soc.*, **110**, 4112 (1988).

21. T. Laube, *Angew. Chem.*, **99**, 580 (1987).

22. (a) G. A. Olah and J. Lukas, *J. Am. Chem. Soc.*, **89**, 2227, 4739 (1967); **90**, 933 (1968); (b) H. Hogeveen and A. F. Bickel, *Chem. Commun.*, 635 (1967); *Rec. Trav. Chim.*, **87**, 319 (1968); **88**, 703 (1969); (c) G. A. Olah and R. H. Schlosberg, *J. Am. Chem. Soc.*, **90**, 2726 (1968); G. A. Olah, G. Klopman and R. H. Schlosberg, *J. Am. Chem. Soc.*, **91**, 3261 (1969).

23. F. H. Field and M. S. B. Munson, *J. Am. Chem. Soc.*, **87**, 3289 (1965) and references therein.

24. G. A. Olah et al., quoted in [15] p. 125.

25. W. J. Hehre, L. Radom, P. v. R. Schleyer and J. A. Pople, *Ab Initio Molecular Orbital Theory*, Wiley Interscience, New York, 1986 and references therein.

26. L. I. Yeh, J. M. Price and Y. T. Lee, *J. Am. Chem. Soc.*, **111**, 5597 (1989).

27. K. Lammertsma, G. A. Olah, M. Barzaghi and M. Simonetta, *J. Am. Chem. Soc.*, **104**, 6851 (1982).

28. K. Lammertsma, M. Barzaghi, G. A. Olah, J. A. Pople, P. v. R. Schleyer and M. Simonetta, *J. Am. Chem. Soc.*, **105**, 5258 (1983).

29. (a) G. A. Olah and M. Simonetta, *J. Am. Chem. Soc.*, **104**, 330 (1982); K. Lammertsma, G. A. Olah, M. Barzaghi and M. Simonetta, *J. Am. Chem. Soc.*, **104**, 6851 (1982); (b) P. v. R. Schleyer, A. Kos, J. Pople and A. T. Balaban, *J. Am. Chem. Soc.*, **104**, 3771 (1982).

30. G. A. Olah and G. Klopman, *Chem. Phys. Lett.* **11**, 604 (1971).

31. M. W. Wang and L. Radom, *J. Am. Chem. Soc.*, **111**, 1155 (1989).

32. (a) A. F. Bickel, C. J. Gaasbeck, H. Hogeveen, J. M. Oelderick and J. C. Plattew, *Chem. Commun.*, 634 (1967); (b) H. Hogeveen and A. F. Bickel, *Rec. Trav. Chim.*, **86**, 1313 (1967).

33. G. A. Olah, *Angew. Chem. Int. Engl. Ed.*, **12**, 173 (1973) and references therein.

34. G. A. Olah, P. S. Iyer and G. K. S. Prakash, *Synthesis*, 513 (1986).

35. W. O. Haag and R. H. Dessau, *Proceedings 8th Cat. Congress,* West-Berlin, Germany, Vol. II, p. 305, Dechema, Frankfurt-am-Main, 1984.

36. G. A. Olah and R. H. Schlosberg, *J. Am. Chem. Soc.*, **90**, 2726 (1968).

37. G. A. Olah, O. Farooq, V. V. Krishnamurthy, G. K. S. Prakash and K. Laali, *J. Am. Chem. Soc.*, **107**, 7541 (1985).

38. (a) G. A. Olah, *Acc. Chem. Res.*, **20**, 422 (1987) and references therein; (b) G. A. Olah, U.S. Patents 4,443,192 (1984); 4,513,164 (1984); 4,465,893 (1984); 4,467,130 (1984); 4,513,164 (1985).

39. G. A. Olah, J. D. Felberg and K. Lammertsma, *J. Am. Chem. Soc.*, **105**, 6259 (1983).

40. J. W. Otvos, D. P. Stevenson, C. D. Wagner and O. Beeck, *J. Am. Chem. Soc.*, **73**, 5741 (1951).

41. (a) H. Hoffman, C. J. Gaasbeck and A. F. Bickel, *Rec. Trav. Chim.*, **88**, 703 (1969); (b) G. A. Olah, Y. Halpern, J. Shen and Y. K. Mo, *J. Am. Chem. Soc.*, **93**, 1251 (1971).

42. M. Siskin, *J. Am. Chem. Soc.*, **98**, 5413 (1976).

43. G. A. Olah and H. C. Lin, *J. Am. Chem. Soc.*, **93**, 1259 (1971).

44. O. Farooq, M. Marcelli, G. K. S. Prakash and G. A. Olah, *J. Am. Chem. Soc.*, **110**, 864 (1988).

45. G. A. Olah, D. G. Parker, N. Yoneda, *Angew. Chem. Int. Engl. Ed.*, **17**, 909 (1978).

46. G. A. Olah, Q. Wang and G. K. S. Prarash, *J. Am. Chem. Soc.*, **112**, 3697 (1990).

47. (a) G. A. Olah and Y. K. Mo, *J. Am. Chem. Soc.*, **94**, 6684 (1972); G. A. Olah, R. Renner, P. Schilling and Y. K. Mo, *J. Am. Chem. Soc.*, **95**, 7686 (1973). (b) G. A. Olah, B. Gupta, M. Farnia, J. D. Felberg, W. M. Ip, A. Husain, R. Karpeles, K. Lammertsma, A. K. Melhotra and N. J. Trivedi, *J. Am. Chem. Soc.*, **107**, 7097 (1985).

48. C. D. Chang, *Cat. Rev. Sci. Eng.*, **25**, 1 (1983) and references therein.

49. (a) G. A. Olah, H. Doggweiler, J. D. Felberg, S. Frohlich, M. J. Grdina, R. Karpeles, T. Keumi, S. Inaba, W. M. Ip, K. Lammertsma, G. Salem and D. C. Tabor, *J. Am. Chem. Soc.*, **106**, 2145 (1984); (b) G. A. Olah, U.S. Patent, 4,373,109 (1983).

50. (a) R. G. Bergman, *Science*, **223**, 902 (1984) and references therein; (b) P. J. Desrosiers, R. S. Shinomoto and T. Flood, *J. Am. Chem. Soc.*, **108**, 7964 (1986); (c) J. Schwartz, *Acc. Chem. Res.*, **18**, 302 (1985); (d) W. D. Jones and F. J. Feher, *J. Am. Chem. Soc.*, **107**, 620 (1983); (e) R. H. Crabtree et al., *Chemtech*, 506 (1982); *Organometallics*, **4**, 519 (1985), **3**, 1727 (1985); *J. Am. Chem. Soc.*, **106**, 2913 (1984), **103**, 1217 (1981), **104**, 107, 6994 (1982); (f) M. A. Green, J. C. Hoffman and K. G. Caulton, *J. Am. Chem. Soc.*, **104**, 2319 (1982); *Organomet. Chem.*, **243**, C78 (1983), **218**, C39 (1981).